The Ecology of Plants

The Ecology of Plants

Editor: Jude Boucher

R CALLISTO REFERENCE

www.callistoreference.com

Callisto Reference,
118-35 Queens Blvd., Suite 400,
Forest Hills, NY 11375, USA

Visit us on the World Wide Web at:
www.callistoreference.com

ISBN: 978-1-64116-221-0 (Hardback)

Cataloging-in-Publication Data

The ecology of plants / edited by Jude Boucher.
    p. cm.
Includes bibliographical references and index.
ISBN 978-1-64116-221-0
1. Plant ecology. 2. Ecology. 3. Botany. I. Boucher, Jude.
QK901 .E26 2019
581.7--dc23

Table of Contents

Preface

The world is advancing at a fast pace like never before. Therefore, the need is to keep up with the latest developments. This book was an idea that came to fruition when the specialists in the area realized the need to coordinate together and document essential themes in the subject. That's when I was requested to be the editor. Editing this book has been an honour as it brings together diverse authors researching on different streams of the field. The book collates essential materials contributed by veterans in the area which can be utilized by students and researchers alike.

Plant ecology is a branch of ecology, which is concerned with the study of the abundance and distribution of plants, the effect of environmental factors and the interactions between plants and other organisms. The plant kingdom ranges in complexity from the single-celled algae to large canopy forming trees. Plant communities are distributed into biomes depending on the dominant plant species present. Some of the important vegetation types are tundra, terrestrial wetlands, temperate grasslands, tropical forests, tropical savannas, etc. The predominant biological interactions occurring in plant communities are competition for resources, mutualism and herbivory. Depending on the level of organization, plant ecology can be divided into plant ecophysiology, community ecology, ecosystem ecology and biosphere ecology, among others. This book studies, analyzes and upholds the pillars of plant ecology and its utmost significance in modern times. It includes some of the vital pieces of work being conducted across the world, on the ecology and diversity of plants. The extensive content herein provides the readers with a thorough understanding of the subject.

Each chapter is a sole-standing publication that reflects each author's interpretation. Thus, the book displays a multi-facetted picture of our current understanding of application, resources and aspects of the field. I would like to thank the contributors of this book and my family for their endless support.

Editor

Belowground advantages in construction cost facilitate a cryptic plant invasion

Joshua S. Caplan[1,2], Christine N. Wheaton[1] and Thomas J. Mozdzer[1,2]*

[1] Department of Biology, Bryn Mawr College, Bryn Mawr, PA, USA
[2] Smithsonian Environmental Research Center, Edgewater, MD, USA

Associate Editor: Dennis F. Whigham

Abstract. The energetic cost of plant organ construction is a functional trait that is useful for understanding carbon investment during growth (e.g. the resource acquisition vs. tissue longevity tradeoff), as well as in response to global change factors like elevated CO_2 and N. Despite the enormous importance of roots and rhizomes in acquiring soil resources and responding to global change, construction costs have been studied almost exclusively in leaves. We sought to determine how construction costs of aboveground and belowground organs differed between native and introduced lineages of a geographically widely dispersed wetland plant species (*Phragmites australis*) under varying levels of CO_2 and N. We grew plants under ambient and elevated atmospheric CO_2, as well as under two levels of soil nitrogen. We determined construction costs for leaves, stems, rhizomes and roots, as well as for whole plants. Across all treatment conditions, the introduced lineage of *Phragmites* had a 4.3 % lower mean rhizome construction cost than the native. Whole-plant construction costs were also smaller for the introduced lineage, with the largest difference in sample means (3.3 %) occurring under ambient conditions. In having lower rhizome and plant-scale construction costs, the introduced lineage can recoup its investment in tissue construction more quickly, enabling it to generate additional biomass with the same energetic investment. Our results suggest that introduced *Phragmites* has had an advantageous tissue investment strategy under historic CO_2 and N levels, which has facilitated key rhizome processes, such as clonal spread. We recommend that construction costs for multiple organ types be included in future studies of plant carbon economy, especially those investigating global change.

Keywords: Carbon dioxide; common reed; construction cost; eutrophication; intraspecific; invasion ecology; *Phragmites*; plant functional traits; rhizomes; wetlands.

Introduction

The energetic requirement of plant tissue biosynthesis, or construction cost (CC), has proven to be a valuable functional trait in investigations of the carbon economy of plants. Research on leaf CC and associated traits has yielded insights into the strategies used by plants for carbon acquisition (investment in leaf longevity, payback time for the investment, light harvesting area, etc.) and has thereby helped to explain patterns in growth at the

* Corresponding author's e-mail address: tmozdzer@brynmawr.edu

individual and population levels (Wright *et al.* 2004; Poorter and Bongers 2006). For instance, a number of studies on invasive species have found lower leaf CCs, higher specific leaf areas (SLAs) and more rapid growth rates relative to co-occurring non-invasive species across life forms (Baruch and Goldstein 1999; Nagel and Griffin 2001; Deng *et al.* 2004; Feng *et al.* 2008; Osunkoya *et al.* 2010; Shen *et al.* 2011). Research on leaf CC has also identified ways in which plants will adjust leaf structure and function as changes in global climate intensify. In prior studies, most species decreased leaf CCs in response to elevated CO_2 (Poorter *et al.* 1997; Lei *et al.* 2012), while leaf CC rose in response to higher nitrogen availability (Griffin *et al.* 1993).

Although functional trait studies that have included CC have almost exclusively used it to gain insight into the carbon economy of leaves, CC is not a trait specific to leaves. The few studies that have addressed CCs of roots, rhizomes or other organs have shown that high investment in one organ does not necessarily correspond to high investment in another (Wullschleger *et al.* 1997; Nagel *et al.* 2005; Osunkoya *et al.* 2008). Given that changes in biomass allocation and tissue composition have been observed in many species following CO_2 and nitrogen manipulation (Poorter *et al.* 1997, 2012; Curtis and Wang 1998; Booker *et al.* 2000; Booker and Maier 2001; Kraus *et al.* 2004), changes in the CCs of organs other than leaves are probably common as well. Research explicitly investigating the CC of belowground organs in response to additions of CO_2 and inorganic nitrogen would be especially useful in understanding how global change will affect the trajectory of plant populations as resource regimes shift.

Phragmites australis, or common reed (hereafter *Phragmites*), is well suited for an investigation of how plants may adjust tissue construction in response to global change. *Phragmites* has a cosmopolitan distribution, with dozens of genetic lineages in the species (Saltonstall 2002; Lambertini *et al.* 2012). It is therefore possible to tightly constrain phylogeny while comparing CCs between lineages that co-occur in natural ecosystems. The most well-studied case is that of two lineages that occur in tidal wetlands along the Atlantic coast of North America. One lineage was introduced from Eurasia to North America in the mid-1800s (haplotype M; *P. australis* subsp. *australis*; hereafter 'introduced *Phragmites*') (Saltonstall 2002). It has invaded wetlands across the Atlantic coast of North America, dramatically changing both ecosystem structure and function (Marks *et al.* 1994; Chambers *et al.* 1999; Kettenring *et al.* 2012). The other lineage present is a haplotype native to the region (haplotype F; *P. australis* subsp. *americanus*; hereafter 'native *Phragmites*') (Saltonstall 2002).

Strong differences in physiology, growth (aboveground and belowground) and abundance have been observed between native and introduced *Phragmites* (Saltonstall 2007; Saltonstall and Stevenson 2007; Park and Blossey 2008; Mozdzer and Zieman 2010; Mozdzer *et al.* 2010, 2013). Further, differences in growth rate between the lineages are known to become exacerbated in response to eutrophication and elevated atmospheric CO_2 (Saltonstall and Stevenson 2007; Holdredge *et al.* 2010; Mozdzer and Megonigal 2012; Tulbure *et al.* 2012; Mozdzer *et al.* 2013). Eutrophication is probably one of the primary drivers of the introduced lineage spreading rapidly in many wetland ecosystems. For instance, its abundance is correlated with shoreline development (King *et al.* 2007), a process that combines elevated nutrient availability, habitat modification and diminished salinity (Silliman and Bertness 2004). Introduced *Phragmites* is able to achieve particularly high rates of seedling establishment and growth in such environments, and also experiences higher rates of outcrossing (rather than self-pollination; McCormick *et al.* 2010). Because outcrossing is associated with greater seedling production, the availability of eutrophied environments is hypothesized to accelerate invasion dramatically (McCormick *et al.* 2010; Hazelton *et al.* 2014). In the context of rising atmospheric CO_2 and intensifying anthropogenic disturbance in wetland systems, information on how introduced *Phragmites* invests in tissue construction, and how it adjusts this investment in response to the environment, could be highly relevant in understanding the ecological processes driving the invasion, as well as in formulating strategies to manage it.

We sought to determine how CCs of plant organs in introduced and native *Phragmites* lineages would vary in response to alterations to CO_2, nitrogen (N) and the combination of these factors. We measured leaf, stem, rhizome and root CCs in greenhouse-grown plants, and compared organ-specific and whole-plant CCs with other functional traits related to growth and morphology. In keeping with prior observations of leaves in invasive species, we hypothesized that CCs of all organ types, as well as whole plants, would be lower for introduced vs. native *Phragmites*. Further, we hypothesized that the difference in CCs between lineages would increase when plants grew under levels of CO_2 or inorganic N expected in the coming century (Hopkinson and Giblin 2008; Meinshausen *et al.* 2011), with the greatest difference being in plants that experienced higher CO_2 and N simultaneously.

Methods

Phragmites australis plant material was originally collected from marshes on the Delmarva Peninsula, USA (38.5°N, 75.5°W); populations of native and introduced

Phragmites were sampled from stands that were located within 50 km of one another. Samples were genetically confirmed to belong to haplotypes F and M, which correspond to North American Atlantic coast native and Eurasian introduced lineages, respectively. Clones from this material were subsequently grown in a common garden at the University of Rhode Island, where they experienced identical abiotic conditions for 3 years (2006–09). We therefore attribute any differences in functional trait expressions between lineages from this experiment strictly to the genetic source. Plants for the experiment described herein were propagated from rhizome fragments at the Smithsonian Environmental Research Center in Edgewater, MD, USA in 2009, where the experiment also took place. Rhizome fragments contained 3–5 intact internodes, which was equivalent to 1.29 ± 0.07 and 1.10 ± 0.70 g (mean $\pm$ SE) dry mass for native and introduced lineages, respectively. Rhizomes were planted individually in plastic pots (15 L; $24 \times 24 \times 33$ cm) that contained reed-sedge peat (Baccto, Houston, TX, USA) on 11–12 June 2009.

The experiment had a three-way factorial design, which included two levels of atmospheric CO_2, two levels of soil N and the two *Phragmites* lineages. Plants from each lineage were randomly distributed among six transparent chambers, in which CO_2 was either not added or elevated to ~330 ppm above ambient air (Mozdzer and Megonigal 2012). This is a conservative estimate of rise in global mean CO_2 concentration by the latter part of the 21st century (Meinshausen *et al.* 2011). Plants were placed in chambers when new growth became visible at the soil surface; the first plant emerged on 19 June 2009. Within each chamber, half of the plants from each lineage received supplemental N at a rate equivalent to 25 g m^{-2} year^{-1}, while the remaining half were unfertilized. The higher N level is typical of those seen in eutrophied tidal marsh ecosystems (Hopkinson and Giblin 2008). Nitrogen was delivered bi-weekly via a solution of NH_4Cl. A sufficient quantity of tapwater to maintain at least 3 cm of standing water was added to each pot daily. To allow for water movement through the potting medium, four macropores were inserted vertically using PVC tubing (1.25 cm i.d.).

Plants were destructively harvested after ~2 months of exposure to treatment conditions (20–27 August 2009). Material from each individual ($N = 52$) was carefully separated into leaf, stem (culm plus leaf sheath), rhizome and root categories. All plant material was oven dried at 60 °C to constant mass, weighed and finely ground. Samples of ground tissue were analysed at the University of Virginia for elemental carbon and nitrogen content (Carlo Erba Instruments, NA2500, Milan, Italy). Tissue mineral content was determined via loss-on-ignition using a separate set of samples; ~0.5 g of each sample was ashed in a muffle furnace for 6 h at 550 °C.

Organ-specific CCs were determined using a method based on the production value of dry matter. Construction cost is defined specifically as the mass of glucose required to synthesize a given mass of plant tissue, but can be determined from the carbon (C_{dm}) and ash (Ash_{dm}) content of dried organic material as follows (Vertregt and Penning de Vries 1987):

$$CC = \frac{5.39C_{dm} + 0.80Ash_{dm} - 1191}{1000}.$$

While estimates of CC are more complicated when the N source available to plants includes NO_3 (Vertregt and Penning de Vries 1987; Poorter *et al.* 1997), NH_4 was the sole N source in this experiment. Further, very little of the NH_4 could have oxidized given that soils were constantly inundated; measurements of redox potential confirmed that soils were predominantly anaerobic (Mozdzer and Megonigal 2013). After calculating CCs for each organ type (CC_{org}, where org is alternately leaf, stem, rhizome or root), we determined the contribution of organ-specific CCs ($Contrib_{org}$) to plant-scale CCs (CC_{plant}) by weighting CC_{org} by the corresponding mass fraction (MF_{org}; organ mass per plant mass) and summing the contributions:

$$Contrib_{org} = CC_{org} \times MF_{org},$$
$$CC_{plant} = Contrib_{leaf} + Contrib_{stem} + Contrib_{root} + Contrib_{rhizome}.$$

Additional functional traits were measured for each plant. Relative growth rate (RGR) was based on the accumulation of dry biomass between planting (M_p) and harvest (M_h):

$$RGR = \frac{(\ln(M_h) - \ln(M_p))}{t},$$

where t is the number of days between emergence and harvest (mean $\pm$ SE: 58 ± 1 days). Masses at the time of planting (M_p) were determined from the fresh masses of rhizome fragments used to propagate plants; the water content of rhizome fragments that were not used in the study was used to estimate dry masses. Dry masses at harvest (M_h) were sums of leaf, stem, rhizome and root masses. Stem heights and diameters were calculated as the mean of all stems in individual pots, with diameters measured at the soil surface. Stem density was a count of the number of stems per pot. Specific leaf area was calculated as the ratio of the area of the leaf blade to the dry mass of the third-most apical, fully developed leaf. Leaf blade areas were measured with an LI-3000 leaf scanner (LI-COR Biosciences, Lincoln, NE, USA). Additional procedural details are provided elsewhere (Mozdzer and Megonigal 2012, 2013).

Differences among experimental factors (CO_2, N and lineage) with respect to CCs (plant scale and organ specific) were evaluated with ANOVA-type linear models in R version 3.0.2. Transformations to response variables (square root or natural log) were made if residuals were not normal and homoscedastic. Models initially contained terms for all main effects and interactions; when F statistics for individual terms (especially interactions) or the model itself were non-significant (using $\alpha = 0.05$), simpler models were sought by sequentially removing non-significant terms. If an interaction term was significant, all lower-order terms were retained regardless of significance. Tukey's honestly significant difference (HSD) tests were used to evaluate pairwise differences among means based on terms in the final models. We assessed the correlation (Pearson coefficient, ρ) between CCs and other functional traits using mean values for each lineage within each combination of treatments. Variables for which $\rho > 0.7$ were considered strongly correlated, as this level of correlation corresponds to ~50 % of the variation in CCs being explained by the functional traits in question (Sokal and Rohlf 1995).

Results

The influence of lineage and environmental manipulations on CCs was strongly organ specific. Aboveground, leaf CCs were influenced by both N and CO_2 treatment, but the magnitude of these effects depended on lineage (Fig. 1A, Table 1). Specifically, N fertilization induced an increase in leaf CC for native *Phragmites*, but this effect was independent of the CO_2 level. In contrast, introduced *Phragmites* only increased its leaf CC with fertilization if CO_2 was elevated as well. When averaging across environmental treatments, leaf CCs were similar for

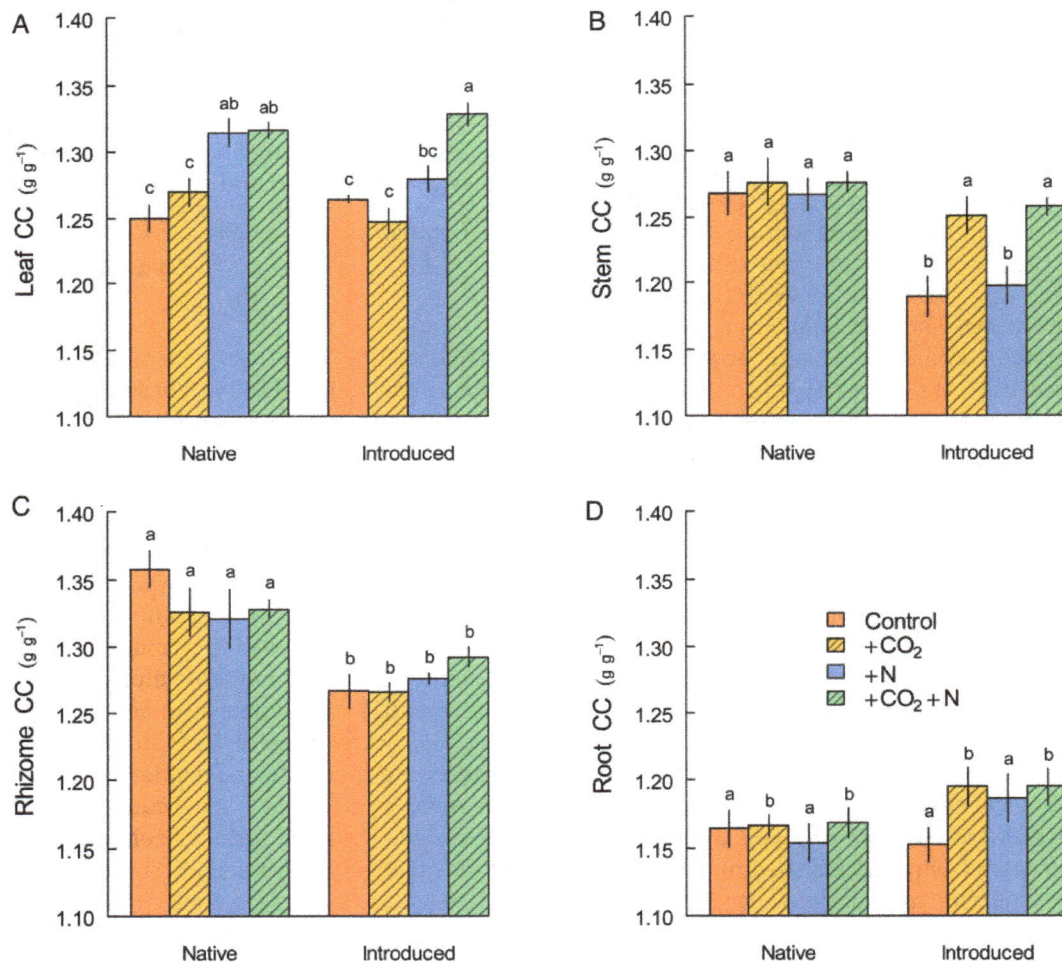

Figure 1. Organ-specific construction costs (CCs) for *Phragmites* lineages native to the North American North Atlantic coast ('Native') and introduced from Eurasia ('Introduced'). Bar heights represent mean ($\pm$ SE) CC for all plants grown in a combination of CO_2 and N fertilization treatments. Within each panel, lowercase letters above bars differ when Tukey's HSD tests for the best-fitting model identified statistically significant differences in means. Units are grams of glucose required per gram of biomass produced.

Table 1. Statistical results for the linear models best describing CCs as a function of *Phragmites* lineage and environmental treatments. Values in the top row for each model correspond to *F*-tests of each model as a whole, while the remaining values correspond to *F*-tests of individual terms.

Model	d.f.	F	P
Plant CC	4	16.07	<0.001
Lineage	1	31.99	<0.001
CO_2	1	13.05	<0.001
N	1	14.30	<0.001
Lineage × CO_2	1	4.96	0.031
Residuals	46		
Leaf CC	7	13.85	<0.001
Lineage	1	1.68	0.20
CO_2	1	6.98	0.011
N	1	74.43	<0.001
Lineage × CO_2	1	0.012	0.91
Lineage × N	1	0.059	0.81
CO_2 × N	1	2.67	0.11
Lineage × CO_2 × N	1	11.11	0.002
Residuals	43		
Stem CC	3	16.88	<0.001
Lineage	1	27.04	<0.001
CO_2	1	14.65	<0.001
Lineage × CO_2	1	8.95	0.004
Residuals	47		
Rhizome CC	1	36.35	<0.001
Lineage	1	36.35	<0.001
Residuals	49		
Root CC	2	3.91	0.027
Lineage	1	3.57	0.065
CO_2	1	4.24	0.045
Residuals	48		

the two lineages. Unlike leaves, the CC of stems was unaffected by N fertilization, but did differ by lineage and CO_2 status (Fig. 1B, Table 1). Specifically, under ambient CO_2 conditions, introduced *Phragmites* generated stems that had 5.8 % lower CCs than did native *Phragmites*.

The largest difference in CC between lineages was seen belowground, specifically in rhizomes. Rhizome CCs were 4.3 % lower for introduced *Phragmites* than for the native, and this difference was not significantly influenced by environmental treatments (Fig. 1C, Table 1). Root CCs were notably lower than they were for any other organ (Fig. 1D). Elevated CO_2 induced slight increases in root

CC for both lineages, while N fertilization had no measurable effect (Table 1). Although there was a trend towards higher root CC for introduced *Phragmites* compared with the native, this effect was not significant.

At the level of the whole plant, CCs differed by lineage and by environmental conditions. Introduced *Phragmites* had a lower mean CC than did the native (Table 1); the magnitude of this effect ranged from 0.6 to 3.3 % depending on the CO_2 and N treatment levels, and was 2.3 % for all treatments pooled (Fig. 2A). Native and introduced *Phragmites* also differed markedly in the size of contribution that each type of organ made to whole-plant CCs. Under all environmental conditions, introduced *Phragmites* had smaller rhizome and root contributions, but larger stem and leaf contributions compared with the native. Relative to unfertilized conditions, elevated nitrogen raised the contribution of belowground organs to whole-plant CCs in both lineages. These differences

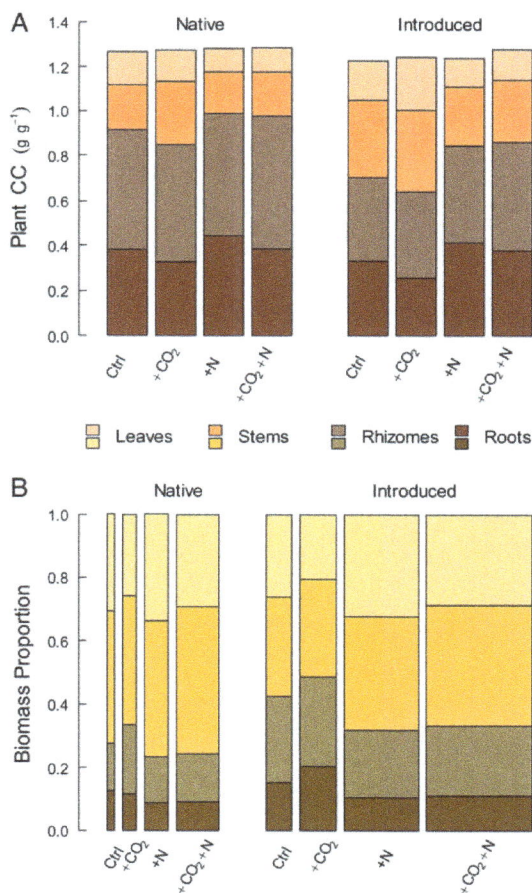

Figure 2. Partitioning of (A) plant-level CCs and (B) plant biomass by organ type for native and introduced *Phragmites* under each of the environmental treatment combinations evaluated in this experiment. Mean values for the replicate individuals within each treatment are shown. The widths of bars in (B) are scaled to the total biomass produced across treatments.

in aboveground vs. belowground contributions were driven by CCs and not biomass distributions, as organ mass fractions were higher belowground for the introduced lineage and under fertilized conditions (Fig. 2B). The addition of CO_2 also raised the contribution of aboveground organs to plant CCs over ambient conditions.

Across environmental treatments, higher leaf, rhizome and whole-plant CCs corresponded to introduced *Phragmites* plants growing faster, more densely and larger (taller, more massive and having wider stems; Table 2, Fig. S1 [see **Supporting Information**]). Correlation coefficients were consistently strongest for rhizome CCs in all of

Table 2. Correlation coefficients between whole-plant or organ-specific CCs and other functional traits. The underlying data were the means for each combination of environmental treatments (Control, $+CO_2$, $+N$ and $+CO_2+N$) within each lineage. Pearson coefficients >0.7 are shown; all coefficients and plots of the underlying data appear in Fig. S1 [see **Supporting Information**]. RGR, aboveground relative growth rate; Height, mean height of stems; Stem diam., mean basal diameter of stems; Biomass, total plant biomass; Density, number of stems per pot; SLA, specific leaf area; N:C, nitrogen to carbon ratio in tissue. All traits were measured at harvest except RGR.

	Plant CC	Leaf CC	Stem CC	Rhizome CC	Root CC
Introduced *Phragmites*					
RGR	0.86	0.87	–	0.94	–
Height	0.93	0.92	–	0.98	–
Stem diam.	0.88	0.97	–	1.00	–
Biomass	0.86	0.91	–	0.97	–
Density	0.79	0.84	–	0.91	–
SLA	−0.87	–	−0.98	–	–
$N:C_{lf}$	–	–	–	–	–
$N:C_{stem}$	–	–	–	–	–
$N:C_{rhiz}$	–	0.89	–	0.91	–
$N:C_{root}$	–	0.78	–	0.79	–
Native *Phragmites*					
RGR	0.99	0.96	–	−0.75	–
Height	0.73	–	0.88	−0.73	–
Stem diam.	0.92	0.80	–	−0.72	–
Biomass	0.94	0.89	–	–	–
Density	0.93	0.94	–	–	–
SLA	0.90	0.86	–	−0.98	–
$N:C_{lf}$	0.76	0.88	–	–	–
$N:C_{stem}$	0.87	0.93	–	–	–
$N:C_{rhiz}$	0.88	0.93	–	–	–
$N:C_{root}$	0.83	0.94	–	–	–

these relationships. Native *Phragmites* likewise grew more rapidly, more densely and larger as plant and leaf CCs increased. In contrast to the introduced lineage, rhizome CC in the native was oppositely, and generally more weakly, correlated to these and other traits than were leaf and whole-plant costs (Table 2). While SLA was negatively correlated with plant and stem CCs for the introduced lineage, it was positively correlated with plant and leaf CCs in the native (Table 2). Finally, for introduced *Phragmites*, leaf and rhizome CCs were strongly and positively associated with the N:C ratio of belowground organs, but only weakly associated with the N:C ratio of aboveground organs ($\rho \approx 0.6$). The native lineage had strong positive associations between leaf and plant CCs and the N:C ratio of all organ types (Table 2).

Discussion

Rhizome construction costs

This study demonstrates that CCs for organs not typically measured (rhizomes, roots and stems) can reveal patterns of plant adaptation well beyond those that can be gleaned from leaf CC alone. Most strikingly, our results identified key advantages in rhizome investment for introduced *Phragmites* over the native lineage that likely contribute to its invasion as a perennial, clonal grass. By maintaining lower CCs under all combinations of CO_2 and nitrogen, introduced *Phragmites* can recoup its investment in tissue construction more quickly (Poorter *et al.* 2006), enabling it to generate additional rhizome biomass and potentially other organs as well. Prior research supports this explanation; under multiple CO_2 and N conditions and in multiple studies, introduced *Phragmites* had greater absolute rhizome mass, higher rhizome mass fractions, higher ramet densities and greater leaf areas compared with the native (League *et al.* 2006; Holdredge *et al.* 2010; Mozdzer and Megonigal 2012).

We suggest that the lower rhizome CC of introduced *Phragmites* ultimately contributes to advantages in belowground dynamics that are known to facilitate its invasion in North American tidal marshes. More specifically, we suggest that lower rhizome CC and shorter payback times allow introduced *Phragmites* to build more extensive rhizome systems (e.g. greater biomass, as seen in this study, as well as greater total length) than it would if CCs were higher. Low rhizome CC may also yield thicker rhizomes (i.e. higher masses per unit length, which could come from greater diameters, as seen by Holdredge *et al.* (2010), and/or from more dense rhizome tissue). Given that clonal expansion occurs by stems emerging from laterally extending rhizomes (Amsberry *et al.* 2000), the favourable carbon economics of rhizome generation may facilitate higher rates of ramet production, spatially and/ or temporally, as reported for introduced vs. native lineages previously (Vasquez et al. 2005; League et al. 2006; Holdredge et al. 2010; Mozdzer and Megonigal 2012). By building and maintaining a network of stems that are connected by rhizomes, Phragmites clones can draw oxygen into belowground organs, aiding respiration and nutrient uptake (Brix et al. 1992; Vretare and Weisner 2000; Tulbure et al. 2012). Introduced Phragmites is able to induce $4\times$ the rate of airflow per unit of pressure differential and stand area than native Phragmites (Tulbure et al. 2012). This efficiency is due, in part, to the higher stem densities of its clones (Rolletschek et al. 1999; Tulbure et al. 2012). In addition, the ability of introduced Phragmites to tolerate substantially higher salinity than native Phragmites contributes to its ability to invade habitats that the native lineage is unable to colonize (Vasquez et al. 2005). Tolerance to relatively high salinity (≤ 0.40 M NaCl) has been attributed to larger rhizome sizes, greater rhizosphere oxygenation and more rapid clonal growth by the introduced lineage (Vretare and Weisner 2000; Bart and Hartman 2003; Vasquez et al. 2005).

The advantages that introduced Phragmites exhibits in connection with low rhizome CC and short payback times, compared with native Phragmites and likely other species, are magnified by its rapid photosynthetic rates. The photosynthetic capacity (A_{max}) of the introduced lineage has been measured as being 12–80 % higher than that of the native (Hansen et al. 2007; Mozdzer and Zieman 2010; Mozdzer et al. 2013). This translated into the introduced lineage producing more than twice as much rhizome biomass as native Phragmites in both field and greenhouse settings (Holdredge et al. 2010; Mozdzer and Zieman 2010; Mozdzer and Megonigal 2012). As past studies of photosynthetic traits did not manipulate CO_2 or N, ecophysiological data collected under predicted future conditions would be extremely valuable in assessing the carbon economy of Phragmites as global change intensifies.

Our results also suggest that introduced Phragmites may avoid a tradeoff in photosynthate allocation between rhizomes and leaves. This is supported by the fact that rhizome CCs were positively correlated with leaf CCs and metrics of plant size in the introduced lineage, but negatively correlated in the native lineage. An ability to make a large investment in rhizomes may lead to greater root production and nutrient acquisition rates for introduced Phragmites (Holdredge et al. 2010), positively feeding back to growth and tissue quality both aboveground and below. We suspect that the native lineage is not sufficiently productive to support the initial investment in rhizome biomass needed to make such feedback possible.

There are several possible changes in rhizome tissue composition that could contribute to the observed differences in CCs between lineages. One possibility is that introduced Phragmites invests in a lower proportion of energetically expensive compounds like lignins, proteins and phenolics in rhizomes. Because they have larger diameters (Holdredge et al. 2010), rhizomes of introduced Phragmites may require less structural support via lignification. As described above, synthesis of fewer expensive compounds in rhizome tissue would lead to lower longevity, but a faster payback time, and a more rapid growth rate (Poorter et al. 2006). It is also possible that introduced Phragmites incorporates a greater proportion of inexpensive compounds than the native, such as nonstructural carbohydrates or organic acids (Poorter et al. 2006). For instance, it may synthesize a larger surplus of starch via photosynthesis, much of which it may allocate belowground for immediate growth or storage (Granéli et al. 1992). Through this mechanism as well, introduced Phragmites would be able to achieve a rapid return on the energetic investment in rhizome tissue, facilitating its further growth.

Response to global change factors

In direct contrast to prior studies (Poorter et al. 1997; Wullschleger et al. 1997; Nagel et al. 2004, 2005), all statistically separable comparisons of mean CCs for ambient vs. elevated CO_2, as well as most of the non-significant comparisons, involved increases in CCs. However, the vast majority of past studies focused specifically on CCs of leaves. As seen in other studies that manipulated CO_2 (Poorter et al. 2012; Langley et al. 2013; Madhu and Hatfield 2013), belowground production increased under elevated CO_2 for both Phragmites lineages. The concomitant rise in root CCs may have been due to shifts in root morphology or architecture, such as larger diameters, higher tissue density or more frequent branching (Madhu and Hatfield 2013). Such shifts allow for increased nutrient uptake, soil penetration ability and resistance to pathogens and herbivores, but require increased synthesis of energetically expensive compounds like lignin and suberin (Vance et al. 1980; Soukup et al. 2002; Baxter et al. 2009). Consistent with this possibility, prior studies have found higher lignin concentrations in fine roots under elevated CO_2 (Booker et al. 2000; George et al. 2003). The strong correlation of stem CCs with plant height in native Phragmites raises the possibility that stems were also more lignified under elevated CO_2. Introduced Phragmites likewise exhibited a positive correlation between these factors, though it was only moderate in strength ($\rho = 0.58$).

Despite the literature's enormous emphasis on leaf CCs, we found no differences in leaf CC between lineages.

While both lineages adjusted leaf CC in response to nitrogen addition, the magnitude of response was similar. Higher N availability probably corresponded to a greater investment in rubisco and other compounds associated with photosynthetic capacity (Griffin *et al.* 1993; Poorter and Bongers 2006). Other studies have also found negative correlations between leaf CC and SLA (e.g. Feng *et al.* 2008), whereas we found a positive correlation. We attribute this discrepancy to the fact that most other studies describe variation among species grown under similar environmental conditions, while our analysis portrays phenotypic plasticity in leaf construction to strongly varying environmental conditions. If we had only investigated leaf CCs for these *Phragmites* lineages, we would have overlooked key differences belowground, and determined little about the carbon economy or differences in invasiveness between lineages.

Our findings on whole-plant CCs suggest that an ability to generate biomass with a relatively short return time on the energetic investment has facilitated introduced *Phragmites* colonizing wetlands in North America over the past century (Saltonstall 2002). Modest differences in CCs, like the 3.3 % difference seen in this study, have previously been linked with large differences in abundance (Nagel and Griffin 2001). In combination with its relatively high photosynthetic rates (Mozdzer and Zieman 2010; Mozdzer *et al.* 2013) and plastic nutrient use efficiency (Mozdzer and Megonigal 2012), introduced *Phragmites* has had an energetic advantage from its establishment to the present day that could have contributed to its invasiveness.

In contrast to our expectations, and unlike most performance metrics measured in introduced *Phragmites* under global change conditions (Holdredge *et al.* 2010; Mozdzer and Megonigal 2012; Eller *et al.* 2014), our plant-scale data suggest that advantages due to CC will diminish with rising atmospheric CO_2 and nutrient proliferation. If efficient tissue construction and short payback time are particularly strong components of introduced *Phragmites* invasiveness, as global change intensifies, the competitive dynamics of these lineages may shift such that introduced *Phragmites* is less able to dominate ecosystems. However, other factors may allow for a continued competitive advantage by introduced *Phragmites*, especially if the increased investment in tissues improves their performance. Such factors include photosynthetic capacity (Mozdzer and Zieman 2010), salinity tolerance (Vasquez *et al.* 2005), production of litter that suppresses competing plants (Holdredge and Bertness 2011) and a propensity to outcross and generate greater numbers of seedlings at eutrophied sites (McCormick *et al.* 2010). In addition, like the processes that are selecting for genotypes well adapted to eutrophied conditions (McCormick

et al. 2010), shifts in CO_2 and N may similarly select for more efficient tissue construction in populations of introduced *Phragmites*.

Conclusions

Leaf CCs alone do not provide an adequate representation of the energy required to produce biomass for *Phragmites*. Accounting for all major plant organs enabled us to identify key patterns in CCs, particularly belowground, that are likely associated with the invasive ability of the introduced lineage. In future studies attempting to address questions of plant carbon economy using CCs, we recommend that organs other than leaves be investigated, especially those belowground. In addition to gaining insight into invasion dynamics associated with rhizome and whole-plant CC patterns, these traits allowed us to identify responses to global change that are not well described in the literature. For instance, we observed greater root and stem CCs under elevated CO_2 and greater leaf CC under high N. Given the critical nature of understanding plant responses to global change, scientists should use the full array of tools available.

Sources of Funding

Funding for J.S.C. was provided by a Bucher-Jackson fellowship through Bryn Mawr College. T.J.M. was supported by a Smithsonian Institution fellowship at the time of the experiment. Additional financial support came from the National Science Foundation (award DEB-0950080), Maryland Sea Grant (award SA7528114-WW) and Bryn Mawr College.

Contributions by the Authors

T.J.M. designed the experiment, T.J.M. and C.N.W. collected data, and J.S.C. analysed the data. All authors contributed to writing, led by J.S.C.

Acknowledgements

The authors thank Patrick Megonigal, Rachel Hager, Tasnim Aziz, an anonymous reviewer and the editors of *AoB PLANTS* for valuable feedback on earlier versions of this manuscript. We also thank Laura Meyerson for providing the source plants used in the experiment.

Literature Cited

Amsberry L, Baker MA, Ewanchuk PJ, Bertness MD. 2000. Clonal integration and the expansion of *Phragmites australis*. *Ecological Applications* **10**:1110–1118.

Bart D, Hartman JM. 2003. The role of large rhizome dispersal and low salinity windows in the establishment of common reed, *Phragmites australis*, in salt marshes: new links to human activities. *Estuaries* **26**:436–443.

Baruch Z, Goldstein G. 1999. Leaf construction cost, nutrient concentration, and net CO_2 assimilation of native and invasive species in Hawaii. *Oecologia* **121**:183–192.

Baxter I, Hosmani PS, Rus A, Lahner B, Borevitz JO, Muthukumar B, Mickelbart MV, Schreiber L, Franke RB, Salt DE. 2009. Root suberin forms an extracellular barrier that affects water relations and mineral nutrition in Arabidopsis. *PLoS Genetics* **5**:e1000492.

Booker FL, Maier CA. 2001. Atmospheric carbon dioxide, irrigation, and fertilization effects on phenolic and nitrogen concentrations in loblolly pine (*Pinus taeda*) needles. *Tree Physiology* **21**: 609–616.

Booker FL, Shafer SR, Wei CM, Horton SJ. 2000. Carbon dioxide enrichment and nitrogen fertilization effects on cotton (*Gossypium hirsutum* L.) plant residue chemistry and decomposition. *Plant and Soil* **220**:89–98.

Brix H, Sorrell BK, Orr PT. 1992. Internal pressurization and convective gas-flow in some emergent fresh-water macrophytes. *Limnology and Oceanography* **37**:1420–1433.

Chambers RM, Meyerson LA, Saltonstall K. 1999. Expansion of *Phragmites australis* into tidal wetlands of North America. *Aquatic Botany* **64**:261–273.

Curtis PS, Wang XZ. 1998. A meta-analysis of elevated CO_2 effects on woody plant mass, form, and physiology. *Oecologia* **113**: 299–313.

Deng X, Ye WH, Feng HL, Yang QH, Cao HL, Hui KY, Zhang Y. 2004. Gas exchange characteristics of the invasive species *Mikania micrantha* and its indigenous congener *M. cordata* (Asteraceae) in South China. *Botanical Bulletin of Academia Sinica* **45**: 213–220.

Eller F, Lambertini C, Nguyen LX, Brix H. 2014. Increased invasive potential of non-native *Phragmites australis*: elevated CO_2 and temperature alleviate salinity effects on photosynthesis and growth. *Global Change Biology* **20**:531–543.

Feng Y-L, Fu G-L, Zheng Y-L. 2008. Specific leaf area relates to the differences in leaf construction cost, photosynthesis, nitrogen allocation, and use efficiencies between invasive and noninvasive alien congeners. *Planta* **228**:383–390.

George K, Norby RJ, Hamilton JG, DeLucia EH. 2003. Fine-root respiration in a loblolly pine and sweetgum forest growing in elevated CO_2. *New Phytologist* **160**:511–522.

Granéli W, Weisner SB, Sytsma M. 1992. Rhizome dynamics and resource storage in *Phragmites australis*. *Wetlands Ecology and Management* **1**:239–247.

Griffin K, Thomas R, Strain B. 1993. Effects of nitrogen supply and elevated carbon dioxide on construction cost in leaves of *Pinus taeda* (L.) seedlings. *Oecologia* **95**:575–580.

Hansen DL, Lambertini C, Jampeetong A, Brix H. 2007. Clone-specific differences in *Phragmites australis*: effects of ploidy level and geographic origin. *Aquatic Botany* **86**:269–279.

Hazelton ELG, Mozdzer TJ, Burdick DM, Kettenring KM, Whigham DF. 2014. *Phragmites australis* management in the United States: 40 years of methods and outcomes. *AoB PLANTS* **6**: plu001; doi: 10.1093/aobpla/plu001.

Holdredge C, Bertness MD. 2011. Litter legacy increases the competitive advantage of invasive *Phragmites australis* in New England wetlands. *Biological Invasions* **13**:423–433.

Holdredge C, Bertness MD, von Wettberg E, Silliman BR. 2010. Nutrient enrichment enhances hidden differences in phenotype to drive a cryptic plant invasion. *Oikos* **119**:1776–1784.

Hopkinson CS, Giblin AE. 2008. Nitrogen dynamics of coastal salt marshes. In: Capone DG, Bronk DA, Mulholland MR, Carpenter EJ, eds. *Nitrogen in the marine environment*. San Diego: Academic Press, 991–1036.

Kettenring KM, de Blois S, Hauber DP. 2012. Moving from a regional to a continental perspective of *Phragmites australis* invasion in North America. *AoB PLANTS* **2012**: pls040; doi:10.1093/aobpla/pls040.

King RS, Deluca WV, Whigham DF, Marra PP. 2007. Threshold effects of coastal urbanization on *Phragmites australis* (common reed) abundance and foliar nitrogen in Chesapeake Bay. *Estuaries and Coasts* **30**:469–481.

Kraus TEC, Zasoski RJ, Dahlgren RA. 2004. Fertility and pH effects on polyphenol and condensed tannin concentrations in foliage and roots. *Plant and Soil* **262**:95–109.

Lambertini C, Sorrell BK, Riis T, Olesen B, Brix H. 2012. Exploring the borders of European *Phragmites* within a cosmopolitan genus. *AoB PLANTS* **2012**: pls020; doi:10.1093/aobpla/pls020.

Langley JA, Mozdzer TJ, Shepard KA, Hagerty SB, Megonigal JP. 2013. Tidal marsh plant responses to elevated CO_2, nitrogen fertilization, and sea level rise. *Global Change Biology* **19**:1495–1503.

League MT, Colbert EP, Seliskar DM, Gallagher JL. 2006. Rhizome growth dynamics of native and exotic haplotypes of *Phragmites australis* (common reed). *Estuaries and Coasts* **29**:269–276.

Lei YB, Wang WB, Feng YL, Zheng YL, Gong HD. 2012. Synergistic interactions of CO_2 enrichment and nitrogen deposition promote growth and ecophysiological advantages of invading *Eupatorium adenophorum* in Southwest China. *Planta* **236**:1205–1213.

Madhu M, Hatfield JL. 2013. Dynamics of plant root growth under increased atmospheric carbon dioxide. *Agronomy Journal* **105**: 657–669.

Marks M, Lapin B, Randall J. 1994. *Phragmites australis* (*P. communis*): threats, management, and monitoring. *Natural Areas Journal* **14**: 285–294.

McCormick MK, Kettenring KM, Baron HM, Whigham DF. 2010. Spread of invasive *Phragmites australis* in estuaries with differing degrees of development: genetic patterns, Allee effects and interpretation. *Journal of Ecology* **98**:1369–1378.

Meinshausen M, Smith SJ, Calvin K, Daniel JS, Kainuma MLT, Lamarque J-F, Matsumoto K, Montzka SA, Raper SCB, Riahi K, Thomson A, Velders GJM, van Vuuren DPP. 2011. The RCP greenhouse gas concentrations and their extensions from 1765 to 2300. *Climatic Change* **109**:213–241.

Mozdzer TJ, Megonigal JP. 2012. Jack-and-Master trait responses to elevated CO_2 and N: a comparison of native and introduced *Phragmites australis*. *PLoS ONE* **7**:e42794.

Mozdzer TJ, Megonigal JP. 2013. Increased methane emissions by an introduced *Phragmites australis* lineage under global change. *Wetlands* **33**:609–615.

Mozdzer TJ, Zieman JC. 2010. Ecophysiological differences between genetic lineages facilitate the invasion of non-native *Phragmites australis* in North American Atlantic coast wetlands. *Journal of Ecology* **98**:451–458.

Mozdzer TJ, Zieman JC, McGlathery KJ. 2010. Nitrogen uptake by native and invasive temperate coastal macrophytes: importance of dissolved organic nitrogen. *Estuaries and Coasts* **33**:784–797.

Mozdzer TJ, Brisson J, Hazelton ELG. 2013. Physiological ecology and functional traits of North American native and Eurasian introduced *Phragmites australis* lineages. *AoB PLANTS* **5**: plt048; doi: 10.1093/aobpla/plt048.

Nagel JM, Griffin KL. 2001. Construction cost and invasive potential: comparing *Lythrum salicaria* (Lythraceae) with co-occurring native species along pond banks. *American Journal of Botany* **88**: 2252–2258.

Nagel JM, Huxman TE, Griffin KL, Smith SD. 2004. CO_2 enrichment reduces the energetic cost of biomass construction in an invasive desert grass. *Ecology* **85**:100–106.

Nagel JM, Wang X, Lewis JD, Fung HA, Tissue DT, Griffin KL. 2005. Atmospheric CO_2 enrichment alters energy assimilation, investment and allocation in *Xanthium strumarium*. *New Phytologist* **166**:513–523.

Osunkoya OO, Daud SD, Wimmer FL. 2008. Longevity, lignin content and construction cost of the assimilatory organs of *Nepenthes* species. *Annals of Botany* **102**:845–853.

Osunkoya OO, Bayliss D, Panetta FD, Vivian-Smith G. 2010. Leaf trait co-ordination in relation to construction cost, carbon gain and resource-use efficiency in exotic invasive and native woody vine species. *Annals of Botany* **106**:371–380.

Park MG, Blossey B. 2008. Importance of plant traits and herbivory for invasiveness of *Phragmites australis* (Poaceae). *American Journal of Botany* **95**:1557–1568.

Poorter H, Van Berkel Y, Baxter R, Den Hertog J, Dijkstra P, Gifford RM, Griffin KL, Roumet C, Roy J, Wong SC. 1997. The effect of elevated CO_2 on the chemical composition and construction costs of leaves of 27 C3 species. *Plant, Cell and Environment* **20**: 472–482.

Poorter H, Pepin S, Rijkers T, de Jong Y, Evans JR, Korner C. 2006. Construction costs, chemical composition and payback time of high- and low-irradiance leaves. *Journal of Experimental Botany* **57**: 355–371.

Poorter H, Niklas KJ, Reich PB, Oleksyn J, Poot P, Mommer L. 2012. Biomass allocation to leaves, stems and roots: meta-analyses of interspecific variation and environmental control. *New Phytologist* **193**:30–50.

Poorter L, Bongers F. 2006. Leaf traits are good predictors of plant performance across 53 rain forest species. *Ecology* **87**:1733–1743.

Rolletschek H, Hartzendorf T, Rolletschek A, Kohl JG. 1999. Biometric variation in *Phragmites australis* affecting convective ventilation and amino acid metabolism. *Aquatic Botany* **64**:291–302.

Saltonstall K. 2002. Cryptic invasion by a non-native genotype of the common reed, *Phragmites australis*, into North America. *Proceedings of the National Academy of Sciences of the USA* **99** 2445–2449.

Saltonstall K. 2007. Comparison of morphological variation indicative of ploidy level in *Phragmites australis* (Poaceae) from eastern North America. *Rhodora* **109**:415–429.

Saltonstall K, Stevenson JC. 2007. The effect of nutrients on seedling growth of native and introduced *Phragmites australis*. *Aquatic Botany* **86**:331–336.

Shen X-Y, Peng S-L, Chen B-M, Pang J-X, Chen L-Y, Xu H-M, Hou Y-P. 2011. Do higher resource capture ability and utilization efficiency facilitate the successful invasion of native plants? *Biological Invasions* **13**:869–881.

Silliman BR, Bertness MD. 2004. Shoreline development drives invasion of *Phragmites australis* and the loss of plant diversity on New England salt marshes. *Conservation Biology* **18**:1424–1434.

Sokal RR, Rohlf FJ. 1995. *Biometry: the principles and practice of statistics in biological research*. 3rd edn. New York: WH Freeman.

Soukup A, Votrubova O, Cizkova H. 2002. Development of anatomical structure of roots of *Phragmites australis*. *New Phytologist* **153**: 277–287.

Tulbure MG, Ghioca-Robrecht DM, Johnston CA, Whigham DF. 2012. Inventory and ventilation efficiency of nonnative and native *Phragmites australis* (common reed) in tidal wetlands of the Chesapeake Bay. *Estuaries and Coasts* **35**:1353–1359.

Vance C, Kirk T, Sherwood R. 1980. Lignification as a mechanism of disease resistance. *Annual Review of Phytopathology* **18**: 259–288.

Vasquez EA, Glenn EP, Brown JJ, Guntenspergen GR, Nelson SG. 2005. Salt tolerance underlies the cryptic invasion of North American salt marshes by an introduced haplotype of the common reed *Phragmites australis* (Poaceae). *Marine Ecology Progress Series* **298**:1–8.

Vertregt N, Penning de Vries FWT. 1987. A rapid method for determining the efficiency of biosynthesis of plant biomass. *Journal of Theoretical Biology* **128**:109–119.

Vretare V, Weisner SEB. 2000. Influence of pressurized ventilation on performance of an emergent macrophyte (*Phragmites australis*). *Journal of Ecology* **88**:978–987.

Wright IJ, Reich PB, Westoby M, Ackerly DD, Baruch Z, Bongers F, Cavender-Bares J, Chapin T, Cornelissen JH, Diemer M, Flexas J, Garnier E, Groom PK, Gulias J, Hikosaka K, Lamont BB, Lee T, Lee W, Lusk C, Midgley JJ, Navas ML, Niinemets U, Oleksyn J, Osada N, Poorter H, Poot P, Prior L, Pyankov VI, Roumet C, Thomas SC, Tjoelker MG, Veneklaas EJ, Villar R. 2004. The worldwide leaf economics spectrum. *Nature* **428**:821–827.

Wullschleger SD, Norby RJ, Love JC, Runck C. 1997. Energetic costs of tissue construction in yellow-poplar and white oak trees exposed to long-term CO_2 enrichment. *Annals of Botany* **80**:289–297.

Responses of sap flow, leaf gas exchange and growth of hybrid aspen to elevated atmospheric humidity under field conditions

Aigar Niglas[1], Priit Kupper[1], Arvo Tullus[1,2] and Arne Sellin[1]*

[1] Institute of Ecology and Earth Sciences, University of Tartu, Lai 40, 51005 Tartu, Estonia
[2] Institute of Forestry and Rural Engineering, Estonian University of Life Sciences, Kreutzwaldi 5, 51014 Tartu, Estonia

Associate Editor: Tim J. Brodribb

Abstract. An increase in average air temperature and frequency of rain events is predicted for higher latitudes by the end of the 21st century, accompanied by a probable rise in air humidity. We currently lack knowledge on how forest trees acclimate to rising air humidity in temperate climates. We analysed the leaf gas exchange, sap flow and growth characteristics of hybrid aspen (*Populus tremula* × *P. tremuloides*) trees growing at ambient and artificially elevated air humidity in an experimental forest plantation situated in the hemiboreal vegetation zone. Humidification manipulation did not affect the photosynthetic capacity of plants, but did affect stomatal responses: trees growing at elevated air humidity had higher stomatal conductance at saturating photosynthetically active radiation ($g_{s\ sat}$) and lower intrinsic water-use efficiency (IWUE). Reduced stomatal limitation of photosynthesis in trees grown at elevated air humidity allowed slightly higher net photosynthesis and relative current-year height increments than in trees at ambient air humidity. Tree responses suggest a mitigating effect of higher air humidity on trees under mild water stress. At the same time, trees at higher air humidity demonstrated a reduced sensitivity of IWUE to factors inducing stomatal closure and a steeper decline in canopy conductance in response to water deficit, implying higher dehydration risk. Despite the mitigating impact of increased air humidity under moderate drought, a future rise in atmospheric humidity at high latitudes may be disadvantageous for trees during weather extremes and represents a potential threat in hemiboreal forest ecosystems.

Keywords: Atmospheric humidity; canopy conductance; climate change; net photosynthesis; photosynthetic capacity; relative stomatal limitation; stomatal conductance; water-use efficiency.

* Corresponding author's e-mail address: arne.sellin@ut.ee

Introduction

With rapid increases in global industrial development, fossil fuel use and changing land-use practices, atmospheric CO_2 concentration ($[CO_2]$) is expected to double within the 21st century. This increase will result in global climate changes: global mean water vapour concentration, evaporation and precipitation rates, as well as global mean surface temperature are projected to increase during the 21st century (IPCC 2007). These changing climate factors along with rising $[CO_2]$ affect the physiological performance of plants: CO_2 assimilation, transpiration, stomatal conductance (g_s) and ultimately plant growth and productivity.

The impact of the most common consequences of climate change—drought, high temperature and high atmospheric vapour pressure deficit (VPD)—on photosynthesis and water use in C_3 plants has been quite well studied, because the occurrence of extreme temperatures, soil water deficit and high VPD, as well as their interactions, alters the physical properties and yield of plants, which are important to agriculture and forestry (Fletcher et al. 2007; Guha et al. 2010; Estrada-Campuzano et al. 2012; Kuster et al. 2013; Li et al. 2013; Sapeta et al. 2013). Considerably less is known of the effect of increasing atmospheric humidity on plants. Increases in precipitation are considered very likely at high latitudes in the long-term perspective (IPCC 2007). Precipitation is predicted to increase in northern Europe, especially in winter, and to decrease in southern and central Europe in summer (Räisänen et al. 2004). There might also be fewer dry days at higher latitudes by the end of the 21st century (IPCC 2007). Increasing rainfall frequency results in higher relative air humidity at local or regional scales.

The leaves of plants grown at high relative humidity (RH) have larger stomata, larger stomatal pore aperture and length, and significantly lower stomatal density due to larger epidermal cells than in plants grown at moderate RH (Torre et al. 2003; Nejad and Van Meeteren 2005; Arve et al. 2013). Therefore, decreasing VPD may lead to increased stomatal conductance and to a consequent increase in transpiration in some plant species grown at high RH (Pospíšilová 1996; Fordham et al. 2001; Nejad and Van Meeteren 2005). Nevertheless, most findings suggest that a decrease in VPD generally leads to decreased steady-state leaf transpiration or sap flux density in a wide range of tree species from different habitats (Pataki et al. 1998; Meinzer 2003; Bovard et al. 2005; Hölscher et al. 2005). Our previous studies have demonstrated decreased sap flux density in response to increased air humidity in silver birch (*Betula pendula*) and hybrid aspen (*Populus tremula* × *P. tremuloides*) trees in moist summers (Kupper et al. 2011).

Growing at high RH not only alters stomatal morphology, but stomatal functioning as well (Fanourakis et al. 2010, 2011). It is known that RH is a key environmental factor mediating changes in stomatal sensitivity to CO_2 (Talbott et al. 2003). Moreover, RH affects stomatal response to water availability and drought. Plants grown at high RH are less hydrosensitive than plants grown at moderate RH: stomata of high-RH-grown leaves are less sensitive to decreases in leaf water potential than moderate-RH-grown leaves, and the homogeneity, speed and degree of stomatal closure are less in high-RH-grown plants (Nejad and Van Meeteren 2005; Rezaei Nejad et al. 2006; Rezaei Nejad and Van Meeteren 2008). Therefore, plants developed under moderate RH are able to retain higher water status due to more efficient stomatal control. Arve et al. (2013) revealed that stomata developed under high RH respond to neither darkness nor drought, but remain open. Thus, high RH may even override the signals given by darkness. The stomata of plants growing in naturally waterlogged soil are also less sensitive to decreasing VPD than those of plants growing in well-drained soil (Sellin 2001).

High RH does not change only the stomatal characteristics of plants. Our previous experiments with silver birch and hybrid aspen have shown that elevated atmospheric RH lowers leaf nutritional status by altering nutrient movement via mass flow in soil and lowering nutrient transfer through xylem flow into leaves (Tullus et al. 2012a; Sellin et al. 2013). The changes in leaf nutrient content and P : N ratio in turn cause a decline in photosynthetic capacity and ultimately changes in tree growth rate.

Experiments on stomatal responses to air humidity and plant stress resistance are typically carried out in greenhouses or growth chambers with seedlings or saplings growing in pots. The objective of the present study was to investigate how artificially increased RH during leaf development affects the sap flow, stomatal responses and photosynthetic parameters of hybrid aspen (*P. tremula* × *P. tremuloides*) under free-air conditions. Hybrid aspen is a fast-growing deciduous tree species suitable for short-rotation forestry in the relatively cold climate of northern Europe (Tullus et al. 2012b). Our aim was to test the following hypotheses. (i) Trees grown at higher atmospheric humidity have higher stomatal conductance and lower water-use efficiency (WUE) than control trees. (ii) The photosynthetic capacity of leaves developed in humid air is lower because of reduced nitrogen uptake due to lower transpirational flux density. (iii) Plants grown in a more humid atmosphere are unable to adjust their WUE quickly because of acclimation to lower VPD or possible stomatal malfunction.

Methods

Study area and sample trees

Studies were performed on hybrid aspen (*P. tremula* × *P. tremuloides*) saplings growing in an experimental forest plantation at the free-air humidity manipulation (FAHM) site, situated at Rõka village (58°24′N, 27°29′E, 40–48 m ASL) in eastern Estonia, representing a hemiboreal vegetation zone. The long-term average annual precipitation in the region is 650 mm and the average temperature is 17.0 °C in July and −6.7 °C in January. In the study year (2011) drought conditions prevailed in June and July (Fig. 1). The growing season lasts 175–180 days from mid-April to October. The soil is a fertile endogenic mollic planosol (WRB) with an A-horizon thickness of 27 cm. Total nitrogen content is 0.11–0.14 %, C/N ratio is 11.4 and pH is 5.7–6.3.

The study site, established on an abandoned agricultural field in 2006–07, is a fenced area of 2.7 ha containing nine circular experimental plots (diameter 14 m) planted with hybrid aspen and silver birch (*B. pendula*) and surrounded by a buffer zone. One-year-old micropropagated hybrid aspen plants were planted in the experimental area in the autumn of 2006. The stand density in the buffer zone is 2500 trees ha^{-1}, and in the experimental plots, 10 000 trees ha^{-1}. The computer-operated FAHM system, based on an integrated approach of two different technologies—a misting technique to atomize/vaporize water and FACE-like technology to mix humidified air inside the plots—enables RH of the air to increase by up to 18 % over the ambient level during the humidification treatment, depending on the wind speed inside the experimental stand. The humidification is applied during daytime 6 days a week throughout the growing period if ambient RH is <75 % and mean wind speed is <4 m s^{-1}. As a long-term average, RH is 7–8 % greater in humidified plots (**H** treatment) than in control plots (**C** treatment). A detailed description of the FAHM site

and technical setup is presented in Kupper *et al.* (2011). The treatment began in June 2008; sap flow and gas exchange were measured in the summer months of 2011. Soil water potential (Ψ_S) was recorded at depths of 15 and 30 cm with EQ2 equitensiometers (Delta-T Devices, Burwell, UK) in eight replications per plot. The daily average Ψ_S varied from June to August and was ~25 % higher in the humidification treatment (Fig. 2). The air temperature (T_a) and RH were measured 1.5–3.5 m above the ground with 2–4 HMP45A sensors (Vaisala, Helsinki, Finland) per plot. Sensor readings were collected every 1 min and stored as 10-min average values with a data logger (DL2e; Delta-T Devices). Air VPD was calculated from T_a, saturated vapour pressure and RH. The daily average VPD in the humidification treatment was 15 % lower than the control in the summer of 2011 (Fig. 3).

Sap flow measurements

Xylem sap flow in the stems of sample trees was measured with FLOW4 sap flow systems (Dynamax Inc.,

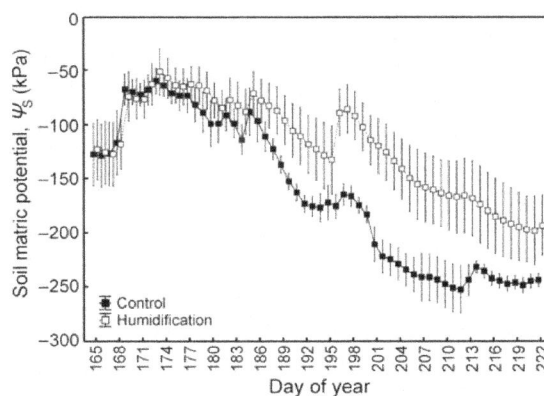

Figure 2. Daily average values of soil water potential at a depth of 15–30 cm in control and humidification plots from June to August in 2011. Scale bars denote SEM.

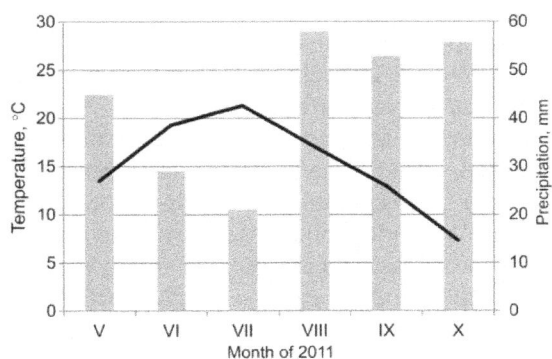

Figure 1. Weather data in the growing period of 2011: the dark line indicates monthly average air temperature, and the grey bars indicate monthly precipitation.

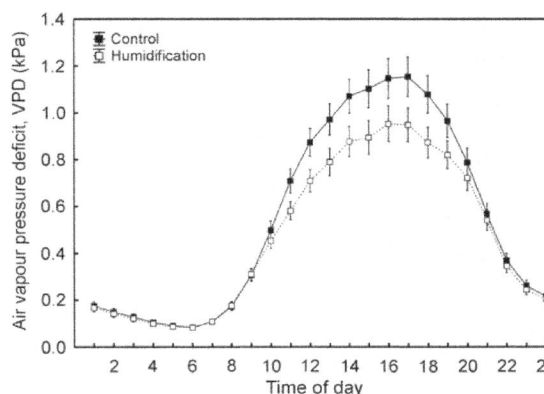

Figure 3. Hourly average values of air VPD in control and humidification plots from June to August in 2011. Scale bars denote SEM.

Houston, TX, USA). Six trees from control plots (C1, C2, C4) and four trees from humidification plots (H1, H2) were fitted with sap flow gauges (SGB35-WS) and sampled episodically from June to August 2011. Sap flow data were recorded every 1 min and stored as 30-min averages. Sap flux density (F; g m^{-2} h^{-1}) was calculated as sap flow divided by whole-tree foliage area estimated by mean sapwood-to-leaf-area ratios (Huber value, HV) measured in nine **C** (3.08×10^{-4} m^2 m^{-2}) and nine **H** (3.09×10^{-4} m^2 m^{-2}) trees using destructive sampling. Foliage area, measured with a LI 3100C optical area meter (LI-COR Biosciences, Lincoln, NE, USA), was on average 31 % greater in **C** trees compared with **H** trees. Whole-tree canopy conductance to water vapour (g_C; mmol m^{-2} s^{-1}) was computed from the data of sap flux density (mmol m^{-2} s^{-1}) using a simplified Penman–Monteith equation (Komatsu *et al.* 2006; Sellin and Lubenets 2010):

$$g_C = \frac{F \times P}{\text{VPD}}, \qquad (1)$$

where P is atmospheric pressure (kPa) and VPD is air vapour pressure deficit (kPa).

Gas exchange measurements

We sampled gasometrically nine trees (mean height 3.8 m) from **C** plots and nine trees (mean height 3.3 m) from **H** plots (i.e. three trees per sample plot) for 1 month, from mid-July to mid-August. Measurements were performed on rainless misting-free days on intact fully expanded leaves *in situ* with a portable photosynthesis system LCpro+ (ADC BioScientific, Great Amwell, UK) at constant air humidity (13 mbar), CO_2 concentration ($C_a = 360$ μmol mol^{-1}) and temperature of the leaf chamber (25 °C). Leaf-to-air vapour pressure difference was relatively similar in the two treatments: on average 2.12 kPa for **C** plants and 1.99 kPa for **H** plants. To generate photosynthetic light response curves (A/Q curves), four leaves per tree were sampled from the middle part of the crown with an instrument equipped with an LED light source. The measurements started with photosynthetically active radiation (PAR) at 1196 μmol m^{-2} s^{-1}, then decreased stepwise to 9 μmol m^{-2} s^{-1} and increased stepwise from 1196 to 1803 μmol m^{-2} s^{-1}. Intrinsic water-use efficiency, expressed as the ratio of net photosynthesis (A_n) to stomatal conductance to water vapour (g_s), was determined at two levels of irradiance: at 400–600 μmol m^{-2} s^{-1} when IWUE was usually at a maximum (IWUE$_{max}$) and at light intensities corresponding to full sunlight (IWUE$_{sat}$; $Q \geq 1400$ μmol m^{-2} s^{-1}).

The response of net photosynthesis to varying intercellular CO_2 concentration (C_i)—A/C_i curves—was also determined on intact leaves (four leaves per tree) *in situ* at constant air humidity (13 mbar), temperature of the leaf chamber (25 °C) and at saturating irradiance (1500 μmol m^{-2} s^{-1}). External CO_2 concentration (C_a) was supplied in 11 steps, decreasing from 360 to 60 μmol mol^{-1} and then increasing from 450 to 1600 μmol mol^{-1}. In addition to IWUE$_{max}$ and IWUE$_{sat}$ calculated from the data of A/Q curves, IWUE$_{in}$ (initial IWUE) was determined using initial values of the A/Q and A/C_i sequences when external [CO_2] was 360 μmol mol^{-1}.

Tree growth assessment

Tree height (H, cm) and stem diameter at 30-cm height (D, mm) of all aspen trees growing at three **C** and three **H** plots were measured before and after the 2011 growing season. H was measured with a telescopic Nedo mEssfix-S measuring rod (Nedo GmbH & Co.KG, Dornstetten, Germany) and stem diameter with a LIMIT digital caliper (Luna AB, Alingsås, Sweden). Current annual increment of the trees (ΔH, ΔD) was estimated as the difference between the two measurements. Relative increment (ΔH_{rel}, ΔD_{rel}) was expressed as the ratio of ΔH and ΔD to their respective characteristics at the beginning of the growing season. The ratio of $H : D$ was defined as tree slenderness (S).

Data analysis

Statistical data analysis was carried out using Statistica, Ver. 7.1 (StatSoft Inc., Tulsa, OK, USA). Repeated-measures analysis of variance (ANOVA) was used to compare the sap flux density (F) and canopy conductance to water vapour (g_C) between trees from the control and the misting treatment. The daily averages of F and g_C were analysed altogether on 31 days from 1000 to 1700 h from 14 June to 7 August 2011 (DOY: 165–176, 197–201, 206–219). Linear regression analysis was carried out to estimate relationships between F, g_C, VPD and Ψ_S. The normality of the regression residuals was checked using the Shapiro–Wilk test.

The gasometric data were analysed with Photosyn Assistant, Ver. 1.2 software (Dundee Scientific, Dundee, UK). The A/Q curves were fitted as a non-rectangular hyperbola expressed as a quadratic equation by Prioul and Chartier (1977). The initial slope of the curve expresses the apparent quantum efficiency (ϕ), whereas the X and Y axes intercepts, respectively, correspond to the light compensation point (Q_{comp}) and apparent dark respiration (R_d), and the upper asymptote approximates the light-saturated rate of photosynthesis (A_{max}). An additional parameter—convexity (θ)—is required to describe the rate of bending between the linear increase and the maximum value. Sub-stomatal cavity CO_2 concentration

(C_i) was calculated using the model of von Caemmerer and Farquhar (1981).

The A/C_i curves were analysed according to the biochemical model proposed by Farquhar et al. (1980), and subsequently modified by Harley and Sharkey (1991) and Harley et al. (1992). This model enables estimation of the CO_2 compensation point (Γ), the maximum rate of carboxylation by Rubisco ($V_{C\ max}$), the PAR-saturated rate of electron transport (J_{max}) and the rate of triose phosphate utilization (V_{TPU}), which indicates the availability of inorganic phosphorus for the Calvin cycle (Sharkey 1985). The relative stomatal limitation on photosynthesis

(L_S), an estimate of the proportion of the reduction in photosynthesis attributable to CO_2 diffusion between the atmosphere and intercellular space, was calculated from the A/C_i curves as follows (Farquhar and Sharkey 1982; Tissue et al. 2005; Huang et al. 2008):

$$L_S = \left(1 - \frac{A_n}{A_0}\right)100, \qquad (2)$$

where A_n is the net photosynthetic rate at normal C_a (360 μmol mol^{-1}) and A_0 is the photosynthetic rate when C_i ($= 360$ μmol mol^{-1}) equals C_a. Under these conditions, A_0 is the rate of photosynthesis that would occur if there were no diffusive limitation to CO_2 transfer through stomatal pores. The effect of humidification on gas exchange parameters was analysed by applying a nested analysis of variance with fixed factors of 'Treatment', 'Experimental plot' and 'Soil water potential' (a continuous variable), the second nested in the first. As plant physiological traits were more strongly related to the soil water potential measured at 30-cm depth (Ψ_{30}), we used this parameter as an index of soil water status. Because of drought development during the measurement period, we divided the datasets of both treatments into two groups according to Ψ_{30} (< -204 kPa for drier

Table 1. Soil water potential (kPa) estimates of the experimental plots: Ψ_{S_mean}, mean across the growing season; Ψ_{S_Q25}, lower quartile; Ψ_{S_Q75}, upper quartile.

Plot	Ψ_{S_mean}	Ψ_{S_Q25}	Ψ_{S_Q75}
C1	−197	−240	−152
C2	−191	−217	−177
C4	−196	−217	−185
H1	−56	−76	−33
H2	−194	−221	−175
H4	−124	−151	−97

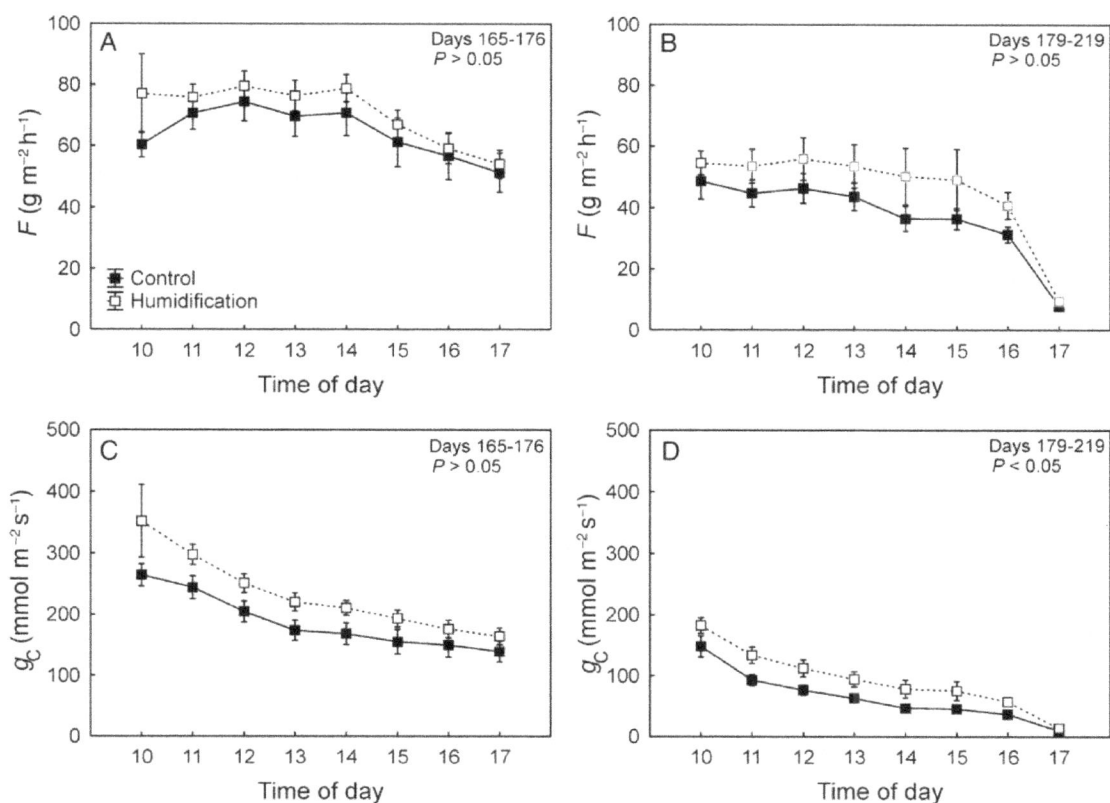

Figure 4. Daily average values of sap flux density (F) and canopy conductance to water vapour (g_C) in control and humidification plots during mist fumigation from June 14 to June 25 (DOY: 165–176; A and C) and July 15 to August 7 (DOY: 197–219; B and D), 2011. Scale bars denote SEM.

soil and ≥ -204 kPa for moister soil in **C** plots; < -163 and ≥ 163 kPa in **H** plots, respectively) and analysed gas exchange data also separately for these conditions.

Student's t-test was applied to estimate the treatment effect on the growth characteristics of individual trees across all experimental plots. Analysis of variance models

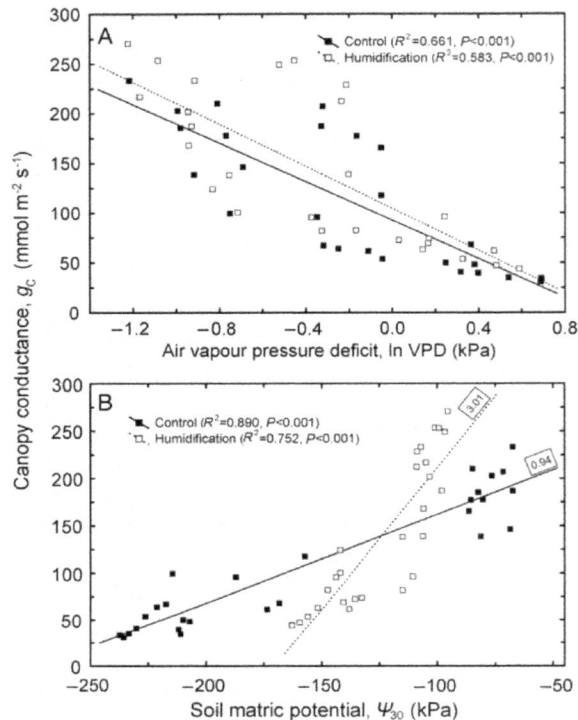

Figure 5. Variation in daily average canopy conductance to water vapour (g_C) depending on atmospheric VPD (A) and bulk soil water potential at a depth of 30 cm (Ψ_{30}; B). The numbers by the regression lines indicate the respective slopes.

were used to study the effects of 'Treatment' and 'Experimental plot' (nested in treatment) or 'Treatment' and 'Soil water potential' as a continuous covariate on the growth characteristics. Means and upper and lower quartiles of daily average soil water potentials (Ψ_{S_mean}, Ψ_{S_Q25}, Ψ_{S_Q75}) across the growing season were used as covariates in separate models (Table 1). When exploring the variance of total and relative growth increment in 2011, tree size (H or D at the end of the previous growing season) was included as a covariate. Type IV sums of squares were used in the calculations; post hoc mean comparisons were conducted using Tukey's HSD test.

Results

Sap flux density and canopy conductance

Although canopy conductance (g_C) was significantly higher (22 %; $P < 0.05$) under humidification across the whole study period, the difference between the treatments was statistically insignificant for days 165–176 (Fig. 4C) when the soil water potential did not differ between the **C** and **H** plots (Fig. 2). Also the sap flux density in the **H** treatment was on average 13 % higher than in the **C** treatment, although the difference was not significant (Fig. 4A and B). g_C decreased with increasing VPD ($P < 0.001$) in both the **C** and **H** plots; the response patterns were completely coincident and the slopes of the respective regression lines did not differ between the treatments (Fig. 5A). g_C also decreased with decreasing Ψ_S ($P < 0.001$), while the treatments demonstrated contrasting sensitivities (dg_C/dΨ_S) to developing soil water deficit—the corresponding slopes were 0.94 and 3.01 for control and humidified trees, respectively (Fig. 5B).

Table 2. Leaf gas exchange characteristics of hybrid aspen growing under control and humidification treatment. Each value is the mean ± SE; the means are compared with Tukey's test. NS, not statistically significant.

Parameter	Treatment		Significance level (P)
	Control ± SE	Humidification ± SE	
A_n (μmol m^{-2} s^{-1})	10.5 ± 0.4	11.3 ± 0.4	NS
g_s (mol m^{-2} s^{-1})	0.22 ± 0.01	0.28 ± 0.01	0.002
$g_{s\,sat}$ (mol m^{-2} s^{-1})	0.19 ± 0.02	0.25 ± 0.02	0.040
IWUE$_{in}$ (μmol mol^{-1})	53.4 ± 1.5	45.1 ± 1.5	<0.001
IWUE$_{max}$ (μmol mol^{-1})	62.3 ± 2.29	54.4 ± 2.19	0.015
IWUE$_{sat}$ (μmol mol^{-1})	56.8 ± 1.88	48.81 ± 2.19	0.008
L_S (%)	41.3 ± 1.01	37.7 ± 1.01	0.016
A_{max} (μmol m^{-2} s^{-1})	12.9 ± 0.6	12.9 ± 0.6	NS
$V_{C\,max}$ (μmol m^{-2} s^{-1})	56.5 ± 2.9	59.9 ± 2.7	NS
J_{max} (μmol m^{-2} s^{-1})	173 ± 10	196 ± 11	NS

Leaf gas exchange

Average net photosynthesis (A_n) tended to be slightly greater in trees growing at elevated atmospheric humidity than those grown at ambient RH, although the treatment means did not differ statistically throughout the experiment (Table 2). Analysis of variance revealed that the humidity treatment affected stomatal response, but not leaf photosynthetic traits (Tables 2 and 3). Specifically, there were significant differences in means of stomatal conductance to water vapour measured at saturating PAR ($g_{s\ sat}$) and IWUE between the treatments: $g_{s\ sat}$ was 32 % higher and $IWUE_{in}$ 16 % lower in the **H** treatment than in **C** trees (Table 2).

The data analysis revealed that soil water availability affected the gas exchange parameters differently within the treatments. A_n and $g_{s\ sat}$ in the **H** treatment were significantly greater under moist soil conditions (12.45 μmol m^{-2} s^{-1} and 0.300 mol m^{-2} s^{-1}, respectively) than under drier conditions (9.78 μmol m^{-2} s^{-1} and 0.192 mol m^{-2} s^{-1}, respectively; Fig. 6A and B). Initial

Table 3. Effects of treatment, plot and soil water potential at a depth of 30 cm (Ψ_{30}) on gas exchange characteristics. NS, not statistically significant.

Characteristic	Factor	Significance level (P)
A_n (μmol m^{-2} s^{-1})	Treatment	NS
	Plot (nested in treatment)	<0.001
	Ψ_{30}	0.013
g_s (mol m^{-2} s^{-1})	Treatment	NS
	Plot (nested in treatment)	0.002
	Ψ_{30}	NS
$g_{s\ sat}$ (mol m^{-2} s^{-1})	Treatment	0.027
	Plot (nested in treatment)	0.022
	Ψ_{30}	NS
$IWUE_{in}$ (μmol mol^{-1})	Treatment	0.013
	Plot (nested in treatment)	0.002
	Ψ_{30}	NS
$IWUE_{max}$ (μmol mol^{-1})	Treatment	<0.001
	Plot (nested in treatment)	0.005
	Ψ_{30}	0.003
$IWUE_{sat}$ (μmol mol^{-1})	Treatment	<0.001
	Plot (nested in treatment)	0.012
	Ψ_{30}	0.007
A_{max}	Treatment	NS
	Plot (nested in treatment)	NS
	Ψ_{30}	NS
$V_{C\ max}$ (μmol m^{-2} s^{-1})	Treatment	NS
	Plot (nested in treatment)	NS
	Ψ_{30}	0.045
J_{max} (μmol m^{-2} s^{-1})	Treatment	NS
	Plot (nested in treatment)	NS
	Ψ_{30}	0.007
L_s (%)	Treatment	NS
	Plot (nested in treatment)	<0.001
	Ψ_{30}	0.015

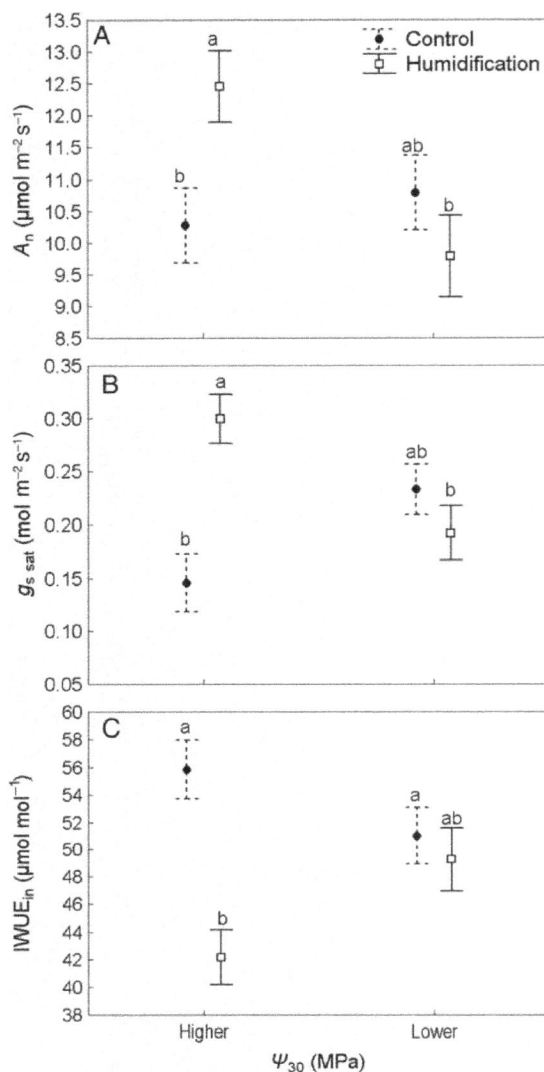

Figure 6. Means of net photosynthesis (A_n; A), stomatal conductance to water vapour at saturating PAR ($g_{s\ sat}$; B) and intrinsic water-use efficiency ($IWUE_{in}$; C) of control (closed circles) and humidified trees (open squares) depending on soil water status. Values are means $\pm$ SE; different letters denote statistically significant ($P < 0.05$) differences.

intrinsic water-use efficiency also differed with respect to soil conditions: it was lower under moist soil conditions than under drier conditions (44.2 and 49.3 $\mu mol\ mol^{-1}$, respectively). There were no differences in A_n, $g_{s\ sat}$ and $IWUE_{in}$ with respect to soil moisture in **C** plots (Fig. 6). It is important to notice that g_s, $g_{s\ sat}$, $IWUE_{in}$ and A_{max} did not depend on Ψ_{30} (as a continuous variable) across the whole dataset (Table 3).

Photosynthesis was strongly associated with g_s in both treatments: $R^2 = 0.84$, $P < 0.001$ in **C** plots and $R^2 = 0.79$, $P < 0.001$ in **H** plots. There was an inverse linear relationship between $IWUE_{in}$ and C_i/C_a ($R^2 = 0.73$, $P < 0.001$),

while the slopes of the corresponding regressions did not differ between the treatments ($P > 0.05$). A_n increased with rising CO_2 concentration (C_a), with significantly ($P < 0.001$) steeper response in the **H** treatment ($\beta = 53.5$; $R^2 = 0.82$, $P < 0.001$) than in the control ($\beta = 45.5$; $R^2 = 0.74$, $P < 0.001$). There were no differences in the $A_n = f(C_i)$ slopes between the treatments. g_s decreased with increasing C_a, but the responses did not differ between the treatments. As a consequence, $IWUE_{in}$ rose with C_a; the slope for control trees was greater than that for humidified trees ($P < 0.001$; Fig. 7).

The maximum rate of carboxylation by Rubisco ($V_{C\ max}$) and the maximum rate of electron transport (J_{max}) did not differ between the treatments across the whole dataset; ANOVA revealed only an effect of soil water status on these parameters (Table 3). When the data were analysed separately in two groups (moist versus dry soil conditions), significant differences between the means of $V_{C\ max}$ and J_{max} became evident only for the humidification treatment—both parameters were higher ($P < 0.001$ for both parameters) in moist soil. No variation with the soil conditions was detected in $V_{C\ max}$ and J_{max} in the control trees (Fig. 8A and B). Regression analysis revealed a positive relationship between $V_{C\ max}$ ($R^2 = 0.168$, $P < 0.001$) and J_{max} ($R^2 = 0.151$, $P < 0.01$) and Ψ_{30}, as well as between $V_{C\ max}$ and J_{max} ($R^2 = 0.85$, $P < 0.001$) across both treatments.

The mean values of relative stomatal limitation of photosynthesis (L_S) were lower in trees grown at elevated RH than in **C** trees—37.9 and 41.3 %, respectively ($P < 0.05$; Table 2), although ANOVA did not establish any significant effect of the treatment (Table 3). Net photosynthesis was negatively correlated with L_S in control trees ($R^2 = 0.15$, $P = 0.03$), but the relationship lacked in the humidification treatment ($P = 0.23$). We found no differences in L_S with respect to soil water status in any treatment separately (Fig. 8C).

Impact on growth rate

Saplings of hybrid aspen growing in **H** plots were significantly shorter and had narrower stems (Table 4), regardless of whether sample plot or soil water potential was included as confounding factors in the models (Table 5). The absolute and relative growth increments in 2011 were either unaffected by treatment or significantly greater in **H** plots (Tables 4 and 5). This was more pronounced when Ψ_{S_mean} or Ψ_{S_Q25} was used as a covariate in ANOVA models, although using Ψ_{S_Q25} yielded slightly better approximations than the two other soil water potential estimates (Table 5). Slenderness (S) was unaffected by treatment, but varied significantly among the experimental plots.

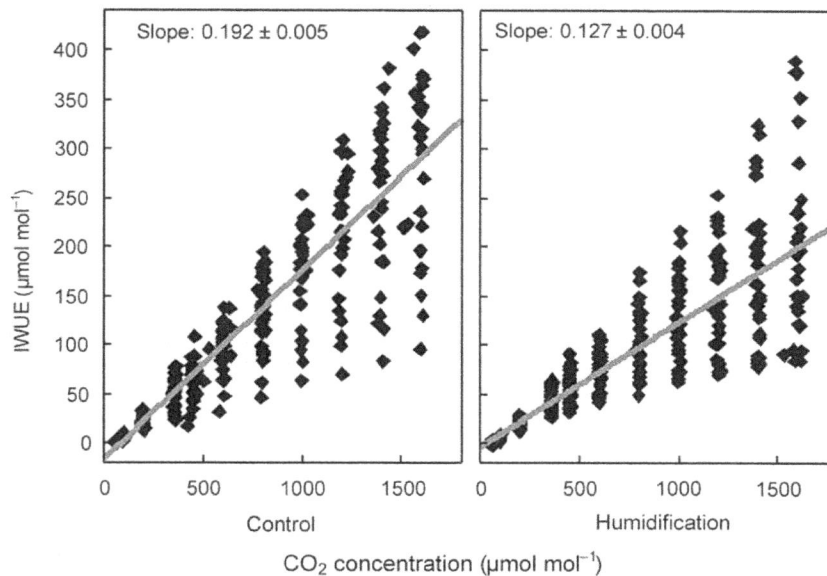

Figure 7. Photosynthetic WUE versus external CO_2 concentration in control and humidification treatment.

Discussion

Effects on sap flow and gas exchange

The sap flux density in hybrid aspen trees changed considerably compared with that at the same experimental site in previous summers (Kupper et al. 2011; Tullus et al. 2012a). F in control plots was significantly greater than that in the humidification treatment during the rainy summer of 2009 (Kupper et al. 2011). The same tendency (not significant) was observed in the drier summer of 2010 (Tullus et al. 2012a). However, our current results demonstrate higher (although not statistically significant) sap flux densities in trees growing in the humidification plots (Fig. 4). The canopy conductance to water vapour in the **H** treatment was greater ($P < 0.05$) across the whole study period. This discrepancy is attributable to relatively low soil water potential in the control treatment owing to the very dry summer: Ψ_S did not rise over -80 kPa in July. The total precipitation during the 2011 growing season (May–October) was 261 mm (Fig. 1), 42 % less than the average of the three previous years (452 mm). As the air humidity manipulation did not affect sapwood-to-leaf-area ratio (HV), the $\sim$30 % greater leaf area of control trees was responsible for greater transpirational water losses causing faster depletion of soil water reserves and a greater decline in Ψ_S despite lower overall sap flux densities in **C** plots (Fig. 4A and B).

The response of canopy conductance to changes in VPD did not vary between the treatments, suggesting that stomatal sensitivity to atmospheric evaporative demand was not affected by the experimental manipulation. In contrast to that, g_C decreased much faster in response to falling Ψ_S in the **H** treatment than in the control (Fig. 5). The differential response of g_C to decreasing soil water availability is probably mediated by plant hydraulic conductance (Cohen and Naor 2002; Domec et al. 2009). Hydraulic measurements performed on aspen trees in 2010 revealed that both soil-to-branch and leaf hydraulic conductances expressed per unit leaf area were smaller in humidified trees, although growing in moister soil (A. Sellin, unpubl. res.). Under conditions of soil water deficit the lower hydraulic capacity probably becomes a crucial factor for **H** trees, limiting leaf water supply and inducing a steep decline in canopy conductance. Furthermore, an experiment with silver birch revealed that a rapid water deficit in **H** plants led to a faster decrease in hydraulic conductance—responsible for liquid water supply—compared with the decrease in g_s, which limits water losses, and exposed plants to a greater risk of dehydration (Sellin et al. 2014).

Lowering A_n and g_s and increasing IWUE are typical responses to water stress in plant species with a drought avoidance strategy. When plants encounter a soil water deficit, abscisic acid (ABA) is synthesized in the roots and translocated to the leaf through the transpiration stream (Assmann and Shimazaki 1999); higher concentrations of ABA in leaves drive mechanisms leading to a decrease in g_s and an increase in WUE (Liu et al. 2005). Pantin et al. (2013) propose that ABA promotes stomatal closure in two ways—via its widely known biochemical effect on guard cells and via an indirect hydraulic effect through a decrease in leaf hydraulic conductance. Maintaining stable gas exchange attributes during drought development means that a plant either possesses a drought

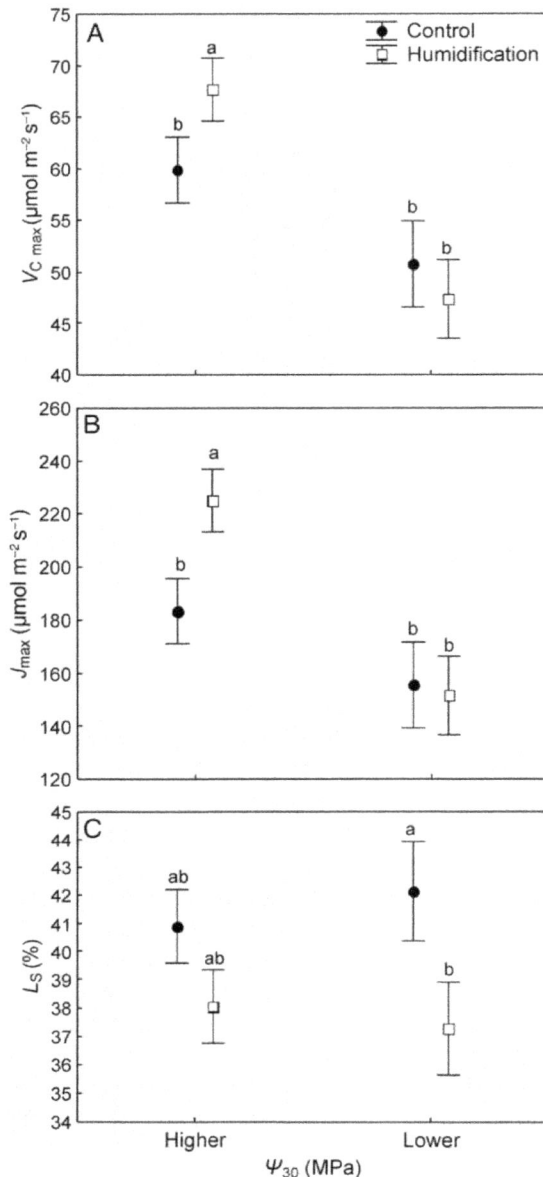

Figure 8. Means of maximum rate of carboxylation by Rubisco ($V_{C\ max}$; A), PAR-saturated rate of electron transport (J_{max}; B) and relative stomatal limitation to photosynthesis (L_S; C) of control (closed circles) and humidified trees (open squares) depending on soil water status. Values are means $\pm$ SE; different letters denote statistically significant ($P < 0.05$) differences.

tolerance strategy or lacks adaptations with respect to drought. *Populus tremula*, one of the parent species, is known to act as a drought avoider (Possen *et al.* 2011). In our case, the lack of variation in gas exchange characteristics in **C** plots with respect to soil water status (Fig. 6) and modification of gas exchange in **H** plots by altering water availability can be explained by the lower Ψ_S values in the **C** treatment (i.e. long-term effects mediated by ABA).

Regardless of the difference in Ψ_S between the treatments, there was still an effect of the humidity manipulation on $g_{s\ sat}$ and IWUE (Table 3). In fact, the differences in leaf gas exchange between the **C** and **H** plots are attributable to the combined effects of soil water availability and increased atmospheric humidity. As such, our first hypothesis is supported by the experiment: growing at higher RH increases stomatal conductance in trees while lowering photosynthetic WUE (Table 2), while the effect is largely mediated by changes in soil water status.

Soil water potential influenced both $V_{C\ max}$ and J_{max} in hybrid aspen (Table 3). Grassi *et al.* (2005) found a positive relationship between $V_{C\ max}$ and Ψ_S in oak and ash trees during summer, as in hybrid aspen in this study. A simulation by Grassi and Magnani (2005) indicated that 30–40 % of the biochemical limitation could be attributed to a reduction in leaf nitrogen content during droughty summers. Plant photosynthetic capacity and leaf N content expressed per leaf area are positively correlated (Grassi *et al.* 2005; van de Weg *et al.* 2012). Previous work at the FAHM experimental site has shown that rising RH reduces the water flux in silver birch (Kupper *et al.* 2011) and alters the nutritional status of leaves, leading to a decline in photosynthetic capacity (Sellin *et al.* 2013). In this experiment, humidification increased rather than decreased water flux through the trees (Fig. 4), which explains why the biochemical capacity of photosynthesis was unaffected by the manipulation, but by changes in Ψ_S (Table 3). The second hypothesis concerning reduced photosynthetic capacity of leaves due to the expectedly smaller N uptake in humidified trees remained unconfirmed. The absence of an impact of air humidity manipulation on photosynthetic machinery of hybrid aspen was also supported by chlorophyll fluorescence measurements performed in the droughty summer of 2012 (A. Niglas, unpubl. res.).

High RH can affect stomatal sensitivity by changing stomatal morphology: plants grown at higher RH have larger stomata that close to a lesser extent when leaves dry (Giday *et al.* 2013). In addition, long-term acclimation to high RH during growth increases heterogeneity in stomatal response characteristics to short-term exposure to stomatal closure-inducing factors (Nejad and Van Meeteren 2005). Experiments showing differences in stomatal sensitivity and morphology between plants grown at high and low RH have been carried out under stable/controlled environmental conditions. The conditions before a leaf is fully expanded are important determinants on whether stomatal closure capacity is affected by leaf dehydration and RH. Moreover, the degree of stomatal adaptation in expanding leaves depends on the duration and timing of the exposure to high RH (Fanourakis *et al.* 2011). The present study was performed under field

Table 4. Comparison (t-test) of mean ($\pm$ SE) growth parameters of individual trees growing in humidified (**H**) and control (**C**) plots. NS, not statistically significant.

Growth characteristic	C	H	t-stat	P
Height (cm)	454 ± 5.8	428 ± 5.8	3.06	0.002
Diameter of stem at 30-cm height (mm)	32.6 ± 0.7	29.5 ± 0.6	3.39	<0.001
Height increment of the current year (cm year^{-1})	111 ± 2.3	117 ± 2.7	-1.70	NS
Relative height increment of the current year	0.34 ± 0.01	0.39 ± 0.01	-3.46	<0.001
Diameter increment of the current year (mm year^{-1})	8.9 ± 0.3	8.5 ± 0.2	1.14	NS
Relative diameter increment of the current year	0.38 ± 0.01	0.41 ± 0.01	-1.65	NS
Slenderness (height-to-diameter ratio)	14.4 ± 0.2	14.8 ± 0.2	-1.97	0.051

Table 5. Results from ANOVA models describing the effect of tree size (i.e. the value of the respective parameter before the start of the growing season, C_{t-1}), treatment (T), plot and soil water potential (Ψ_{S_mean}, Ψ_{S_Q25}, Ψ_{S_Q75}) on the growth parameters.

Factors		Response variables						
		H	D	ΔH	ΔH_{rel}	ΔD	ΔD_{rel}	S
Model 1								
C_{t-1}	P	–	–	0.824	<0.001	<0.001	0.766	–
T	P	<0.001	<0.001	0.074	0.247	0.069	0.086	0.051
Plot (T)	P	<0.001	<0.001	<0.001	<0.001	<0.001	<0.001	0.015
	Adj. R^2	0.29	0.13	0.09	0.37	0.37	0.16	0.03
Model 2								
C_{t-1}	P	–	–	0.802	<0.001	<0.001	0.006	–
T	P	<0.001	<0.001	0.019	0.013	0.041	0.017	0.668
Ψ_{S_mean}	P	<0.001	0.157	<0.001	<0.001	<0.001	<0.001	0.005
	Adj. R^2	0.08	0.03	0.09	0.36	0.32	0.09	0.03
Model 3								
C_{t-1}	P	–	–	0.880	<0.001	<0.001	0.004	–
T	P	<0.001	<0.001	0.013	<0.001	0.049	0.018	0.800
Ψ_{S_Q25}	P	<0.001	0.081	<0.001	<0.001	<0.001	<0.001	0.010
	Adj. R^2	0.09	0.04	0.10	0.37	0.32	0.09	0.03
Model 4								
C_{t-1}	P	–	–	0.684	<0.001	<0.001	0.008	–
T	P	<0.001	0.002	0.083	0.077	0.075	0.052	0.622
Ψ_{S_Q75}	P	<0.001	0.351	<0.001	<0.001	<0.001	<0.001	0.002
	Adj. R^2	0.07	0.03	0.07	0.35	0. 32	0.08	0.04

conditions with natural diurnal fluctuations of RH; misting was applied when ambient RH was <75 % and could be increased to as much as 18 % (versus 60 and 95 % of RH in Fanourakis *et al.* 2011). Our data suggest that stomatal sensitivity to atmospheric VPD remained unaffected in saplings of hybrid aspen. Although we did not explore stomatal dimensions, we presume that differences in morphology and the putative morphological effect on stomatal sensitivity were rather minor as our trees grew *in natura*, in both diurnally and seasonally variable environments, under conditions requiring flexible stomatal adjustment.

The findings that stomatal conductance decreases and photosynthesis increases with rising external CO_2 level are well-known phenomena (reviewed by Araújo et al. 2011). A steeper A_n response to C_a in **H** plots is attributable to higher stomatal conductance (evidenced by $g_{s\ max}$) due to leaf development under lower VPD (Table 2). This is indirectly confirmed also by the negative correlation between A_n and L_S in the **C** treatment.

Initial intrinsic water-use efficiency responded more sensitively to C_a in **C** plots than in **H** plots (Fig. 7), testifying once more to the effect of elevated atmospheric humidity on leaf gas exchange. Because high IWUE is advantageous to plants under drought conditions, a slower response in high-RH-grown trees to changing ambient conditions may be disadvantageous in the case of abrupt climatic fluctuations becoming more frequent in the future (Easterling et al. 2000); these plants are not able to adjust their water use as quickly as plants grown in drier air and experience greater water loss. Thus, our results support the third hypothesis on the capacity of plants to modify WUE under changing environmental conditions, albeit not directly tested with respect to air humidity.

Consequences on tree growth

The above-ground growth response of aspen trees to humidification in 2011 demonstrated some trends in contrast to those observed in previous years (Tullus et al. 2012a). However, the positive effect of humidification was detectable only in current-year height increments, while overall dimensions remained smaller in **H** plots, where hybrid aspen trees had grown slower than in **C** plots in the two previous experimental years (Tullus et al. 2012a). The inverse growth response is also attributable to dry weather conditions prevailing in summer 2011. Generally Ψ_S or experimental plot was the more significant factor influencing tree growth response than humidity manipulation. One must take into account that the two factors—Ψ_S and treatment—are partly interrelated, as transpirational flux through trees was lower in **H** plots (Fig. 4A; see also Kupper et al. 2011 and Tullus et al. 2012a) and more water was retained in the soil (Fig. 2). However, the humidity manipulation also had an impact on growth when considering the effect of Ψ_S; thus, the humidification effect on tree growth was clearly not due solely to altered soil water availability. In average or rainy years, when soil water does not limit growth, lowered transpiration hinders nutrient uptake by trees in **H** plots (Tullus et al. 2012a), especially for nutrients migrating to the roots with mass flow in soil. Under these conditions increased atmospheric humidity does not improve the growth rate of hybrid aspen. Sellin et al. (2013) also showed that humidification treatment lowers the photosynthetic capacity and growth rate of silver birch in moist summers. In dry years, when soil water availability limits growth, the impact of this mechanism is obviously irrelevant.

Conclusions

The current study demonstrates that higher air humidity mitigates the effect of low soil water availability on broadleaved trees during dry years by reducing stomatal limitation to photosynthesis, allowing higher net photosynthetic rates and supporting higher growth rates (relative height growth). At the same time, higher RH increases the sensitivity of canopy conductance to water deficit and reduces the responsiveness of IWUE to factors inducing stomatal closure. The present and our earlier results (Tullus et al. 2012a; Sellin et al. 2013, 2014) imply that a future rise in atmospheric humidity at high latitudes may be disadvantageous in evenly rainy/humid years and expose trees to a higher dehydration risk during weather extremes, although mitigating the impact of soil water deficit under moderate drought.

Sources of Funding

This study was supported by the Estonian Science Foundation (Grant no. 8333), by the Estonian Ministry of Education and Research (target financing project SF0180025s12), and by the EU through the European Social Fund (Mobilitas postdoctoral grant MJD 257) and the European Regional Development Fund (Project No. 3.2.0802.11-0043 'BioAtmos' and Centre of Excellence in Environmental Adaptation).

Contributions by the Authors

A.N. and A.S. designed and performed the experiment, and wrote the manuscript. P.K. and A.T. performed the experiment, analysed the data and revised the paper. All authors read and approved the final manuscript.

Acknowledgements

We are grateful to Kristina Lubenets for installing sap flow gauges, Jaak Sõber for operating the FAHM humidification system and providing weather data, and Robert Szava-Kovats for language revision.

Literature Cited

Araújo WL, Fernie AR, Nunes-Nesi A. 2011. Control of stomatal aperture: a renaissance of the old guard. *Plant Signaling & Behavior* 6: 1305–1311.

Arve LE, Terfa MT, Gislerød HR, Olsen JE, Torre S. 2013. High relative air humidity and continuous light reduce stomata functionality by affecting the ABA regulation in rose leaves. *Plant, Cell and Environment* **36**:382–392.

Assmann SM, Shimazaki KI. 1999. The multisensory guard cell. Stomatal responses to blue light and abscisic acid. *Plant Physiology* **119**:809–815.

Bovard BD, Curtis PS, Vogel CS, Su HB, Schmid HP. 2005. Environmental controls on sap flow in a northern hardwood forest. *Tree Physiology* **25**:31–38.

Cohen S, Naor A. 2002. The effect of three rootstocks on water use, canopy conductance and hydraulic parameters of apple trees and predicting canopy from hydraulic conductance. *Plant, Cell and Environment* **25**:17–28.

Domec J-C, Noormets A, King JS, Sun G, Mcnulty SG, Gavazzi MJ, Boggs JL, Treasure EA. 2009. Decoupling the influence of leaf and root hydraulic conductances on stomatal conductance and its sensitivity to vapour pressure deficit as soil dries in a drained loblolly pine plantation. *Plant, Cell and Environment* **32**:980–991.

Easterling DR, Meehl GA, Parmesan C, Changnon SA, Karl TR, Mearns LO. 2000. Climate extremes: observations, modeling, and impacts. *Science* **289**:2068–2074.

Estrada-Campuzano G, Slafer GA, Miralles DJ. 2012. Differences in yield, biomass and their components between triticale and wheat grown under contrasting water and nitrogen environments. *Field Crops Research* **128**:167–179.

Fanourakis D, Matkaris N, Heuvelink E, Carvalho SMP. 2010. Effect of relative air humidity on the stomatal functionality in fully developed leaves. *Acta Horticulturae* **870**:83–88.

Fanourakis D, Carvalho SMP, Almeida DPF, Heuvelink E. 2011. Avoiding high relative air humidity during critical stages of leaf ontogeny is decisive for stomatal functioning. *Physiologia Plantarum* **142**:274–286.

Farquhar GD, Sharkey TD. 1982. Stomatal conductance and photosynthesis. *Annual Review of Plant Physiology* **33**:317–345.

Farquhar GD, von Caemmerer S, Berry JA. 1980. A biochemical model of photosynthetic CO_2 assimilation in leaves of C_3 species. *Planta* **149**:78–90.

Fletcher AL, Sinclair TR, Allen LH Jr. 2007. Transpiration responses to vapor pressure deficit in well watered 'slow-wilting' and commercial soybean. *Environmental and Experimental Botany* **61**:145–151.

Fordham MC, Harrison-Murray RS, Knight L, Evered CE. 2001. Effects of leaf wetting and high humidity on stomatal function in leafy cuttings and intact plants of *Corylus maxima*. *Physiologia Plantarum* **113**:233–240.

Giday H, Kjaer KH, Fanourakis D, Ottosen CO. 2013. Smaller stomata require less severe leaf drying to close: a case study in *Rosa hydrida*. *Journal of Plant Physiology* **170**:1309–1316.

Grassi G, Magnani F. 2005. Stomatal, mesophyll conductance and biochemical limitations to photosynthesis as affected by drought and leaf ontogeny in ash and oak trees. *Plant, Cell and Environment* **28**:834–849.

Grassi G, Vicinelli E, Ponti F, Cantoni L, Magnani F. 2005. Seasonal and interannual variability of photosynthetic capacity in relation to leaf nitrogen in a deciduous forest plantation in northern Italy. *Tree Physiology* **25**:349–360.

Guha A, Rasineni GK, Reddy AR. 2010. Drought tolerance in mulberry (*Morus* spp.): a physiological approach with insights into growth dynamics and leaf yield production. *Experimental Agriculture* **46**:471–488.

Harley PC, Sharkey TD. 1991. An improved model of C_3 photosynthesis at high CO_2: reversed O_2 sensitivity explained by lack of glycerate reentry into the chloroplast. *Photosynthesis Research* **27**:169–178.

Harley PC, Loreto F, Marco GD, Sharkey TD. 1992. Theoretical considerations when estimating the mesophyll conductance to CO_2 flux by analysis of the response of photosynthesis to CO_2. *Plant Physiology* **98**:1429–1436.

Hölscher D, Koch O, Korn S, Leuschner C. 2005. Sap flux of five co-occurring tree species in a temperate broad-leaved forest during seasonal soil drought. *Trees* **19**:628–637.

Huang Z, Xu Z, Blumfield TJ, Bubb K. 2008. Variations in relative stomatal and biochemical limitations to photosynthesis in a young blackbutt (*Eucalyptus pilularis*) plantation subjected to different weed control regimes. *Tree Physiology* **28**:997–1005.

IPCC. 2007. *Climate change 2007: the physical science basis.* Contribution of Working Group I to the Fourth Assessment Report of the Intergovernmental Panel on Climate Change. Solomon S, Qin D, Manning M, Chen Z, Marquis M, Averyt KB, Tignor M, Miller HL, eds. Cambridge: Cambridge University Press.

Komatsu H, Kang Y, Kume T, Yoshifuji N, Hotta N. 2006. Transpiration from a *Cryptomeria japonica* plantation, Part 2: Responses of canopy conductance to meteorological factors. *Hydrological Processes* **20**:1321–1334.

Kupper P, Sõber J, Sellin A, Lõhmus K, Tullus A, Räim O, Lubenets K, Tulva I, Uri V, Zobel M, Kull O, Sõber A. 2011. An experimental facility for free air humidity manipulation (FAHM) can alter water flux through deciduous tree canopy. *Environmental and Experimental Botany* **72**:432–438.

Kuster TM, Schleppi P, Hu B, Schulin R, Günthardt-Goerg MS. 2013. Nitrogen dynamics in oak model ecosystems subjected to air warming and drought on two different soils. *Plant Biology* **15**(Suppl. 1):220–229.

Li D, Liu H, Qiao Y, Wang Y, Cai Z, Dong B, Shi C, Liu Y, Li X, Liu M. 2013. Effects of elevated CO_2 on the growth, seed yield, and water use efficiency of soybean (*Glycine max* (L.) Merr.) under drought stress. *Agricultural Water Management* **129**:105–112.

Liu F, Jensen CR, Shahanzari A, Andersen MN, Jacobsen SE. 2005. ABA regulated stomatal control and photosynthetic water use efficiency of potato (*Solanum tuberosum* L.) during progressive soil drying. *Plant Science* **168**:831–836.

Meinzer FC. 2003. Functional convergence in plant responses to the environment. *Oecologia* **134**:1–11.

Nejad AR, Van Meeteren U. 2005. Stomatal response characteristics of *Tradescantia virginiana* grown at high relative air humidity. *Physiologia Plantarum* **125**:324–332.

Pantin F, Monnet F, Jannaud D, Costa JM, Renaud J, Muller B, Simonneau T, Genty B. 2013. The dual effect of abscisic acid on stomata. *New Phytologist* **197**:65–72.

Pataki DE, Oren R, Katul G, Sigmon J. 1998. Canopy conductance of *Pinus taeda, Liquidambar styraciflua* and *Quercus phellos* under varying atmospheric and soil water conditions. *Tree Physiology* **18**:307–315.

Pospíšilová J. 1996. Effect of air humidity on the development of functional stomatal apparatus. *Biologia Plantarum* **38**:197–204.

Possen BJHM, Oksanen E, Rousi M, Ruhanen H, Ahonen V, Tervahauta A, Heinonen J, Heiskanen J, Kärenlampi S, Vapaavuori E. 2011. Adaptability of birch (*Betula pendula* Roth) and aspen (*Populus tremula* L.)

genotypes to different soil moisture conditions. *Forest Ecology and Management* **262**:1387–1399.

Prioul JL, Chartier P. 1977. Partitioning of transfer and carboxylation components of intracellular resistance to photosynthetic CO_2 fixation: a critical analysis of the methods used. *Annals of Botany* **41**:789–800.

Räisänen J, Hansson U, Ullerstig A, Döscher R, Graham LP, Jones C, Meier HEM, Samuelsson P, Willén U. 2004. European climate in the late twenty-first century: regional simulations with two driving global models and two forcing scenarios. *Climate Dynamics* **22**:13–31.

Rezaei Nejad A, Van Meeteren U. 2008. Dynamics of adaptation of stomatal behaviour to moderate or high relative air humidity in *Tradescantia virginiana*. *Journal of Experimental Botany* **59**:289–301.

Rezaei Nejad A, Harbinson J, Van Meeteren U. 2006. Dynamics of spatial heterogeneity of stomatal closure in *Tradescantia virginiana* altered by growth at high relative air humidity. *Journal of Experimental Botany* **57**:3669–3678.

Sapeta H, Costa JM, Lourenço T, Maroco J, van der Linde P, Oliveira MM. 2013. Drought stress response in *Jatropha curcas*: growth and physiology. *Environmental and Experimental Botany* **85**:76–84.

Sellin A. 2001. Hydraulic and stomatal adjustment of Norway spruce trees to environmental stress. *Tree Physiology* **21**:879–888.

Sellin A, Lubenets K. 2010. Variation of transpiration within a canopy of silver birch: effect of canopy position and daily versus nightly water loss. *Ecohydrology* **3**:467–477.

Sellin A, Tullus A, Niglas A, Õunapuu E, Karusion A, Lõhmus K. 2013. Humidity-driven changes in growth rate, photosynthetic capacity, hydraulic properties and other functional traits in silver birch (*Betula pendula*). *Ecological Research* **28**:523–535.

Sellin A, Niglas A, Õunapuu-Pikas E, Kupper P. 2014. Rapid and long-term effects of water deficit on gas exchange and hydraulic conductance of silver birch trees grown under varying atmospheric humidity. *BMC Plant Biology* **14**:72.

Sharkey TD. 1985. Photosynthesis in intact leaves of C3 plants: physics, physiology and rate limitations. *Botanical Review* **51**:53–105.

Talbott LD, Rahveh E, Zeiger E. 2003. Relative humidity is a key factor in the acclimation of the stomatal response to CO_2. *Journal of Experimental Botany* **54**:2141–2147.

Tissue DT, Griffin KL, Turnbull MH, Whitehead D. 2005. Stomatal and non-stomatal limitations to photosynthesis in four tree species in a temperate rainforest dominated by *Dacrydium cupressinum* in New Zealand. *Tree Physiology* **25**:447–456.

Torre S, Fjeld T, Gislerød HR, Moe R. 2003. Leaf anatomy and stomatal morphology of greenhouse roses grown at moderate or high air humidity. *Journal of the American Society for Horticultural Science* **128**:598–602.

Tullus A, Kupper P, Sellin A, Parts L, Sõber J, Tullus T, Lõhmus K, Sõber A, Tullus H. 2012*a*. Climate change at Northern latitudes: rising atmospheric humidity decreases transpiration, N-uptake and growth rate of hybrid aspen. *PLoS ONE* **7**:e42648.

Tullus A, Rytter L, Tullus T, Weih M, Tullus H. 2012*b*. Short-rotation forestry with hybrid aspen (*Populus tremula* L. × *P. tremuloides* Michx.) in Northern Europe. *Scandinavian Journal of Forest Research* **27**:10–29.

van de Weg MJ, Meir P, Grace J, Ramos GD. 2012. Photosynthetic parameters, dark respiration and leaf traits in the canopy of a Peruvian tropical montane cloud forest. *Oecologia* **168**:23–34.

von Caemmerer S, Farquhar GD. 1981. Some relationships between the biochemistry of photosynthesis and the gas exchange of leaves. *Planta* **153**:376–387.

3

When Michaelis and Menten met Holling: towards a mechanistic theory of plant nutrient foraging behaviour

Gordon G. McNickle[1]* and Joel S. Brown[2]

[1] Department of Biology, Wilfrid Laurier University, 75 University Avenue West, Waterloo, ON N2L 3C5, USA
[2] Department of Biological Sciences, University of Illinois at Chicago, 845 W. Taylor St. (MC066), Chicago, IL 6060, USA

Associate Editor: James F. Cahill

Abstract. Plants are adept at assessing and responding to nutrients in soil, and generally proliferate roots into nutrient-rich patches. An analogy between this growth response and animal foraging movement is often drawn, but because of differences between plants and animals it has not always been clear how to directly apply existing foraging theory to plants. Here we suggest one way to unite pre-existing ideas in plant nutrient uptake with foraging theory. First, we show that the Michaelis–Menten equation used by botanists and the Holling disc equation used by zoologists are actually just rearrangements of the same functional response. This mathematical unity permits the translation of existing knowledge about the nutrient uptake physiology of plants into the language of foraging behaviour, and as a result gives botanists direct access to foraging theory. Second, we developed a model of root foraging precision based on the Holling disc equation and the marginal value theorem, and parameterize it from the literature. The model predicts (i) generally plants should invest in higher quality patches compared to lower quality patches, and as patch background–contrast increases; (ii) low encounter rates between roots and nutrients result in high root foraging precision; and (iii) low handling times for nutrients should result in high root foraging precision. The available data qualitatively support these predictions. Third, to parameterize the model above we undertook a review of the literature. From that review we obtained parameter estimates for nitrate and/or ammonium uptake for 45 plant species from 38 studies. We observe that the parameters ranged over six orders of magnitude, there was no trade-off in foraging ability for nitrate versus ammonium: plants that were efficient foragers for one form of nitrogen were efficient foragers for the other, and there was also no phylogenetic signal in the parameter estimates.

Keywords: Encounter rate; handling time; Holling's disc equation; Michaelis–Menten kinetics; nutrient foraging; plant foraging behaviour; root foraging precision.

* Corresponding author's e-mail address: gmcnickle@wlu.ca

Introduction

Nutrients are typically distributed heterogeneously throughout the soil (Jackson and Caldwell 1993; Hutchings and de Kroon 1994; Hodge 2004) and plants are adept at assessing and responding to this nutrient heterogeneity (Robinson 1994; de Kroon and Hutchings 1995; Hodge 2004; Kembel and Cahill 2005; Cahill and McNickle 2011; Tian and Doerner 2013). Generally, plants respond to nutrient-rich patches by preferentially proliferating roots into those patches. This growth response results in an increased absorptive surface area inside nutrient-rich patches relative to lower quality regions of the average background soil and is generally considered to be an adaptive response. Increasingly, there has been a trend towards viewing this plasticity in root growth with respect to nutrients through a lens of behavioural ecology (Sutherland and Stillman 1988; Silvertown and Gordon 1989; Kelly 1992; Hutchings and de Kroon 1994; de Kroon and Hutchings 1995; Gersani et al. 2001; Dudley and File 2007; Hodge 2009; Karban 2008; McNickle et al. 2009; Cahill and McNickle 2011; Tian and Doerner 2013). This gradual shift in perspectives on how plants acquire nutrients has been driven by data demonstrating that plants are not passively following pre-determined growth trajectories, but instead plant growth is based on actively assessing and responding to cues from the soil nutrient environment (Silvertown and Gordon 1989; Hutchings and de Kroon 1994; Hodge 2009; Karban 2008; Cahill and McNickle 2011; Tian and Doerner 2013).

However, despite substantial gains in our knowledge of the range of plant nutrient foraging behaviours, we are still a long way from incorporating plant foraging behaviour as a united sub-field of behavioural ecology (McNickle et al. 2009; Cahill and McNickle 2011). Indeed, many questions remain about plant root foraging behaviour. For example, the average plant appears to strongly proliferate roots into nutrient-rich patches; however, some species do not strongly proliferate roots into patches (Hodge 2004; Kembel and Cahill 2005; Kembel et al. 2008; Cahill and McNickle 2011). Additionally, some species of plants appear to discriminate among patches of varying quality by proliferating higher root mass into more nutrient-rich patches, while other species show little discrimination among patches that differ in nutrient availability (Gleeson and Fry 1997; Hutchings et al. 2003; McNickle and Cahill 2009). There are also unresolved questions about how plants should invest in patches based on the contrast between nutrient availability in rich patches versus poorer background soil (Lamb et al. 2004). Logically, we would expect all plant species to benefit from nutrient-rich patches (when they are nutrient limited) and so we lack a first principles explanation

that can permit an understanding of why species differ so much in their foraging responses (Kembel et al. 2008; McNickle and Cahill 2009; McNickle et al. 2009).

In a perfect world plant foraging theory would not reinvent the wheel, but integrate existing ideas about plant ecology, plant nutrient uptake physiology and behavioural ecology. Here, we attempt such a synthesis by exploring the previously recognized fact that the Holling disc equation used by foraging ecologists to model resource intake (Holling 1959; Stephens and Krebs 1986; Vincent et al. 1996; Stephens et al. 2007) and the Michaelis–Menten equation used by plant physiologists to model nutrient uptake (Michaelis and Menten 1913; Lineweaver and Burk 1934; Epstein and Hagen 1951; Johnson and Goody 2011) are actually rearranged forms of the same functional response (Real 1977). As we show, this identity in functional response permits the translation of plant nutrient uptake physiology into the language of foraging behaviour. We have three main objectives: first, we compare the models used by biologists to describe resource capture to show that the models used by plant physiologists and animal behaviourists are mathematically identical. Second, we derive a simple example model to predict the root foraging precision of plants that is based on the well-described functional response of plants (Epstein and Hagen 1951; Bassirirad 2000) and the marginal value theorem (Charnov 1976; McNickle and Cahill 2009). Third, we parameterize the foraging model with a realistic range of plant foraging traits obtained from a literature review of existing studies of plant uptake kinetics for nitrate and ammonium and recast these results from 'enzyme-kinetics' into 'foraging kinetics'. We also present a summary of these data with three sub-objectives: (i) we describe the range and central tendency within the observed patterns of nutrient uptake traits; (ii) we ask whether there is any relationship in the ability of plants to capture the substitutable resources of nitrate and ammonium and (iii) we ask whether there is any phylogenetic signal in these uptake parameters. We conclude with a discussion of the value of rethinking plant uptake of nutrients as a process of enzyme kinetics to a process of foraging behaviour.

Methods

Identical models, different packaging

In the broader ecological literature on foraging and forager functional responses, Holling's disc equation (Holling 1959) provides one commonly used framework for modelling resource capture. In the plant literature on nutrient uptake kinetics, the Michaelis–Menten equation (Michaelis and Menten 1913; Lineweaver and Burk 1934; Johnson and Goody 2011) provides the framework for modelling nutrient capture (Epstein and Hagen 1951). Mathematically, these are simply different arrangements of the same functional response, but the arrangements produce distinct interpretations of parameters, and each arrangement naturally lends itself to different predictive objectives (Real 1977). Both equations produce a Type II functional response (*sensu* Holling 1959), where the resource harvest rate, dH/dt, increases up to an asymptote with resource availability, N (Fig. 1), and both have two parameters.

The Michaelis–Menten equation (Michaelis and Menten 1913; Lineweaver and Burk 1934; Johnson and Goody 2011) for nutrient uptake at the level of the whole root system in plants is

$$\frac{dH}{dt} = \frac{rV_{max}N}{K_m + N} \qquad (1)$$

where dH/dt is the resource harvest rate (units of resource uptake per time per gram of root), r is the biomass of roots possessed by the plant, N is the available nutrient concentration in the environment (units of resources per unit volume of soil), V_{max} is the maximum influx rate (units of resources per time per gram of root) and K_m is the half saturation constant (units of resources per unit volume), representing the resource concentration where the harvest rate is half of the theoretical maximum. Readers should note that V_{max} and K_m, simply describe the shape of the functional response (Fig. 1A); the asymptote on the y-axis is given by V_{max}, and the resource concentration on the x-axis where the harvest rate is halfway to the asymptote is given by K_m (Fig. 1A).

Holling's disc equation (Holling 1959) is written as

$$\frac{dH}{dt} = \frac{aN}{1 + ahN}. \qquad (2a)$$

As above, dH/dt is the harvest rate (units of resource per time per individual forager); N is the concentration of prey in the environment (more typically referred to as prey abundance, but abundance per area or volume is mathematically identical to the concept of nutrient concentration); a is the rate that prey are encountered by the forager (here in units of per time per individual forager, often called search efficiency) and h is the time required by the forager to handle each encountered prey item (units of individual forager × time per prey). Note here that instead of describing the shape of the functional response, the parameters a and h describe activities relevant to the process of foraging.

Typically, Holling's disc equation describes the harvest rate of one animal with one mouth, and so most typically ecologists do not need to clarify that the equation is

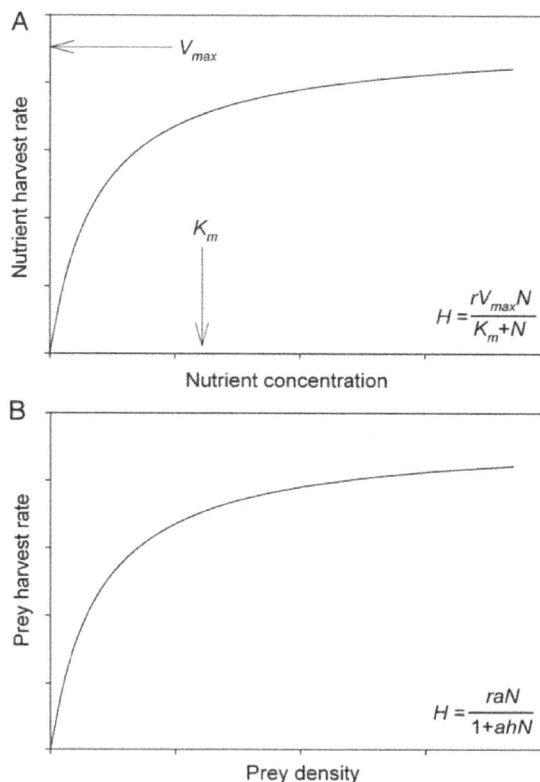

Figure 1. Graphical comparison of (A) the Michaelis–Menten equation that relates resource harvest rates to resource abundance and (B) the Holling disc equation that relates resource harvest to resource abundance. The parameters of the Michaelis–Menten equation describe the shape of the curve where V_{max} is the maximum resource harvest rate and K_m is the concentration of nutrients that produces half of the maximum resource harvest rate (A). The parameters of the Holling disc equation describe traits of the organisms, and though they produce the same curve, these parameters cannot be placed in the figure.

parameterized on a 'per-individual forager' basis as we have done above (McNickle et al. 2009). But, recognizing the 'per unit of forager' aspect of the equation becomes important when using Holling's disc equation to understand the foraging behaviour of modular plants that are more like colonial animals than solitary animals (see McNickle et al. 2009 for discussion). As above, taking into account the per-root foraging effect in plants, equation (2a) becomes

$$\frac{dH}{dt} = \frac{raN}{1 + ahN}. \qquad (2b)$$

In the root foraging form of the Holling disc equation, a unit of root (r, units of mass or length) substitutes for the individual forager, and the plant can effectively be many foragers at once by proliferating many units of root into a volume of soil (McNickle et al. 2009).

The parameters of the Michelis–Menten equation can be recast into Holling's disc equation with a simple rearrangement of equation (1). Dividing both sides of the Michaelis–Menten equation by K_m and setting equal to the disc equation we find

$$\frac{r(V_{max}/K_m)N}{1+(1/K_m)N} = \frac{raN}{1+ahN}. \tag{3}$$

Equation (3) shows how to convert the enzyme-kinetic parameters into the foraging parameters where the encounter rate between a unit of root and a nutrient molecule is given by $a = V_{max}/K_m$, and the cost in time associated with handling a given amount of nutrient molecules is given by $h = 1/V_{max}$. This translation produces estimates of plant foraging parameters for the Holling disc equation that are in the correct units and maintain the correct theoretical interpretation for plant foraging (Table 1).

From functional responses to root behaviour

From the Holling equation, where parameters map directly to functional behavioural traits, many aspects of foraging behaviour can be intuitively derived as a direct consequence of search and handling (reviewed in Stephens et al. 2007). Here we advance a simple nutrient foraging model for plants which is based on Charnov's (1976) marginal value theorem and the Holling disc equation as one example of the value of translating Michaelis–Menten kinetics into Holling's foraging kinetics. The marginal value theorem hypothesizes that foragers should invest effort (here effort is root biomass) into patches until the nutrient uptake rate inside the patch balances the rate in the background soil (Charnov 1976; Gleeson and Fry 1997), and several species of plants have been shown to follow this prediction (Kelly 1990; Kelly 1992; McNickle and Cahill 2009). For plants that can place foraging roots in multiple locations, this prediction is also similar to the ideal free distribution (Fretwell

and Lucas 1969; McNickle and Brown 2014). However, plant foraging is sufficiently different from animal foraging that one further modification is necessary.

Plant foraging is often measured as root foraging precision, which compares the investment of root biomass or root length inside a patch with other locations in the soil. This plant foraging behaviour differs from animal foraging where questions are typically about understanding time investment or energy requirements (McNickle et al. 2009). Thus, the model we develop predicts root foraging precision, which we define as the ratio of root production inside a nutrient-rich patch of some volume to the root production in the poorer quality background soil of equal volume (e.g. Rajaniemi and Reynolds 2004; James et al. 2009; McNickle and Cahill 2009). It is important to note that many other definitions of root foraging precision have been used by empiricists. However, all of these definitions of precision are conceptually similar in that they attempt to estimate the relative investment of root biomass inside a nutrient-rich patch relative to the investment in average background habitat quality. These other definitions are not easily predicted from the functional response without more complicated treatments of root : shoot growth or spatial dimensions of soil. For example, some authors defined precision as the mass of roots inside a patch as a fraction of total body mass (Campbell et al. 1991) but a model for this type of foraging precision would require significantly more complex treatments of root growth relative to nutrients and shoot growth relative to photosynthetically active radiation that are beyond the scope of this manuscript. Other authors have defined precision as the relative proportion of total root system inside a patch (e.g. Kembel and Cahill 2005) or the relative root mass difference between patch and background as a fraction of total root biomass (e.g. Einsmann et al. 1999). These are also difficult to solve without complex and explicit treatments of space at the scale of the entire root system that are beyond the goals of this manuscript.

Table 1. Comparison of the parameters of the Michaelis–Menten equation, and the Holling disc equation. The two models are built from an equation of the same general form (Real 1977) and the two equations model identical processes in plants and animals. This produces interchangeable parameters, where the units of measurement also perfectly translate.

	Parameters	Units	Biological meaning
Holling disc equation	$a = V_{max}/K_m$	L g^{-1} min^{-1}	Effective encounter rate or search efficiency; the maximum volume of nutrients of concentration R (μmol/L) that are encountered per gram of root per minute.
	$h = 1/V_{max}$	min g μmol^{-1}	Handling time: the time taken for 1 μmol of nutrient to be captured by a gram of root
Michaelis–Menten equation	$V_{max} = 1/h$	μmol g^{-1} min^{-1}	Maximum theoretical rate of nutrient uptake, per gram of root.
	$K_m = 1/ah$	μmol L^{-1}	Half saturation constant or the nutrient concentration where the rate of uptake is half of V_{max}. This is sometimes called the enzyme affinity for the substrate.

Consider a plant searching for j forms of nitrogen (N_{ij}) spread across i patches throughout the soil. The total amount of nitrogen j encountered is given by the search efficiency for nitrogen types j to n ($a_j \ldots a_n$), the concentration of nitrogen types j to n at location i in the soil ($N_{ij} \ldots N_{in}$) and the amount of roots that are searching in each location i ($r_i \ldots r_k$). Uptake rate is discounted by the rate at which resources are encountered while searching ($a_j \ldots a_n$), and by the time lags associated with handling nitrogen type j to n ($h_j \ldots h_n$). Assuming that N_{ij} is experimentally held constant over the course of the experiment then with no depletion, (e.g. Campbell and Grime 1989; Shemesh et al. 2010), and also assuming that the concentrations of nutrients other than nitrogen are experimentally held constant among locations (e.g. Drew and Saker 1975; McNickle et al. 2013), then the harvest rate of nitrogen types j to n at location i is given by the multiple resource form of the Holling disc equation, with root biomass:

$$\frac{dH_i}{dt} = \frac{r_i\left(\sum_{j=1}^{n} a_j N_{ij}\right)}{1 + \left(\sum_{j=1}^{n} a_j h_j N_{ij}\right)}. \qquad (4)$$

Equation (4) predicts the uptake rate of all nitrogen types from any location i. Consider a simple experiment where plants are grown with one spatially discrete nutrient-rich patch (location p), surrounded by nutrient-poor background soil (location b). Comparing one patch with a similar volume of background soil, we expect that, all else equal, plants would produce roots in the patch (r_p) and in the background (r_b), such that the amount of roots combine with the traits of the forager (a and h) to produce equal rates of nutrient uptake (Charnov 1976), given by

$$\frac{r_p\left(\sum_{j=1}^{n} a_j N_{pj}\right)}{1 + \left(\sum_{j=1}^{n} a_j h_j N_{pj}\right)} = \frac{r_b\left(\sum_{j=1}^{n} a_j N_{bj}\right)}{1 + \left(\sum_{j=1}^{n} a_j h_j N_{bj}\right)}. \qquad (5a)$$

Here we are interested in root foraging precision, which is the optimal ratio of roots inside the patch relative to the roots inside the background soil ($P^* = r_p^*/r_b^*$) given by:

$$P^* = \frac{r_p^*}{r_b^*} = \frac{\sum_{j=1}^{n} N_{pj}}{1 + \sum_{j=1}^{n} a_j h_j N_{pj}} \frac{1 + \sum_{j=1}^{n} a_j h_j N_{bj}}{\sum_{j=1}^{n} N_{bj}}. \qquad (5b)$$

Equation (5b) thus represents a simple approximation of foraging precision in plants based on plant nutrient uptake physiology and foraging theory. A key assumption of this model is that plants are nitrogen limited and not limited by other resources, particularly carbon. When plants are carbon limited they may shift allocation away from roots and towards shoot production. Additionally, this model assumes that the roots are the sole source of nitrogen uptake. For example, root production may not be as important for nitrogen acquisition in nitrogen-fixing plants or mycorrhizal species. These assumptions can be easily met in controlled manipulative experiments and by choosing appropriate model species, but may not apply to all species and contexts.

Literature review: range of foraging traits

To estimate the range of behavioural foraging traits in plants and parameterize our model, we broadly searched the literature for estimates of V_{max} and K_m for nitrate and ammonium and translated the reported V_{max} and K_m into encounter rates and handling times (Table 1). In February 2011, we searched the ISI Web of Science for the topic 'root uptake kinetics' which returned 870 papers. To make search results more manageable, we filtered the results to the Web of Science Category 'Plant Sciences'. This produced 509 papers. We then inspected titles and abstracts to reduce the search to only papers that reported parameters for nitrate and/or ammonium. From the remaining 219 papers we read each manuscript to collect parameter estimates. We limited our data collection to papers that estimated parameters based on either fresh or dry weight of roots and that estimated both V_{max} and K_m using the Michaelis–Menten equation. Despite the fact that all plant papers we reviewed used the two-parameter Michaelis–Menten equation to fit their data, a surprisingly large number of papers only reported V_{max}, while failing to report the second parameter, K_m. We excluded these papers. Additionally, we limited the data to only plant species with areal shoots so that nutrient capture was achieved exclusively through roots. Fully aquatic plants and algae were therefore excluded, but wetland plants were included. A small number of studies (<10) were not in English, were unavailable after an exhaustive physical and online search or did not report the units of measurement, and these were excluded. When different studies reported estimates of V_{max} and K_m for the same species, we report these as separate data points [see Supporting Information]. When multiple treatments were employed we used only the control treatment or the equivalent 'no manipulation' treatment. These search criteria resulted in a final set of 38 studies, with parameter estimates for nitrate and/or ammonium for 45 distinct plant species, and three species that had been studied more than once [see Supporting Information].

Parameter estimates were adjusted uniformly to $\mu mol\ g^{-1}\ min^{-1}$ for V_{max} and μM for K_m. Parameter estimates from fresh weight and dry weight of roots were plotted and interpreted separately. We used linear regressions to compare foraging ability for nitrate and ammonium (R Statistical environment, R Development Core Team 2009). To summarize the taxonomic diversity and patterns in these parameter estimates, we performed a phylogenetic analysis. The hypothesized phylogenetic relationships among species were constructed using the online phylogenetic database and assembly tool, Phylomatic (Webb and Donoghue 2005), with Phylomatic tree version R20120829 as the backbone for our phylogenetic hypotheses. The Phylomatic tree is well resolved up to the level of family, but the tool places all genera as polytomies within family and all species as polytomies within genera (Kembel and Cahill 2005; Webb and Donoghue 2005). We tested for a phylogenetic signal in the foraging trait data (a and h) by calculating a K statistic using the R library 'Picante' (Kembel *et al.* 2010). The K statistic compares the observed phylogenetic signal in the trait with the signal that would be expected under the Brownian motion model of trait evolution. K values >1 imply a strong phylogenetic signal, K values equal to 1 imply the Brownian motion model and K values <1

imply a random or convergent pattern of evolution. Traits were mapped onto the phylogeny for visualization using the 'plotBranchbyTrait' tool in the R library 'phytools'. For several species there were multiple independent estimates of traits. In these cases, we took the average trait value. Traits were $\ln(x + 1)$ transformed for normality and to control for differences between fresh and dry weight estimates. There was considerable variation in the methods used among studies to estimate nitrogen foraging parameters [**see Supporting Information**]. Thus, we envision the phylogenies as a way of summarizing the data with respect to taxonomy, but urge caution in interpreting the phylogenetic signal from these data.

Results

Range of reported uptake parameters

Parameter estimates ranged over six orders of magnitude (Table 2), but were relatively evenly spaced along this range (Fig. 2). Examining the Holling parameters, the minimum value for per gram of root *encounter rate* for nitrate

Table 2. Summary of observed parameter estimates from the literature review for plant uptake of nitrate and ammonium. Authors sometimes calculated based on dry or fresh weight of roots, these are summarized independently. Note that a and h were calculated from V_{max} and K_m according to Table 1.

	Statistic	Dry weight		Fresh weight	
		a	h	a	h
NO_3	Min	7.57E−06	0.062	1.11E−06	5.455
	Max	0.360	54.545	0.074	952.381
	Mean	0.051	11.241	0.005	94.288
	Median	0.011	1.901	0.001	16.300
	CV	0.133	0.135	0.062	0.087
NH_4	Min	2.16E−04	0.178	1.12E−04	0.902
	Max	0.368	19.690	0.081	30.000
	Mean	0.049	3.191	0.009	10.874
	Median	0.011	2.150	0.003	5.454
	CV	0.109	0.157	0.111	0.259
	Statistic	**Dry weight**		**Fresh weight**	
		K_m	V_{max}	K_m	V_{max}
NO_3	Min	1.45	0.0183	1.4	0.001
	Max	2422	16.001	2140	0.183
	Mean	480.3	2.206	205.8	0.0696
	Median	44.0	0.526	75.6	0.062
	CV	0.662	0.545	0.435	1.365
NH_4	Min	2.3	0.0508	8.3	0.033
	Max	1908	5.633	3930	1.108
	Mean	293.1	1.052	332.03	0.246
	Median	27.8	0.465	72	0.183
	CV	0.507	0.778	0.333	0.904

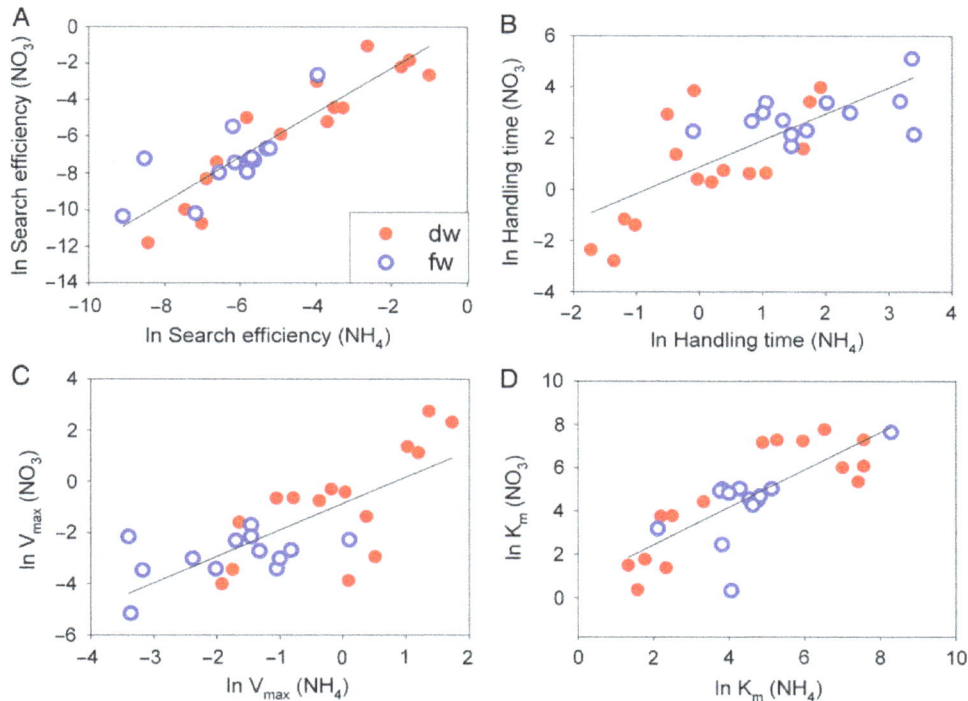

Figure 2. Scatter plots and linear regressions of the observed relationship between (A) the search time for nitrate and ammonium ($F_{1,26} = 100.4$, $P < 0.0001$, $R^2 = 0.79$), (B) the handling time for nitrate and ammonium ($F_{1,26} = 31.2$, $P < 0.0001$, $R^2 = 0.53$), (C) V_{max} for nitrate and ammonium ($F_{1,26} = 31.2$, $P < 0.0001$, $R^2 = 0.53$) and (D) K_m for nitrate and ammonium ($F_{1,26} = 43.2$, $P < 0.0001$, $R^2 = 0.61$). Fresh (fw) and dry (dw) weights are plotted separately, but the patterns were qualitatively similar and so they were pooled for regression fits. Data were $\ln(x + 1)$ transformed.

was two orders of magnitude lower than the minimum value for ammonium, while the maximum, mean, mode and coefficient of variation were all similar between nitrate and ammonium (Table 2). The pattern was retained whether the estimate was based on fresh or dry weight of tissue. Again, note that this is the encounter rate between active uptake sites and not the encounter rate of nutrients and the surface of the root.

For *handling time* the maximum and mean estimates were generally larger for nitrate compared with ammonium, while other statistics were relatively similar, or showed no clear pattern (Table 2). Again the pattern was retained whether the estimates were based on fresh or dry tissue. The range of parameter values suggests that nitrate may be more difficult or costly to transport across root membranes compared with ammonium given the higher average handling costs. It also suggests that at the extreme, the number of encounters which turn into effective encounters (i.e. uptake) may be lower for nitrate.

We also present the range of estimates of parameters for the Michaelis–Menten equation. The parameter V_{max} is simply the inverse of handling time, and so the patterns in V_{max} were the same as for *h* above, but inverted. That is, where handling times were larger for nitrate compared

with ammonium, maximum influx rates (V_{max}) tended to be lower for nitrate compared with ammonium. For the half saturation constant, K_m, there seemed to be no obvious differences between ammonium and nitrate (Table 2). If nutrient uptake in plants is a foraging process, then this may not be surprising since K_m is actually a combination of search and handling time, and the patterns described above for each of these are cancelled by confounding them within this parameter (i.e. $K_m = 1/ah$).

We also explored the relationships in nutrient uptake ability for nitrate and ammonium within a species to examine whether plants might specialize in one type of nitrogen over the other (Fig. 2). Here, all four parameters (*a*, *h*, V_{max} and K_m) tell a similar story: plants that are efficient foragers for nitrate are also efficient foragers for ammonium. Interestingly, this suggests that there are no general trade-offs in foraging ability for these two common forms of nitrogen, and instead species are either efficient or inefficient foragers. However, note that it is not possible to be simultaneously good at searching and handling because of the way that these parameters are conceptualized (Table 1).

In our phylogenetic analysis of trait values, we found no evidence of any phylogenetic signal for any of the foraging traits (Fig. 3). The foraging traits for nitrate

Figure 3. Phylogeny of species for which we have foraging parameters for nitrate (A and B) or ammonium (C and D). Species come from three major clades including conifers, monocots and eudicots. Colour on the branch tips represent ln(x + 1) transformed trait values for search efficiency (A and C) and handling time (B and D). The species lists for nitrate and ammonium foraging parameters were not identical and so the upper and lower phylogenies are not identical.

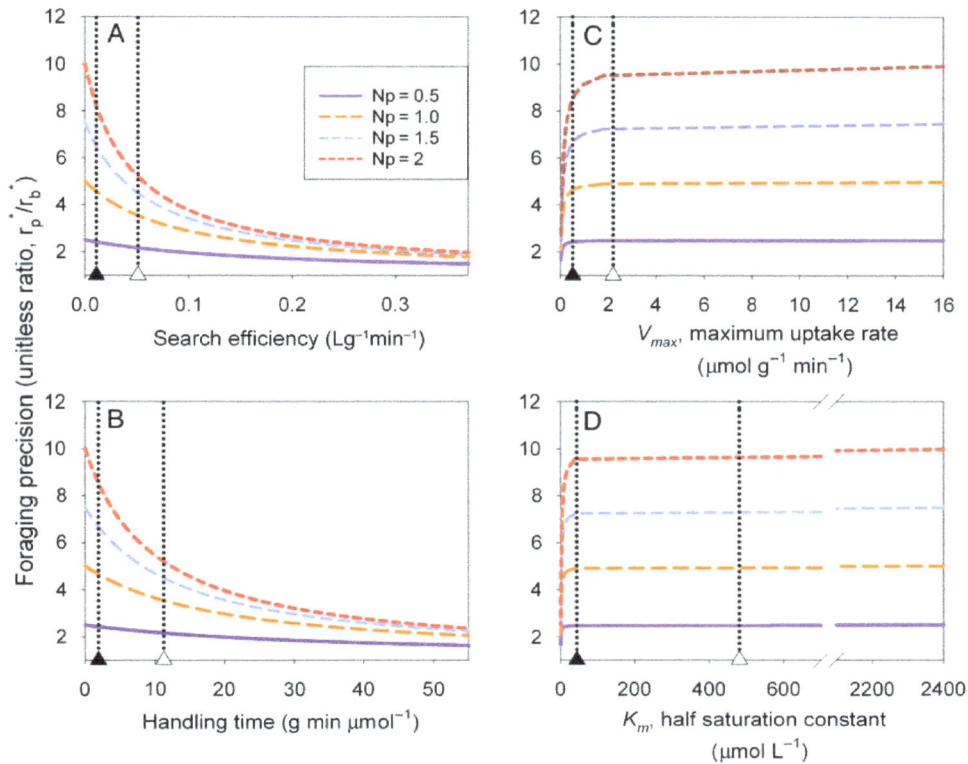

Figure 4. Predicted root foraging precision over the observed range of search efficiency (A), handling time (B), maximum uptake rate (C) and half saturation constant (D) among plants. In each case the value of the background soil was $N_b = 0.2$ μmol L^{-1}, and the value of the patch associated with each curve is shown in the figure legend (0.5 μmol $L^{-1} < N_p < 2$ μmol L^{-1}). In each panel, the x-axis parameter was varied, while the second parameter was held constant at the mean observed value for nitrate calculated from dry weight (Table 2). In each panel, the mean (open triangle) and median (closed triangle) observed values of parameters are indicated by dotted vertical lines.

encounter rates ($K = 0.23$, $P = 0.032$; Fig. 3A) or handling ($K = 0.17$, $P = 0.011$; Fig. 3B) and ammonium encounter rates ($K = 0.15$, $P = 0.161$; Fig. 3C) or handling ($K = 0.09$, $P = 0.625$; Fig. 3D) all had a K statistic <1 implying random evolution of traits, or convergent evolution towards relatively uniform foraging traits across all taxa. At this time we are unable to determine whether this result reflects a true lack of historical relationships, or merely the variability in methodologies used to estimate parameters and urge caution in interpreting this signal.

Root foraging precision

Using the observed range of parameter values for nitrate (Table 2), we can generate hypotheses concerning the corresponding hypothesized range of foraging precision among species foraging for nitrate and how these relate to foraging traits of plants (Equation 5b). The model predicts precision between positive infinity (i.e. all roots inside the patch) and one (i.e. no discrimination between patch and background). In most regions of parameter space, and regardless of which uptake model is used (Michaelis–Menten or Holling), root foraging precision is

hypothesized to increase with increasing patch quality (Fig. 4). The model predicts that plants should allocate more roots to increasingly nutrient-rich patches relative to the average poorer quality background soil; that is with increasing patch–background contrast. Each parameter has specific links to predicted foraging behaviour (Fig. 4), and in the following paragraphs we examine each of the model parameters and their hypothesized effect on root foraging precision individually.

Within the range of a_j observed for plants (Table 2), the marginal value theorem hypothesizes that plants possessing the ability to encounter nitrogen at a high rate should discriminate less among patches of differing quality compared with plants with low encounter rates. Theoretically this occurs because encounter rate acts like a scaling parameter for the effectiveness of each unit of root produced. Plants with high encounter rates between uptake sites and nutrient ions will be able to gather more resources, with a low investment in root surface area compared with plants with low encounter rates that require high amounts of surface area to effectively encounter nutrients. We observe from our examination of the literature that the mean and median plant species in our dataset of 45 species possess relatively low values of this parameter compared with the maximum range that is observed in the literature (Table 2, Fig. 4A). Thus, based on our literature review parameterized model the marginal value theorem would predict that the average plant should have relatively high root foraging precision, and should discriminate among patches of different quality by putting more roots in higher quality patches. However, the literature review also reveals that species exist with extremely efficient encounter rates, and these species are predicted to exhibit low root foraging precision in any patch and exhibit similar root foraging precision regardless of patch quality.

In the context of handling time (Fig. 4B), plants with low handling times have low time lags associated with nutrient uptake, meaning that within a prescribed amount of time spent foraging for nutrients (e.g. the duration of a foraging experiment), plants with low handling times are able to actually acquire more of the available nitrogen compared with plants with high handling time. Logically, plants with low handling times for nitrogen are hypothesized to have high nutrient foraging precision because plants with low handling times are able to capitalize on high quality patches more than plants with high handling costs. As above, we observe from our literature review that the mean and median plant species in our dataset of 45 species possess relatively low handling time compared with the range observed in the literature (Table 2). This means that as above, the average plant in our dataset of 45 species has relatively high foraging precision, and discriminates a great deal among patches of differing quality. However, the observed range for handling time is large, suggesting that some species are hypothesized to have relatively low foraging precision, and not discriminate among patches of variable quality (Fig. 4B).

Finally, in the context of either parameter of the Michaelis–Menten equation (Fig. 4C and D), only the lowest values of V_{max} or K_m lead to much discrimination among species with respect to root foraging precision. For either parameter, we see that a plateau is reached quite quickly, and then root foraging precision changes only slightly. Mathematically, this happens because V_{max} is the inverse of handling time, and K_m is the inverse and the product of both search and handling. Roughly, this reverses the patterns observed for the Holling parameters. Biologically, it suggests that the concepts of maximum uptake rate (V_{max}) and the half saturation constant (K_m) are simply not concepts that are particularly informative about processes important for root proliferation and nutrient foraging. Instead, as we have shown, ecologists interested in nutrient foraging behaviour of

plants will be able to discriminate more clearly among the behaviours and traits of plants by translating the Michaelis–Menten parameters into Holling parameters (Fig. 4, Table 1).

Discussion

We had three major objectives in this paper: First, we showed that the Michaelis–Menten and Holling equations are mathematically identical, and how to translate parameter estimates for each model back and forth (Table 1). The most important insight from this exercise is that under a foraging interpretation of nutrient uptake, the parameter K_m turns out to be a confounded mixture of search and handling that is not particularly useful for predicting plant foraging behaviour (Fig. 4). The equality of the Michaelis–Menten and Holling equations is mathematically straightforward, and we do not discuss this further.

Empirically derived parameter estimates

The analysis of the parameter estimates themselves yielded interesting insights. Interestingly, within species, their ability to capture ammonium and nitrate was positively correlated for all parameters (Fig. 2). This means that there are no trade-offs in the ability of species to capture these two important nitrogen types; instead there are 'super-foragers' where some species are extremely efficient either encountering or handling both nitrate and ammonium simultaneously, while other species are extremely inefficient. Given that these parameters ranged over six orders of magnitude, this 'super-forager' effect is very large indeed (Table 2). However, it should be noted that there is a trade-off (not shown) between encounter rate and handling time. We do not show this because it is a necessary mathematical condition on the way a and h are calculated (see Table 1). But it is important to note that while plants can be simultaneously efficient at either encountering or handling nitrate and ammonium, they cannot be simultaneously efficient at both encountering and handling. Thus, species must specialize on one or the other of these foraging processes, and 'super-foragers' should actually be unpacked into 'super-encounterers' versus 'super-handlers' where species cannot be both. It remains unclear what forces cause this specialization, and our phylogenetic analysis did not shed any light on this problem.

We analysed the parameter data in the context of historical phylogenetic relationships among species, and the results of the phylogenetic analysis were consistent with random or convergent evolution on foraging traits. However, we suggest that convergent evolution is unlikely since the parameter values varied over six orders of magnitude instead of converging on a single value. We are reluctant to draw too many conclusions about this result because of the large differences in methods among studies from which we obtained parameter estimates. Currently we are inclined to suspect that the observed lack of phylogenetic signal may simply reflect this diversity of methods. We suggest caution in interpreting this result at this time, but this is certainly a pattern that demands further investigation using a common set of methods and a large taxonomic sample.

A model of root foraging precision

The foraging model generated three predictions: (i) on average, plants should invest more root biomass into higher quality patches relative to lower quality patches but not all species should be expected to discriminate among patches of differing quality depending on their foraging traits (Fig. 4); (ii) root foraging precision, and discrimination among patches of variable quality, should be lowest in species with high encounter rates between nutrients and active uptake sites (Fig. 4A); and (iii) root foraging precision and discrimination among patches of variable quality should be lowest in plants with the highest handling times for nutrients (Fig. 4B).

Empirically testing these hypotheses will require studies that include a large number of taxonomically diverse plant species and produce paired estimates of physiological uptake parameters and root foraging precision (e.g. to generate an empirically derived version of Fig. 4). Kembel and Cahill (2005) assembled a dataset of root foraging precision for ~120 species. Unfortunately, there are only five species in common between the precision dataset from Kembel and Cahill (2005) and the assembled dataset of uptake parameters presented here. A second problem is that Kembel and Cahill (2005) defined precision as the percentage of total roots inside a patch which is not predicted from the model presented here. Thus, currently there are not enough data available to quantitatively test the foraging model. However, we note that, qualitatively, the available data support the model. For example, the model predicts based on data we assembled from the literature that the average plant should have relatively high root foraging precision, and indeed high root foraging precision appears to be the behaviour of the average plant species (Hutchings and de Kroon 1994; Hodge 2004; Kembel and Cahill 2005; Cahill and McNickle 2011). Further, the prediction that some species should strongly discriminate among patches of variable quality while others should not discriminate among patches is also supported by the available data (Fransen and de Kroon 2001; Hutchings et al. 2003; Lamb et al. 2004; McNickle and Cahill 2009).

There has been no clear explanation for why some plants should exhibit high root foraging precision while others should not (Robinson 1994; Robinson 1996; Kembel et al. 2008), or why some species discriminate among patches of variable quality while others do not (McNickle and Cahill 2009; McNickle et al. 2009). Indeed, some authors have even gone so far as to suggest that the large range of behaviour of root proliferation is illogical (Robinson 1994, 1996) or that certain results suggest the behaviour might even be maladaptive (Fransen and de Kroon 2001) when explained using previous conceptual frameworks. We suggest that recasting enzyme kinetics as a foraging process of search and handling provides one clear first principles hypothesis for why so many plant species exhibit high root foraging precision (Fig. 4). Low encounter rates and low handling times intuitively lead to high root foraging precision by virtue of the marginal value theorem (Fig. 4). However, there is a theoretical trade-off between search and handling: foragers cannot do both simultaneously leading to differences in adaptation and ultimately foraging behaviour (Holling 1959).

Switching from the Michaelis–Menten enzyme kinetics view of nutrient uptake to Holling's functional response view of nutrient uptake as foraging behaviour will require integration of some new concepts into our understanding of plant nutrient foraging. Since handling time is just the inverse of the maximum uptake rate ($h = 1/V_{max}$) then handling time, as a concept, is already in common use by plant biologists. Plants with high influx rates necessarily have low handling times. However, as we have shown, switching to the inverse of V_{max} allows a more subtle discrimination between the foraging behaviour of different species with differing uptake abilities (Fig. 4B versus C) and as a result is a more ecologically valuable parameter.

The concept of encounter rate, which can also be thought of as search efficiency (Stephens et al. 2007), is a relatively new idea for plant ecologists that was confounded, along with handling time, inside the half saturation constant ($K_m = 1/ah$). The concept of encounter rates is critically important in the foraging literature, and is important for understanding the patch-use behaviour of foragers (Vincent et al. 1996; Stephens et al. 2007). Just as the inverse of influx rate is a more informative parameter for root foraging behaviour, unpacking encounter rate from within K_m provides better insights into root foraging behaviour of plants that was obscured inside of K_m (Fig. 4A versus D). Encounters will be influenced by any factor that influences the rate at which nutrient ions are encountered by active uptake sites on a plant root and can include behavioural responses of the plant such as changing the number of active uptake sites (Lauter et al. 1996; McNickle et al. 2009), or by changing total root biomass/length and therefore the number of active uptake sites per volume of soil (Hutchings and de Kroon 1994; Cahill and McNickle 2011). Encounter rate will also be influenced by physical properties of the soil and physical properties of the nutrient molecules that might limit ion movement in soil solution. For example, most uptake studies are conducted in nutrient solutions within laboratories, which likely have quite high mobility of ions leading to artificially high encounter rates. However, physical factors that limit diffusion rates in field soil will also limit the rate at which plants can encounter nutrients and should have an influence on plant foraging behaviour. An experimental test of the root foraging precision model could manipulate encounter rates by manipulating properties of the soil environment. For example, soils with high clay content have lower mobility of cations, and this would limit the ability of plants to encounter positively charged ions such as nitrate.

With any model there are caveats around the assumptions made. We assumed that nitrogen was the only limiting resource. This is unlikely to be true in many contexts, but can and has been achieved in controlled experiments (e.g. Drew and Saker 1975; McNickle et al. 2013). Mathematically, the model could be extended to include foraging for multiple essential resources by the use of a minimum function, where foraging decisions were based on Liebig's law of the minimum. This would require a more complex model, but foraging theory exists for this problem (See Vincent et al. 1996; Simpson et al. 2004). Additionally, we assumed that nitrogen levels were not depleted over the course of the experiment. Again, controlled manipulative experiments can and have met this assumption (e.g. Campbell and Grime 1989; Shemesh et al. 2010). This assumption could be relaxed by allowing nutrients to have their own dynamics in the model (see Vincent et al. 1996). Relaxing both of these assumptions would change the quantitative predictions of our model, but we do not believe they would change the qualitative predictions. Specifically, that encounter rates and handling times are important predictors of foraging precision. The assumptions that plants are limited by nitrogen more than carbon can be more easily met by ensuring adequate light supplies. Similarly, the assumption that uptake is achieved through roots alone can be met by the selection of model species, or by sterilizing soil prior to experimentation. Finally, this model assumes that foraging is all that matters to plants. There are many other problems such as mutualisms, enemy attack and competition that plants must solve (De Deyn and Van der Putten 2005; McNickle and Dybzinski 2013), and trade-offs required to solve these problems may cause undermatching in foraging behaviour as plants direct resources away from nutrient foraging and towards solving these other problems (Brown 1988; Nonacs 2001).

The model of root foraging precision presented here is just one example of how the application of a Holling functional response to plant nutrient uptake could enhance our understanding of plant nutrient foraging behaviour, and we hope this work will lead to further advances. For example, much of the existing foraging theory in the animal literature is based upon the Holling disc equation (Real 1977; Stephens and Krebs 1986; Vincent et al. 1996; Stephens et al. 2007; Abrams 2010a, 2010b) and a diversity of models can be derived from this functional response. We suggest that interested plant ecologists can now begin to take full advantage of the existing foraging literature by using the Holling equation to interpret nutrient uptake instead of the Michaelis–Menten equation. It is beyond the scope of this manuscript to review the existing animal models (see Stephens and Krebs 1986; Stephens et al. 2007), but we believe that there is considerable room for enhanced linkages between processes of interest to plant ecologists and plant physiologists mediated through pre-existing understanding of foraging ecology.

Conclusions

We have argued that a switch from the phenomenological view of plants as passive *enzyme-like* entities that are largely governed by chemical fluxes to a more mechanistic view of plants as active foragers that assess and respond to their environment fits with the trend towards viewing plant plasticity as a behavioural process. We believe that the ability to translate existing plant physiological data into information relevant to foraging behaviour and theory will be valuable for plant ecologists. Our model has the potential to generate improved ecological understanding by uniting traditionally separate fields of ecology, while still preserving our existing knowledge and understanding.

Sources of Funding

Our work was funded by a Natural Sciences and Engineering Research Council (Canada) Post-Doctoral Fellowship and a Banting Post-Doctoral Fellowship to G.G.M.

Contributions by the Authors

G.G.M. and J.S.B. conceived the study and developed the theoretical approach. J.S.B. developed the comparison between the Holling and Michaelis–Menten equations. G.G.M. developed the root foraging model, performed the literature review and analysed the models. Both authors contributed to the writing of the manuscript.

Acknowledgements

The authors thank R. Julia Kilgour, Paul Orlando and Christopher J. Whelan for their helpful discussion during the development of these ideas. They thank Steve W. Kembel for advice on phylogenetic analysis and Liam Revell for his blog (http://blog.phytools.org) containing detailed advice about how to use the 'phytools' library in R.

Supporting Information

The following Supporting Information is available in the online version of this article –

File S1. Text with two figures. The Holling disc equation and Michaelis–Menton equation were fit to the same randomly generated data to demonstrate that they are mathematically identical.

File S2. Supporting data. The Michaelis–Menten parameter estimates collected from the literature for (i) nitrate, (ii) ammonium including (iii) metadata, and (iv) reference list.

Literature Cited

Abrams PA. 2010a. Implications of flexible foraging for interspecific interactions: lessons from simple models. *Functional Ecology* **24**:7–17.

Abrams PA. 2010b. Quantitative descriptions of resource choice in ecological models. *Population Ecology* **52**:47–58.

Bassirirad H. 2000. Kinetics of nutrient uptake by roots: responses to global change. *New Phytologist* **147**:155–169.

Brown JS. 1988. Patch use as an indicator of habitat preference, predation risk, and competition. *Behavioral Ecology and Sociobiology* **22**:37–47.

Cahill JF Jr, McNickle GG. 2011. The behavioral ecology of nutrient foraging by plants. *Annual Review of Ecology, Evolution and Systematics* **42**:289–311.

Campbell BD, Grime JP. 1989. A comparative-study of plant responsiveness to the duration of episodes of mineral nutrient enrichment. *New Phytologist* **112**:261–267.

Campbell BD, Grime JP, Mackey JML. 1991. A trade-off between scale and precision in resource foraging. *Oecologia* **87**:532–538.

Charnov EL. 1976. Optimal foraging, marginal value theorem. *Theoretical Population Biology* **9**:129–136.

De Deyn GB, Van der Putten WH. 2005. Linking aboveground and belowground diversity. *Trends in Ecology and Evolution* **20**:625–633.

de Kroon H, Hutchings MJ. 1995. Morphological plasticity in clonal plants—the foraging concept reconsidered. *Journal of Ecology* **83**:143–152.

Drew MC, Saker LR. 1975. Nutrient supply and growth of seminal root system in Barley. II. Localized, compensatory increases in lateral root growth and rates of nitrate uptake when nitrate supply is restricted to only part of root system. *Journal of Experimental Botany* **26**:79–90.

Dudley SA, File AL. 2007. Kin recognition in an annual plant. *Biology Letters* **3**:435–438.

Einsmann JC, Jones RH, Pu M, Mitchell RJ. 1999. Nutrient foraging traits in 10 co-occurring plant species of contrasting life forms. *Journal of Ecology* **87**:609–619.

Epstein E, Hagen CE. 1951. A kinetic study of the absorption of alkali cations by barley roots. *Plant Physiology* **27**:457–474.

Fransen B, de Kroon H. 2001. Long-term disadvantages of selective root placement: root proliferation and shoot biomass of two perennial grass species in a 2-year experiment. *Journal of Ecology* **89**:711–722.

Fretwell SD, Lucas HL. 1969. On territorial behavior and other factors influencing habitat distribution in birds. *Acta Biotheoretica* **19**:16–36.

Gersani M, Brown JS, O'Brien EE, Maina GM, Abramsky Z. 2001. Tragedy of the commons as a result of root competition. *Journal of Ecology* **89**:660–669.

Gleeson SK, Fry JE. 1997. Root proliferation and marginal patch value. *Oikos* **79**:387–393.

Hodge A. 2004. The plastic plant: root responses to heterogeneous supplies of nutrients. *New Phytologist* **162**:9–24.

Hodge A. 2009. Root decisions. *Plant, Cell and Environment* **32**:628–640.

Holling CS. 1959. The components of predation as revealed by a study of small mammal predation of the European pine sawfly. *The Canadian Entomologist* **91**:293–320.

Hutchings MJ, de Kroon H. 1994. Foraging in plants—the role of morphological plasticity in resource acquisition. *Advances in Ecological Research* **25**:159–238.

Hutchings MJ, John EA, Wijesinghe DK. 2003. Toward understanding the consequences of soil heterogeneity for plant populations and communities. *Ecology* **84**:2322–2334.

Jackson RB, Caldwell MM. 1993. The scale of nutrient heterogeneity around individual plants and its quantification with geostatistics. *Ecology* **74**:612–614.

James JJ, Mangold JM, Sheley RL, Svejcar T. 2009. Root plasticity of native and invasive Great Basin species in response to soil nitrogen heterogeneity. *Plant Ecology* **202**:211–220.

Johnson KA, Goody RS. 2011. The original Michaelis constant: translation of the 1913 Michaelis–Menten paper. *Biochemistry* **50**:8264–8269.

Karban R. 2008. Plant behaviour and communication. *Ecology Letters* **11**:727–739.

Kelly CK. 1990. Plant foraging—a marginal value model and coiling response in *Cuscuta subinclusa*. *Ecology* **71**:1916–1925.

Kelly CK. 1992. Resource choice in *Cuscuta europaea*. *Proceedings of the National Academy of Sciences of the USA* **89**:12194–12197.

Kembel SW, Cahill JF. 2005. Plant phenotypic plasticity belowground: a phylogenetic perspective on root foraging trade-offs. *The American Naturalist* **166**:216–230.

Kembel SW, de Kroon H, Cahill JF, Mommer L. 2008. Improving the scale and precision of hypotheses to explain root foraging ability. *Annals of Botany* **101**:1295–1301.

Kembel SW, Cowan PD, Helmus MR, Cornwell WK, Morlon H, Ackerly DD, Blomberg SP, Webb CO. 2010. Picante: R tools for integrating phylogenies and ecology. *Bioinformatics* **26**:1463–1464.

Lamb EG, Haag JJ, Cahill JF. 2004. Patch-background contrast and patch density have limited effects on root proliferation and plant performance in *Abutilon theophrasti*. *Functional Ecology* **18**:836–843.

Lauter FR, Ninnemann O, Bucher M, Riesmeier JW, Frommer WB. 1996. Preferential expression of an ammonium transporter and of two putative nitrate transporters in root hairs of tomato. *Proceedings* of the *National Academy of Sciences of the USA* **93**:8139–8144.

Lineweaver H, Burk D. 1934. The determination of enzyme dissociation constants. *Journal of the American Chemical Society* **56**:658–666.

McNickle GG, Brown JS. 2014. An ideal free distribution explains the root production of plants that do not engage in a tragedy of the commons game. *Journal of Ecology* **102**:963–971.

McNickle GG, Cahill JF. 2009. Plant root growth and the marginal value theorem. *Proceedings of the National Academy of Sciences of the USA* **106**:4747–4751.

McNickle GG, Dybzinski R. 2013. Game theory and plant ecology. *Ecology Letters* **16**:545–555.

McNickle GG, Cahill JFJ, St Clair CC. 2009. Focusing the metaphor: plant foraging behaviour. *Trends in Ecology and Evolution* **24**:419–426.

McNickle GG, Deyholos MK, Cahill JF. 2013. Ecological implications of single and mixed nitrogen nutrition in *Arabidopsis thaliana*. *BMC Ecology* **13**:28. doi:10.1186/1472-6785-13-28.

Michaelis L, Menten ML. 1913. Die Kinetik der Invertinwirkung. *Biochemische Zeitschrift* **49**:333–369.

Nonacs P. 2001. State dependent behavior and the marginal value theorem. *Behavioral Ecology* **12**:71–83.

Rajaniemi TK, Reynolds HL. 2004. Root foraging for patchy resources in eight herbaceous plant species. *Oecologia* **141**:519–525.

R Development Core Team. 2009. *R: A language and environment for statistical computing*. R Foundation for Statistical Computing, Vienna, Austria, ISBN 3-900051-07-0, http://www.R-project.org/.

Real LA. 1977. Kinetics of functional response. *The American Naturalist* **111**:289–300.

Robinson D. 1994. The responses of plants to nonuniform supplies of nutrients. *New Phytologist* **127**:635–674.

Robinson D. 1996. Resource capture by localized root proliferation: why do plants bother? *Annals of Botany* **77**:179–185.

Shemesh H, Arbiv A, Gersani M, Ovadia O, Novoplansky A. 2010. The effects of nutrient dynamics on root patch choice. *Plos One* **5**:e10824. doi:10.1371/journal.pone.0010824.

Silvertown J, Gordon DM. 1989. A framework for plant behavior. *Annual Review of Ecology and Systematics* **20**:349–366.

Simpson SJ, Sibly RM, Lee KP, Behmer ST, Raubenheimer D. 2004. Optimal foraging when regulating intake of multiple nutrients. *Animal Behaviour* **68**:1299–1311.

Stephens DW, Krebs JR. 1986. *Foraging theory*. Princeton, NJ, Princeton University Press.

Stephens DW, Brown JS, Ydenberg RC. 2007. *Foraging: behavior and ecology*, 1st edn. Chicago: The University of Chicago Press.

Sutherland WJ, Stillman RA. 1988. The foraging tactics of plants. *Oikos* **52**:239–244.

Tian X, Doerner P. 2013. Root resource foraging: does it matter? *Frontiers in Plant Science* **4**:303. doi:10.3389/fpls.2013.00303.

Vincent TLS, Scheel D, Brown JS, Vincent TL. 1996. Trade-offs and coexistence in consumer-resource models: it all depends on what and where you eat. *The American Naturalist* **148**:1038–1058.

Webb CO, Donoghue MJ. 2005. Phylomatic: tree assembly for applied phylogenetics. *Molecular Ecology Notes* **5**:181–183.

4

Competition and soil resource environment alter plant–soil feedbacks for native and exotic grasses

Loralee Larios[1,2]* and Katharine N. Suding[1,3]

[1] Department of Environmental Science, Policy & Management, University of California Berkeley, 137 Mulford Hall, Berkeley, CA 94720-3114, USA
[2] Present address: Division of Biological Sciences, University of Montana, 32 Campus Dr HS104, Missoula, MT 59812, USA
[3] Present address: EBIO, University of Colorado, Ramaley N122, Campus Box 334, Boulder, CO 80309-0334, USA

Associate Editor: Inderjit

Abstract. Feedbacks between plants and soil biota are increasingly identified as key determinants of species abundance patterns within plant communities. However, our understanding of how plant–soil feedbacks (PSFs) may contribute to invasions is limited by our understanding of how feedbacks may shift in the light of other ecological processes. Here we assess how the strength of PSFs may shift as soil microbial communities change along a gradient of soil nitrogen (N) availability and how these dynamics may be further altered by the presence of a competitor. We conducted a greenhouse experiment where we grew native *Stipa pulchra* and exotic *Avena fatua*, alone and in competition, in soils inoculated with conspecific and heterospecific soil microbial communities conditioned in low, ambient and high N environments. *Stipa pulchra* decreased in heterospecific soil and in the presence of a competitor, while the performance of the exotic *A. fatua* shifted with soil microbial communities from altered N environments. Moreover, competition and soil microbial communities from the high N environment eliminated the positive PSFs of *Stipa*. Our results highlight the importance of examining how individual PSFs may interact in a broader community context and contribute to the establishment, spread and dominance of invaders.

Keywords: *Avena fatua*; California grasslands; competition; exotic species; native species; nitrogen enrichment; plant–soil feedbacks; *Stipa pulchra*.

Introduction

Increasingly, feedbacks between plants and soil biota are being identified as key determinants of the abundance and composition of plant communities (Wardle *et al.* 2004; van der Putten *et al.* 2013). Negative feedbacks, where plant species are less productive in their 'home' soil biota, are thought to be important in the maintenance of plant diversity (Reynolds *et al.* 2003; Vogelsang *et al.* 2006) and promote species coexistence at small scales. Positive feedbacks, where species are more productive in 'home' soil biota, can contribute to species dominance and patch dynamics on a landscape scale

* Corresponding author's e-mail address: loralee.larios@mso.umt.edu

(Chase and Leibold 2003; Shurin 2007). Introduced species seem to be exceptions to the rule, as soil biota is often found to have little impact on invasion success (Callaway et al. 2004; Inderjit and van der Putten 2010; Suding et al. 2013). However within the introduced range, the positive effects of 'home' soil biota may contribute to exotic dominance (Grman and Suding 2010). Translating demonstrated plant–soil feedbacks (PSFs) to abundance patterns has had varied results (Klironomos 2002; Yelenik and Levine 2011), as these effects are often considered in isolation from other ecological processes. Environmental factors can affect the dependency of plants on soil biota (Johnson et al. 2003) and the composition of the soil communities (Zeglin et al. 2007). However, the relative strength of these feedbacks may be small compared with interactions such as plant competition (Shannon et al. 2012). Addressing this context dependency of PSFs is key to our understanding of the role of PSFs in plant invasions and exotic dominance.

Soil nitrogen (N) enrichment, via fertilization, atmospheric deposition or other anthropogenic inputs, can alter soil microbial communities (Bissett et al. 2013) and facilitate plant invasions (Vitousek et al. 1997; Brooks 2003). Despite this evidence, our understanding of how feedbacks may shift in light of these changes to impact plant performance and subsequent invasion dynamics is limited. Under elevated soil N, microbial composition can shift towards a more bacterial dominated community (Bardgett et al. 1999; Allison 2002; Bradley et al. 2006; Zeglin et al. 2007) and can experience a loss of arbuscular mycorrhizal fungal (AMF) species within soil microbial communities (Egerton-Warburton et al. 2007; Liu et al. 2012). However, the net effect of these soil microbial community shifts on PSFs and invasions is unclear. In addition to these changes in soil microbial communities, host plant identity, which plays a significant role in dictating soil microbial community composition and feedback strength (Bardgett and Cook 1998; Hausmann and Hawkes 2009), can also shift in tandem with resources. For example, native and exotic species loss has been observed with increasing resource availability across multiple grassland systems, but resident natives had a greater likelihood of loss than exotics (Suding et al. 2005). Synergistic interactions between shifts in soil microbial communities due to altered resources and shifts in exotic abundance may result in enhanced PSFs that benefit the exotic vs. the native, contributing to invasion; yet these interactive effects are seldom studied.

Plant–soil feedbacks are often assessed at the individual plant level in isolation of other ecological processes such as plant–plant interactions, although they can jointly operate in regulating community diversity and abundance (Hodge and Fitter 2013). Plants can actively secrete compounds within their rhizosphere to promote the acquisition of resources (Hartmann et al. 2009), but the presence of the competitor can cause resources to be more limiting and potentially alter the magnitude of PSFs, either intensifying the PSF (Van der Putten and Peters 1997) or eliminating them (Casper and Castelli 2007). Scaling up individual plant responses to soil communities to the community level requires an understanding of how competitive hierarchies may interact with existing PSFs; however, only a handful of studies have investigated both (Van der Putten and Peters 1997; Casper and Castelli 2007; Hol et al. 2013) and rarely in the context of invasion (Yelenik and Levine 2011; Shannon et al. 2012).

Here, we propose that (i) soil microbial communities from differing resource environments and host plants and (ii) the interaction between plant competition and microbial community can influence the magnitude and direction of PSFs. We focus our study on California grasslands, which have experienced a large-scale shift from native perennial grasses mixed with annual forbs to exotic annual grasses over the last century (Jackson 1985), as well as an increase in atmospheric N deposition (Fenn et al. 2003). In this system, annual exotic grasses can shift the composition of soil microbial communities (Hawkes et al. 2005, 2006) and can alter the community of AMF colonizing roots of native grasses (Hausmann and Hawkes 2009, 2010), reducing the growth of native species (Vogelsang and Bever 2009).

We conducted a greenhouse experiment where we grew a native, *Stipa pulchra,* and exotic, *Avena fatua* (hereafter, *Stipa* and *Avena,* respectively), in soils inoculated with conspecific ('home') and heterospecific ('away') soil communities. To examine the interactive effects of resource environment and plant species identity on microbial communities, soil inocula were collected from a field experiment where *Avena* and *Stipa* plots had been treated with either carbon or N addition to alter soil resource availability. To examine the interaction between competitive interactions and microbial function on plant species performance, we grew plants individually or with a neighbour. We hypothesized that if positive PSFs contributed to invasion, then *Avena* would grow better in its 'home' soil than 'away' soil communities (note: we refer to 'home' soil as soils conditioned by the exotic in the introduced range vs. in its native range). Conversely, if *Stipa* were to grow better in its 'home' soil compared with 'away', positive PSFs would prevent invasion. Moreover, we hypothesized that soil communities from different soil resource environments would contribute to invasion if *Avena* were to grow better with soil communities from high N sites. Lastly, we hypothesized that plant–plant interactions would contribute to invasion if the presence of a competitor weakened the benefit that *Stipa* has when grown in its 'home' soil communities.

Methods

Study species and soil

We focused on two grass species common to southern California grasslands: the native perennial, *S. pulchra*, and the exotic annual, *A. fatua* (nomenclature follows Baldwin *et al.* 2012). Soils for the experiment were collected from Loma Ridge in Irvine, CA within the Irvine Ranch Land Reserve (N: 33.7501, W: −117.71787)—a grassland largely dominated by a mixture of exotic annual grasses and native perennial grasses (Larios *et al.* 2013). Background soil was collected from this site and upon collection the soil was air dried, sieved through a 2-mm sieve to remove rocks and debris and steam sterilized at 120 °C. This soil was then mixed 1 : 1 with sterile coarse sand and used as the sterile background soil to fill 164 mL cone-tainers for the greenhouse experiment described below.

To test how soil communities from varying N environments affected the strength of PSFs on plant performance, we collected soil inocula in March 2010 from a field experiment where native and exotic plants had been grown separately under low, ambient and high soil N (L. Larios and K. N. Suding, unpubl. data). Within the experiment, N was increased at a rate of 6 g N m^{-2} year^{-1}, which we applied in the form of slow-release calcium nitrate (Florikan®, Sarasota, FL), and was decreased using table sugar at a rate of 421 g C m^{-2} year^{-1}. In similar sites, this level of carbon addition decreased N by ~30 % (Cleland *et al.* 2013). Soil amendments were applied three times over each growing season, beginning in the 2009 growing season (i.e. 2009 growing season is defined as October 2008 to June 2009) until the end of the 2011 growing season. In total, the experiment consisted of 30 plots (5 replicate blocks × 2 neighbourhood types × 3 soil N). Within each of the five experimental blocks, we collected soils from both the native and exotic plots. Within the native plots, soils were collected directly under a *Stipa* individual and for the exotics, under a stand of *Avena*, ensuring that roots were collected with each soil sample. This soil was kept cool (~4–6 °C) and shipped to the University of California, Berkeley. Within 3 weeks of collection, the soils from each block were bulked to form the soil inocula used in the experiment. Spatial variation can contribute to high variability in microbial communities within a site (Pereira e Silva *et al.* 2012). Our goal was not to assess this spatial variability by testing the effects of the field soil resource additions on soil microbial communities *per se*, but to ask how soil communities from different resource environments impact plant growth and feedbacks. Therefore, we composited the soils from each block to form the soil inocula used in our soil treatments to ensure that we inoculated with the entire microbial taxa found across a resource environment. We additionally included a sterile soil treatment with no inoculum. Therefore, we had a total of seven soil-community treatments: *Stipa*-conditioned, (i) low N, (ii) ambient N, (iii) high N; *Avena*-conditioned, (iv) low N, (v) ambient N, (vi) high N and (vii) sterile control. The inoculum was added to the cone-tainers at a ratio of 30 : 1, sterile background soil (described above) to inoculum (Bever 1994).

Experimental design

To assess the interaction between soil communities from different resource environments and plant host on plant–soil interactions in the absence of competitive interactions, we planted three individual seeds of each species by themselves into cone-tainers with the soil inoculated with either conspecific or heterospecific soil communities from low N, ambient and high N sites. To examine the effect of competitive interactions on plant–soil interactions, we also planted species mixtures (consisting of one *Stipa* and one *Avena*) with the seven soil-community treatments described above. After initial germination we removed individuals from all cone-tainers so that each cone had a single individual for the no-competition *Stipa* and *Avena* treatments and one individual of each species for the competitive mixtures. We transplanted seedlings into the cones if no seeds germinated. The transplanted seedlings were planted at the same time as the other seeds so that they were comparable in size upon transplant. Thus we had a total of 420 cone-tainers (7 soil-community inocula × 3 species plantings × 10 blocks × 2 replicates within each block). The multiple replicates within a single block were averaged so that only block means were used in subsequent analyses.

The plants were grown at the Oxford Tract Greenhouse at the University of California, Berkeley, and were watered regularly with distilled water, without supplemental lighting or fertilizer. The blocks were rotated every week to minimize any differential effects of lighting and temperature within the greenhouse. Additionally, the cone-tainers were spaced such that there were never two cone-tainers adjacent to each other, to minimize any potential cross-contamination of soil inocula with watering. All above- and below-ground biomass was harvested 10 weeks after initial planting. Transplanted individuals were harvested 10 weeks after transplanting. The biomass was sorted to species for the competition treatment, and all biomass was dried for 48 h at 60 °C.

Statistical analysis

To evaluate how plant growth varied across the experiment, we analysed total biomass (sum of above- and below-ground biomass) with a three-way ANOVA, specifying block as a random factor, using the Proc Mixed module (SAS Institute, v 9.1).

We calculated the effect of the soil inoculum pairwise between the sterile soil treatment and the other soil inocula within each block with a natural log-response ratio, '$\ln(B_i/B_c)$', where B was the total biomass of the plant in either an inoculated soil treatment ('i') or sterile soil ('c'). We assessed the directionality of the response ratio using t-tests, where a value >0 indicated a significant positive response and a value <0 indicated a significant negative response. To assess whether the effect of simply adding soil inocula changed with culturing species or soil resource site, we ran a mixed effects model using the Proc Mixed module separately for each species with the inoculum response ratio as the response variable, soil-community sources (plant species, soil resource site) as two fixed factors and block as a random effect.

To assess whether soil communities from varying soil resources affect plant performance, we calculated for each species a natural log-response ratio (i.e. $\ln(B_{alteredN}/B_{ambN})$), separately for the conspecific and heterospecific soil communities. We then analysed this soil resource response ratio in a mixed model with soil-community sources (i.e. species and soil resource environment) as fixed effects and block as a random effect. We assessed directionality where a positive value would indicate that the individual grew better in the altered soil communities, while a negative value would indicate that it grew worse using t-tests as described above. A significant effect of soil resource environment for *Avena* would indicate that the changes in soil communities due to resource environment do alter performance, supporting our second hypothesis. A significant effect of the species soil inocula would indicate whether the effect of the soil communities from varied resourced environments varied between conspecific and heterospecific soil inocula.

Plant–soil feedback strength was calculated as '$\ln(B_{home}/B_{away})$', where B_{home} is the total biomass of an individual when grown in their conspecific soil communities and B_{away} is the total biomass when grown in heterospecific soil communities. Plant–soil feedback strength was calculated within each soil resource soil microbial community and competition treatment (i.e. *Avena* feedback for no-competition and low N would be the comparison of *Avena* biomass when grown alone, between conspecific (home) and heterospecific (away) cultured soils at low N sites). For blocks where individuals of a specific treatment died, we averaged biomass across the other blocks for that species as a substitute. We did this five times for *Stipa* when grown alone. For the competition treatments, we replaced the biomass of both the species nine times. However, we dropped any blocks that had lost replicates for three or more soil inocula

treatments, resulting in a loss of one block for the no-competition treatment and three for the competition treatments.

To assess how PSF responses changed with competition or across soil communities from different soil N environments, we ran a mixed effects model with PSF as the response variable and soil N inocula sources, target species identity and competition as fixed factors. Block was included as a random factor and any significant interactions were evaluated with post-hoc Tukey pairwise difference tests. A significant culturing species–target species interaction would indicate that PSFs could facilitate invasion, if *Avena* experienced no feedbacks when grown in 'away' soil communities, but would indicate invasion resistance if *Stipa* experienced positive feedbacks when grown in 'home' soil communities. A significant competition–species interaction would indicate that PSFs changed in the presence of a competitor, where a negative shift in feedbacks for *Stipa* when grown in competition would support our third hypothesis.

Results

Stipa pulchra response

Soil inocula and competitive environment both affected *Stipa* growth. *Stipa* total biomass was affected by soil microbial inoculum from *Avena* and from different soil N environments (culturing species × soil N interaction: $F_{2,76} = 8.22$, $P < 0.001$; [see Supporting Information]). Competition decreased *Stipa* biomass by almost 90 % (0.327 vs. 0.036 g, $F_{1,76} = 595.9$, $P < 0.0001$). Additionally, the competitive environment influenced the effect of soil inoculum on *Stipa* (competition × culturing species interaction: $F_{1,76} = 9.72$, $P < 0.01$; Fig. 1, square symbols). Comparisons of growth in sterilized soil indicate that *Avena*-cultured soil communities decreased *Stipa* growth while conspecific-cultured soils had a combination of neutral and negative effects compared with sterilized conditions (culturing species: $F_{1,40} = 14.18$, $P < 0.0001$; soil N: $F_{2,40} = 0.90$, $P = 0.41$; Fig. 2A).

When grown alone, *Stipa* grew better with conspecific-cultured soil communities compared with heterospecific (better in home vs. away soils), resulting in positive feedbacks when *Stipa* was grown alone (Fig. 3A, dark grey bars). These positive feedbacks diminished when *Stipa* was grown with *Avena* (Spp × Comp, $F_{1,76} = 7.45$, $P < 0.01$; Fig. 3A, light grey bars) and with high N soil communities (soil N × Spp, $F_{2,76} = 6.24$, $P < 0.01$, low and ambient N vs. high N Tukey HSD $P < 0.01$ and $P < 0.05$, respectively), resulting in the development of a strong negative feedback when in competition with *Avena* and in high N soil communities (Fig. 3).

Soil microbial communities from different N environments did not alter *Stipa* growth; however, *Stipa* grew better with soil communities cultured by the heterospecific, *Avena* (culturing species: $F_{1,24} = 4.25$, $P = 0.05$; soil N: $F_{1,24} = 0.23$, $P = 0.63$; Spp × soil N: $F_{1,24} = 0.95$, $P = 0.34$; Fig. 4).

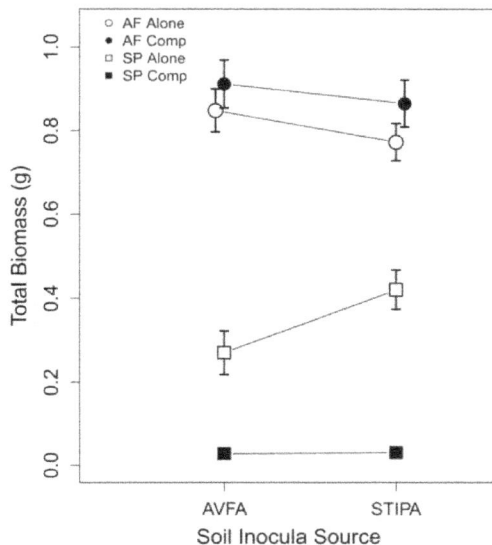

Figure 1. Total (above- and below-ground) biomass for *S. pulchra* (squares) and *A. fatua* (circles) when grown alone (open symbols) or with a competitor (filled symbols) with soil inocula cultured under ambient resources by conspecifics and heterospecifics. Competition decreased *Stipa* biomass, regardless of which soil community *Stipa* was grown. *Avena* grew similarly in both conspecific (*Stipa*) and heterospecific (*Avena*) soils regardless of the presence of a competitor. Mean ± 1 SE. Error bars for *Stipa* with competitors are hidden by symbols.

Avena fatua response

Avena exhibited little response to different soil communities (Fig. 3). The only exception to this pattern was a negative feedback at low N, where it grew worse in 'home' low N soil communities (soil N × Spp, low vs. ambient N: Tukey HSD, $P < 0.05$). Interactions with *Stipa* did not alter *Avena* growth ($F_{1,76} = 0.01$, $P = 0.91$; Fig. 1, circles) nor change PSFs (Fig. 3). Additionally, *Avena* growth was greater in 'away', low N soil communities than under sterile soil conditions (Fig. 2B, culturing species × soil N: $F_{2,40} = 3.36$, $P < 0.05$).

The soil resource environment did not alter the soil community in a way that altered *Avena* biomass. Much like *Stipa*'s response, *Avena* grew better in soils conditioned by heterospecifics compared with conspecifics (culturing species: $F_{1,24} = 10.22$, $P < 0.01$; soil N: $F_{1,24} = 1.87$, $P = 0.18$; Spp × soil N: $F_{1,24} = 1.45$, $P = 0.23$).

Discussion

Plant–soil feedbacks involve two effects: soil-community effects on plant growth and plant species effects on soil communities (Bever 1994). As such, these feedbacks have most often been studied by isolating these two factors (Kulmatiski *et al.* 2008). However, many other factors can affect the composition of microbial communities (Waldrop *et al.* 2006; Bissett *et al.* 2013), as well as the growth of plant species (Chase and Leibold 2003), leading us to expect that PSFs may be dependent on the broader environmental context (Kardol *et al.* 2013). Indeed, we find that two of these additional factors (soil resource environment effects on soil microbial communities and

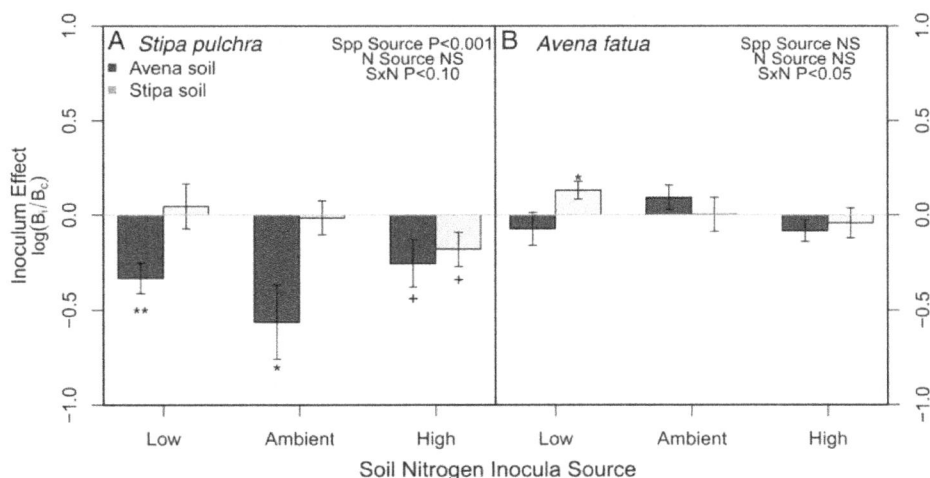

Figure 2. Effect of inoculating soil on plant performance for *S. pulchra* (A) and *A. fatua* (B). *Stipa* experienced negative effects (i.e. grew worse in the inoculated soil treatments compared with sterile) when grown in heterospecific (*Avena*) soil inoculum. Soil inocula affected *Avena* growth only when grown with inoculum from the heterospecific (*Stipa*) grown under low N environments. Mean ± 1 SE. Significantly different from zero: +P < 0.07, *P < 0.05, **P < 0.01.

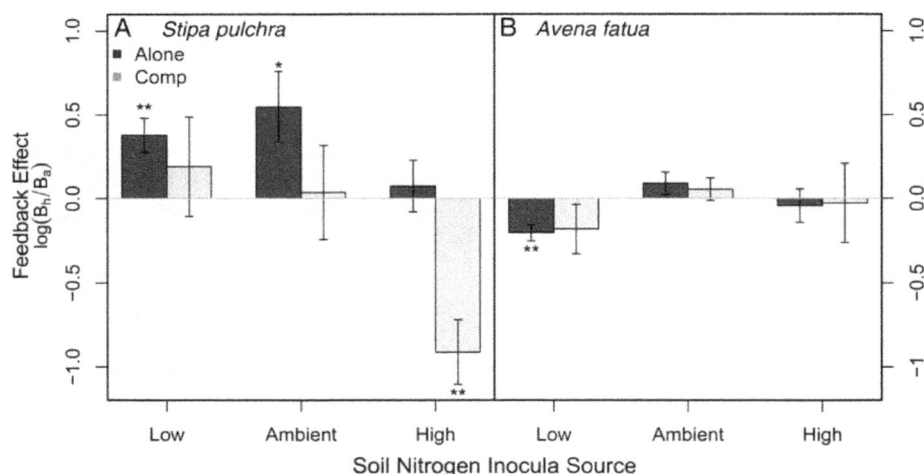

Figure 3. Plant–soil feedbacks for *S. pulchra* (A) and *A. fatua* (B) grown alone or with a competitor, across soils cultured by conspecifics or heterospecifics under varying resources. *Stipa* experienced positive feedbacks (i.e. grew better with its home soil communities) when grown alone in low N and ambient N soil communities, but these feedbacks became negative when grown in high N soil communities. *Avena* grew worse in its conspecific soil compared with heterospecific low N soil communities, resulting in a negative feedback. Means ± 1 SE. Significantly different from zero: *P < 0.05, **P < 0.01.

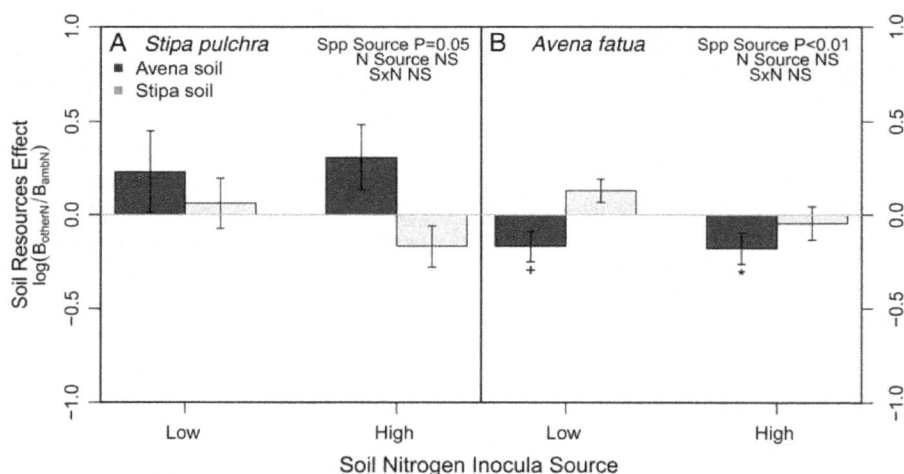

Figure 4. Effect of changes in soil community due to changes in soil nitrogen (N) resources on *S. pulchra* (A) and *A. fatua* (B) growth. *Stipa* grew better in soil communities from ambient N availability compared with low or high N availability when these soils were also cultured by heterospecifics, *Avena*. Conversely, *Avena* grew better in soil communities from ambient N availability when these soils were cultured by conspecifics. Means ± 1 SE. Significantly different from zero: +P < 0.07, *P < 0.05.

competitive effects on plant growth) strongly impact the strength and even direction of PSFs.

Plant species effects on microbial communities can strongly regulate species establishment and performance (Bever *et al.* 2010) and the presence or lack of these effects may have strong implications for plant invasions (Inderjit and van der Putten 2010). Here, we observed that the native *Stipa* responded to culturing plant identity, where it grew less in soils conditioned by *Avena*, suggesting that *Avena* is able to culture a distinct soil community that negatively affects the native *Stipa*. On the other hand, we found that *Avena* was not responsive to culturing plant

species identity as it grew similarly in soil conditioned by either conspecifics (*Avena*-conditioned) or heterospecifics (*Stipa*-conditioned) compared with sterile soil. While recent reviews have suggested that sterilized and unsterilized comparisons can be biased towards detecting negative responses to soil inocula (Kulmatiski *et al.* 2008; Brinkman *et al.* 2010), the strong response of *Stipa* to soil conditioned by *Avena* suggests that *Avena* may foster soil pathogens at a high enough density to affect *Stipa* growth. Interestingly, we observed an interaction between culturing plant host and soil N environment for both species, but the directionality varied for the native and exotic. *Stipa* grew worse in home soils compared with sterile when the soils came from the high N environment, and *Avena* grew better in heterospecific soils that were cultured at low N compared with sterile soil.

Our results support the idea that resource-induced changes to soil communities can impact PSFs, but the response may be species specific (Manning *et al.* 2008). Across the resource environments, we observed that neither *Stipa* nor *Avena* responded to changes in soil communities conditioned by *Stipa*. However both species responded to shifts in the *Avena*-conditioned soil communities, regardless of whether the conditioning was in low or high N environments, where *Stipa*'s performance improved, while *Avena*'s worsened (Fig. 4, dark grey bars). These results support previous findings that *Stipa* is able to foster a more diverse assemblage of soil biota compared with exotic annual grasses (Hausmann and Hawkes 2009), and thus, resource-induced shifts in soil communities may not have a large impact on plant growth. The positive response of *Stipa* to *Avena*-conditioned soil communities in different resource environments has interesting applications for management efforts aimed at native recovery. Soil N reduction activities are traditionally used to alter competitive interactions in favour of the natives (Blumenthal *et al.* 2003), and our results suggest that these soil N reductions may also minimize some of the negative effects on native species' growth that result from the soil conditioning of an exotic species like *Avena*. The small amount of inocula that we used may have resulted in lower densities of harmful pathogens and beneficial symbionts and contributed to the positive/neutral feedbacks that we observed for *Stipa* and *Avena*, respectively (Brinkman *et al.* 2010). However by assessing both the inocula effects and feedback effects, our results suggest that *Stipa*'s positive feedback is likely a result of *Avena* culturing a microbial community that negatively impacts *Stipa*. Additional experiments that explore the spatial variability in the soil community and partition the members of the community to assess the groups driving this pattern are needed to further our understanding of how consistent this response will be across a landscape.

Integrating PSFs into other ecological processes such as competition is key to scaling the impact of PSFs observed at the individual plant level up to the community level (Hodge and Fitter 2013; Kardol *et al.* 2013). Competition had no impact on *Avena* growth, either independently or through a PSF interaction. Independently we observed: (i) when grown alone, *Stipa* grew better in its home soil compared with *Avena*-conditioned soil and (ii) *Stipa* had a strong negative response to competition by *Avena*. However, when we assessed the potential interactive effects of competition and feedbacks, we observed that *Stipa*'s positive feedback was eliminated under competition. While this result is consistent with the competitive hierarchy previously observed between *Avena* and *Stipa* seedlings (Dyer and Rice 1997, 1999), this study does not allow us to decipher whether this result is also due to the strong control that *Avena* species may have on the soil community (Hausmann and Hawkes 2009). The strong effect of *Avena* on *Stipa* performance suggests that restoration efforts should continue to focus on ways to reduce the abundance of exotics in order to promote native species recovery.

Our approach also allowed us to examine how feedbacks may change in the presence of a competitor and soil communities conditioned in different soil N environments. We observed that soils from high N environments eliminated *Stipa*'s positive feedback and interacted strongly with competition such that *Stipa* grew worse in its 'home' soil compared with 'away' soils. Similarly to the individual effects of soil communities from different resource environments, we observed that *Avena* grew worse in its 'home' soil compared with 'away' soils. Our results highlight the importance of future studies to explore how PSFs may interact with ongoing environmental change such as atmospheric N deposition to influence the resilience of existing native communities to invasion.

Conclusions

In conclusion, we found that both plant host and soil resource environment effects on soil communities may alter plant growth and that these impacts can shift in the presence of a competitor. Although the relationships of plant host and soil microbial communities are often assessed in isolation, our ability to understand how they may contribute to observed abundance patterns require us to investigate them in light of other key ecological processes. This more integrated assessment is key to our improved understanding of how plant–soil interactions may contribute to invader establishment, spread and dominance.

Sources of Funding

This work was supported by the NSF Graduate Research Fellowship Program (DEB 1106400 to L.L.) and NSF (DEB 09-19569 to K.N.S.).

Contributions by the Authors

Both L.L. and K.N.S. designed the experiment and edited the manuscript. L.L. conducted the data collection and statistical analyses and wrote the first draft of the manuscript.

Acknowledgements

We thank L. August-Schmidt, J. Butler, A. Carlson, H. Gao and J. Martinez for help in the greenhouse and H. Bueno, E. Stone for help in the lab. We also thank K. Baer, J. Maron, M. Spasojevic, L. Waller and two anonymous reviewers for their helpful comments on this manuscript. Lastly, we thank the Irvine Ranch Conservancy for access to our research sites.

Supporting Information

The following Supporting Information is available in the online version of this article –

Figure S1. The average individual total biomass for *Stipa pulchra* (A) and *Avena fatua* (B) as above- and below-ground biomass across soil inocula and competition treatments.

Literature Cited

Allison VJ. 2002. Nutrients, arbuscular mycorrhizas and competition interact to influence seed production and germination success in *Achillea millefolium*. *Functional Ecology* 16:742–749.

Baldwin BG, Goldman DH, Keil DJ, Patterson R, Rosatti TJ, Wilken DH (eds) 2012. *The Jepson manual: vascular plants of California*, 2nd edn. Berkeley, CA: University of California Press.

Bardgett RD, Cook R. 1998. Functional aspects of soil animal diversity in agricultural grasslands. *Applied Soil Ecology* 10:263–276.

Bardgett RD, Mawdsely JL, Edwards S, Hobbs PJ, Rodwell JS, Davies WJ. 1999. Plant species and nitrogen effects on soil biological properties of temperate upland grasslands. *Functional Ecology* 13:650–660.

Bever JD. 1994. Feedback between plants and their soil communities in an old field community. *Ecology* 75:1965–1977.

Bever JD, Dickie IA, Facelli E, Facelli JM, Klironomos J, Moora M, Rillig MC, Stock WD, Tibbett M, Zobel M. 2010. Rooting theories of plant community ecology in microbial interactions. *Trends in Ecology and Evolution* 25:468–478.

Bissett A, Brown MV, Siciliano SD, Thrall PH. 2013. Microbial community responses to anthropogenically induced environmental change: towards a systems approach. *Ecology Letters* 16:128–139.

Blumenthal DM, Jordan NR, Russelle MP. 2003. Soil carbon addition controls weeds and facilitates prairie restoration. *Ecological Applications* 13:605–615.

Bradley K, Drijber RA, Knops J. 2006. Increased N availability in grassland soils modifies their microbial communities and decreases the abundance of arbuscular mycorrhizal fungi. *Soil Biology and Biochemistry* 38:1583–1595.

Brinkman EP, Van der Putten WH, Bakker EJ, Verhoeven KJF. 2010. Plant–soil feedback: experimental approaches, statistical analyses and ecological interpretations. *Journal of Ecology* 98: 1063–1073.

Brooks ML. 2003. Effects of increased soil nitrogen on the dominance of alien annual plants in the Mojave desert. *Journal of Applied Ecology* 40:344–353.

Callaway RM, Thelen GC, Barth S, Ramsey PW, Gannon JE. 2004. Soil fungi alter interactions between the invader *Centaurea maculosa* and North American natives. *Ecology* 85:1062–1071.

Casper BB, Castelli JP. 2007. Evaluating plant–soil feedback together with competition in a serpentine grassland. *Ecology Letters* 10: 394–400.

Chase JM, Leibold MA. 2003. *Ecological niches: linking classical and contemporary approaches*. Chicago, IL: University of Chicago Press.

Cleland EE, Larios L, Suding KN. 2013. Strengthening invasion filters to re-assemble native plant communities: soil resources and phenological overlap. *Restoration Ecology* 21:390–398.

Dyer AR, Rice KJ. 1997. Intraspecific and diffuse competition: the response of *Nassella pulchra* in a California grassland. *Ecological Applications* 7:484–492.

Dyer AR, Rice KJ. 1999. Effects of competition on resource availability and growth of a California bunchgrass. *Ecology* 80: 2697–2710.

Egerton-Warburton LM, Johnson NC, Allen EB. 2007. Mycorrhizal community dynamics following nitrogen fertilization: a cross-site test in five grasslands. *Ecological Monographs* 77: 527–544.

Fenn ME, Haeuber R, Tonnesen GS, Baron JS, Grossman-Clarke S, Hope D, Jaffe DA, Copeland S, Geiser L, Rueth HM, Sickman JO. 2003. Nitrogen emissions, deposition, and monitoring in the western United States. *Bioscience* 53:391–403.

Grman E, Suding KN. 2010. Within-year soil legacies contribute to strong priority effects of exotics on native California grassland communities. *Restoration Ecology* 18:664–670.

Hartmann A, Schmid M, van Tuinen D, Berg G. 2009. Plant-driven selection of microbes. *Plant and Soil* 321:235–257.

Hausmann NT, Hawkes CV. 2009. Plant neighborhood control of arbuscular mycorrhizal community composition. *New Phytologist* 183:1188–1200.

Hausmann NT, Hawkes CV. 2010. Order of plant host establishment alters the composition of arbuscular mycorrhizal communities. *Ecology* 91:2333–2343.

Hawkes CV, Wren IF, Herman DJ, Firestone MK. 2005. Plant invasion alters nitrogen cycling by modifying the soil nitrifying community. *Ecology Letters* 8:976–985.

Hawkes CV, Belnap J, D'Antonio C, Firestone MK. 2006. Arbuscular mycorrhizal assemblages in native plant roots change in the presence of invasive exotic grasses. *Plant and Soil* 281:369–380.

Hodge A, Fitter AH. 2013. Microbial mediation of plant competition and community structure. *Functional Ecology* 27:865–875.

Hol WHG, de Boer W, ten Hooven F, van der Putten WH. 2013. Competition increases sensitivity of wheat (*Triticum aestivum*) to biotic plant–soil feedback. *PLoS ONE* 8:6.

Inderjit, van der Putten WH. 2010. Impacts of soil microbial communities on exotic plant invasions. *Trends in Ecology and Evolution* 25:512–519.

Jackson LE. 1985. Ecological origins of California's Mediterranean grasses. *Journal of Biogeography* 12:349–361.

Johnson NC, Rowland DL, Corkidi L, Egerton-Warburton LM, Allen EB. 2003. Nitrogen enrichment alters mycorrhizal allocation at five mesic to semiarid grasslands. *Ecology* 84:1895–1908.

Kardol P, De Deyn GB, Laliberte E, Mariotte P, Hawkes CV. 2013. Biotic plant–soil feedbacks across temporal scales. *Journal of Ecology* 101:309–315.

Klironomos JN. 2002. Feedback with soil biota contributes to plant rarity and invasiveness in communities. *Nature* **417**:67–70.

Kulmatiski A, Beard KH, Stevens JR, Cobbold SM. 2008. Plant–soil feedbacks: a meta-analytical review. *Ecology Letters* **11**: 980–992.

Larios L, Aicher RJ, Suding KN. 2013. Effect of propagule pressure on recovery of a California grassland after an extreme disturbance. *Journal of Vegetation Science* **24**:1043–1052.

Liu YJ, Shi GX, Mao L, Cheng G, Jiang SJ, Ma XJ, An LZ, Du GZ, Johnson NC, Feng HY. 2012. Direct and indirect influences of 8 yr of nitrogen and phosphorus fertilization on Glomeromycota in an alpine meadow ecosystem. *New Phytologist* **194**: 523–535.

Manning P, Morrison SA, Bonkowski M, Bardgett RD. 2008. Nitrogen enrichment modifies plant community structure via changes to plant–soil feedback. *Oecologia* **157**:661–673.

Pereira e Silva MC, Dias ACF, van Elsas JD, Salles JF. 2012. Spatial and temporal variation of archaeal, bacterial and fungal communities in agricultural soils. *PLoS ONE* **7**:e51554. doi:10.1371/journal. pone.0051554.

Reynolds HL, Packer A, Bever JD, Clay K. 2003. Grassroots ecology: plant–microbe–soil interactions as drivers of plant community structure and dynamics. *Ecology* **84**:2281–2291.

Shannon S, Flory SL, Reynolds H. 2012. Competitive context alters plant–soil feedback in an experimental woodland community. *Oecologia* **169**:235–243.

Shurin JB. 2007. How is diversity related to species turnover through time? *Oikos* **116**:957–965.

Suding KN, Collins SL, Gough L, Clark C, Cleland EE, Gross KL, Milchunas DG, Pennings S. 2005. Functional- and abundance-based mechanisms explain diversity loss due to N fertilization. *Proceedings of the National Academy of Sciences of the USA* **102**:4387–4392.

Suding KN, Harpole WS, Fukami T, Kulmatiski A, MacDougall AS,

Stein C, van der Putten WH. 2013. Consequences of plant–soil feedbacks in invasion. *Journal of Ecology* **101**:298–308.

Van der Putten WH, Peters BAM. 1997. How soil-borne pathogens may affect plant competition. *Ecology* **78**:1785–1795.

van der Putten WH, Bardgett RD, Bever JD, Bezemer TM, Casper BB, Fukami T, Kardol P, Klironomos JN, Kulmatiski A, Schweitzer JA, Suding KN, Van de Voorde TFJ, Wardle DA. 2013. Plant–soil feedbacks: the past, the present and future challenges. *Journal of Ecology* **101**:265–276.

Vitousek PM, Aber JD, Howarth RW, Likens GE, Matson PA, Schindler DW, Schlesinger WH, Tilman DG. 1997. Human alteration of the global nitrogen cycle: sources and consequences. *Ecological Applications* **7**:737–750.

Vogelsang KM, Bever JD. 2009. Mycorrhizal densities decline in association with nonnative plants and contribute to plant invasion. *Ecology* **90**:399–407.

Vogelsang KM, Reynolds HL, Bever JD. 2006. Mycorrhizal fungal identity and richness determine the diversity and productivity of a tallgrass prairie system. *New Phytologist* **172**: 554–562.

Waldrop MP, Zak DR, Blackwood CB, Curtis CD, Tilman D. 2006. Resource availability controls fungal diversity across a plant diversity gradient. *Ecology Letters* **9**:1127–1135.

Wardle DA, Bardgett RD, Klironomos JN, Setala H, van der Putten WH, Wall DH. 2004. Ecological linkages between aboveground and belowground biota. *Science* **304**:1629–1633.

Yelenik SG, Levine JM. 2011. The role of plant–soil feedbacks in driving native-species recovery. *Ecology* **92**:66–74.

Zeglin LH, Stursova M, Sinsabaugh RL, Collins SL. 2007. Microbial responses to nitrogen addition in three contrasting grassland ecosystems. *Oecologia* **154**:349–359.

5

Extractable nitrogen and microbial community structure respond to grassland restoration regardless of historical context and soil composition

Sara Jo M. Dickens[1]*, Edith B. Allen[1], Louis S. Santiago[1] and David Crowley[2]

[1] Department of Botany and Plant Sciences, University of California Riverside, Riverside, CA 92521, USA
[2] Department of Environmental Sciences, University of California Riverside, Riverside, CA 92521, USA

Associate Editor: Inderjit

Abstract. Grasslands have a long history of invasion by exotic annuals, which may alter microbial communities and nutrient cycling through changes in litter quality and biomass turnover rates. We compared plant community composition, soil chemical and microbial community composition, potential soil respiration and nitrogen (N) turnover rates between invaded and restored plots in inland and coastal grasslands. Restoration increased microbial biomass and fungal : bacterial (F : B) ratios, but sampling season had a greater influence on the F : B ratio than did restoration. Microbial community composition assessed by phospholipid fatty acid was altered by restoration, but also varied by season and by site. Total soil carbon (C) and N and potential soil respiration did not differ between treatments, but N mineralization decreased while extractable nitrate and nitrification and N immobilization rate increased in restored compared with unrestored sites. The differences in soil chemistry and microbial community composition between unrestored and restored sites indicate that these soils are responsive, and therefore not resistant to feedbacks caused by changes in vegetation type. The resilience, or recovery, of these soils is difficult to assess in the absence of uninvaded control grasslands. However, the rapid changes in microbial and N cycling characteristics following removal of invasives in both grassland sites suggest that the soils are resilient to invasion. The lack of change in total C and N pools may provide a buffer that promotes resilience of labile pools and microbial community structure.

Keywords: Carbon; exotic grasses; exotic plants; phospholipid fatty acid; resilience.

Introduction

The effects of exotic plant invasions on terrestrial ecosystems vary temporally and spatially and span scales ranging from the plant rhizosphere to changes in nutrient flux that occur at the ecosystem level (Ehrenfeld 2003; Potthoff *et al.* 2009). Previously, the impacts of exotic invasive plants on soil microbial communities and nutrient fluxes have received considerable attention (e.g. Jackson *et al.* 1988, 1989; Bever *et al.* 1997; Hawkes *et al.* 2005, 2006; Wolfe and Klironomos 2005). However, belowground

* Corresponding author's e-mail address: ecologybridge@gmail.com

responses to restoration practices and studies on the legacy effects of plant invasions are relatively new areas of research (Potthoff et al. 2006, 2009; Kulmatiski and Beard 2008, 2011; Dickens 2010; Dickens and Allen 2014). The capacity of invaded systems to recover from short-term and legacy effects of exotic plants is unknown. In addition, the role of legacy effects of exotic invasion and exotic species identity in the success of restoration is unclear, greatly limiting the knowledge base needed for strategic restoration of invaded lands.

One mechanism by which exotic plants impact ecosystems is by decoupling plant–soil feedback loops that previously functioned in soils under native vegetation. We define decoupling here and the interruption of interactions between plants and soil via soil inputs and microbial community responses (Bardgett et al. 2013). Feedback loops describe how plants, soils and microorganisms interact through resources. For example, a plant species may produce particular soil inputs via senescent biomass and exudates that become resources for soil microbes. Microbes that use these resources determine rates of nutrient cycling and thus nutrient availability to plants. Through this feedback loop, plants and microbes may exert selective pressure on one another (Wardle 2002; Eviner and Chapin 2003; Santiago et al. 2005; Santiago 2007). In the case of plant invasion, a new species' arrival may alter the microbial community, leading to further modifications of belowground processes such as nutrient turnover or the introduction of microbial species associated with this novel plant. The end result can be inhibition of native plant species and/or the facilitation of the invading, exotic plant species (Bever et al. 1997; Ehrenfeld 2003; Wolfe and Klironomos 2005).

Introductions of plant species that differ in litter quality, phenology and relative distribution of above and belowground biomass may result in especially strong plant–soil feedbacks. Exotic species may introduce novel nutrient uptake or litter deposition traits that could create positive feedbacks with the soil microbial community (Grayston et al. 1998; Eviner 2004; Batten et al. 2006). Exotics may shift the seasonal availability of extractable nitrogen (N) by introducing phenologies with earlier germination and growth rates (Jackson et al. 1988; Dickens 2010; Dickens and Allen 2014) and changes in soil properties that drive the selection and composition of microbial communities (Ehrenfeld 2003; Wardle et al. 2004; Berg and Smalla 2009; Potthoff et al. 2009). Additionally, exotic plant invasion can change cycling and availability of C, N and other nutrients (Christian and Wilson 1999; Ehrenfeld 2003; Yoshida and Allen 2004). Litter with high C : N promotes immobilization of N by microbes resulting in reduced available N (Brady and Weil 1996; Grayston et al. 1998; Cione et al. 2002; Potthoff et al. 2009). Invasion of exotic, annual grasses into a perennial bunchgrass grassland would be expected to introduce litter of lower C : N compared with native perennials which would increase decomposition and N cycling rates (Eviner and Firestone 2007; Potthoff et al. 2009).

California grasslands are highly invaded by exotic annuals and undergoing restoration in many locations, and thus an ideal system for studying plant–soil feedbacks through decoupling exotic plant species' plant–soil feedbacks using restoration. Plant biomass in grasslands turns over annually (Jackson et al. 1988) so grassland soils are likely to respond to altered plant inputs over a relatively short time scale. Due to the almost complete conversion of native perennial grasslands with annual forbs to exotic annual grassland with annual forbs (Biswell 1956; D'Antonio 2007; Minnich 2008), native California grasslands are a system of high conservation value and concern. Annual plant invasions began >200 years ago (Minnich 2008), and invasion is so widespread that there are no true relic grasslands to use as reference sites. However, even without relic grasslands, differences in soil microbial community structure, soil chemistry and nutrient flux rates between unrestored and restored soils can be used to evaluate the capacity of grassland soils to respond to changes in vegetation type.

Few studies have observed soil recovery after removal of invasives and native species restoration (but see Potthoff et al. 2006; Kulmatiski and Beard 2008, 2011; Dickens and Allen 2014). Shifts in microbial community structures can occur within a few years of plant species community compositional changes and microbial abundances may remain affected by land-use legacies for 50 years (Kulmatiski and Beard 2008). Further studies are necessary to determine which system responses are capable of rapid recovery or slower re-establishment of native feedback loops and whether patterns of responses are similar across differing environments. The objective of this study was to assess the capacity of southern California grassland soils to diverge from their invaded condition following the decoupling of long-term exotic plant–soil feedbacks. Invasion has likely led to the establishment of exotic plant–soil feedbacks that overwhelm feedbacks produced by the limited native plant population. Through restoration there are two possible, successful restoration scenarios. The first is successful removal of exotics and their associated plant–soil feedbacks leaving the restored grassland with limited native cover and bare ground initially. The second is a partially restored grassland that is dominated by native plant–soil feedbacks but still experiences some exotic plant–soil feedbacks due to constant, but limited, reinvasion (Fig. 1). We hypothesized that (i) restoration by removing exotic annual grasses will lead to shifts in the microbial community,

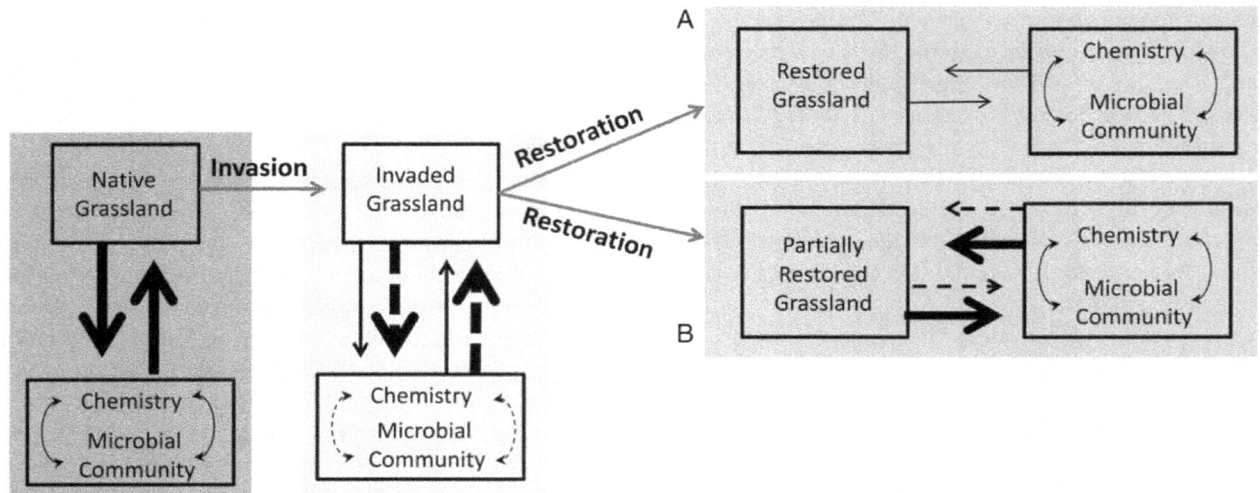

Figure 1. Invasion of original, native grasslands introduced new, exotic plant–soil feedbacks (solid black arrows represent native feedbacks and dotted black arrows represent plant soil feedbacks of exotics). Through restoration efforts plant–soil feedbacks can be altered leading to (A) a restored grassland experiencing no exotic feedbacks and moderate native feedbacks from the re-establishing native community or (B) a partially restored grassland experiencing a much higher proportion of native plant–soil feedbacks than exotic. Thickness of arrows indicates the degree to which feedbacks are influencing the system.

reflecting the development of new plant–soil feedbacks; (ii) exotic annuals often have higher quality litter than native perennial grasses, so microbial community shifts would translate into reduced carbon (C) and N cycling rates following restoration; and (iii) soil responses to restoration are sensitive to environmental conditions, which will result in different magnitudes of shifts in both microbial community and nutrient cycling at sites with different land-use/management histories and environmental contexts.

Methods

We investigated inland and coastal grassland sites in southern California that have been invaded by Mediterranean annual grasses and forbs, but that still support sparse native bunchgrasses and forbs. No uninvaded reference grasslands occur in southern California (Minnich 2008). An important contextual factor in this study is the difference in site histories, soils and current management strategies applied between our two study sites (Table 1). The inland grassland is located within the 4000-ha Santa Rosa Plateau Ecological Reserve in Murrieta, CA (33°31′N, 117°15′E). Soils at this location are basalts of the Vallecitos loam, thick solum variant (USDA NCSS SoilWeb Network), and restoration consisted of exotic grass control through prescribed spring burns but no reseeding (Gillespie and Allen 2004). The 120-ha White Point Preserve coastal grassland is located in San Pedro, Los Angeles County (33°43′N, 118°18′W), and soils are classified as a clay loam of the Diablo Clay

Table 1. Comparison of site abiotic properties, land-use history and restoration methodologies.

	Inland	Coastal
Mean annual precipitation (cm)	48	30
Annual temperature range (°C)	1–37	8–26
Soil clay (%)	12	36
Soil silt (%)	57	35
Soil sand (%)	31	29
Elevation (m)	579	47
Land-use history	Grazing	Defence missile facility
Year of restoration	1997	2000
Restoration method	Prescribed burn	Mowing, hand seeding, irrigation

Adobe series (Nelson *et al.* 1919). Restoration consisted of hand weeding and mowing of invasive plant species and reseeding of native species. To examine the effects of restoration on the structure and function of the invaded grassland, we used nine previously established 1 m^2 plots within areas that had undergone long-term restorations (9 years) and an additional nine plots in an adjacent unrestored grassland at each location. Restored areas were defined as those having experienced active restoration that had an exotic plant species cover of <40 % and were dominated by native species, while unrestored areas had ≥50 % cover by exotic plant species with sparse native species.

Measurements of ecosystem structure

Plant species richness, per cent cover by individual species and per cent of litter cover were measured annually by visual estimation in gridded 1 m² frames in each treatment at the peak of the growing season (March) in 2007–09. Net annual productivity of annuals was determined by harvesting biomass in four functional groups (native forb, native grass, exotic forb and exotic grass) clipped at the soil level from 0.25 m² sub-plots and scaled up to the 1 m² plot size using regression of plant biomass and per cent cover in 0.25 m² and per cent cover of 1 m² plots. Additional biomass was collected for chemical analysis of the vegetative plant tissue at peak plant growth. All biomass was oven dried at 60 °C and weighed. Biomass for tissue analysis was ground and analysed for total C and N on a soil combustion analyser system (Flash AlllZ, Thermo-Finnigan).

To determine the effects of restoration on soil biological and chemical characteristics, three soil cores of 2.5 cm diameter and 10 cm depth were collected per plot, composited to the plot level and then transported on ice to the laboratory where a portion of each sample was stored at −20 °C until processed for chemical analyses and the remaining portion of the sample at −80 °C for microbial analyses. Soils were analysed for total C and N by combustion, KCl-extractable NO_3 and NH_4, and bicarbonate-extractable phosphorus (Olsen P) by the University of California Analytical Laboratory at UC Davis (anlab. ucdavis.edu). Soil pH was measured using a 2:1 soil:water slurry. Soil cores were collected once annually in 2007 and 2009 at peak growth, and three times annually (at germination, peak plant growth and plant senescence) during 2007–08 for analysis of KCl-extractable N (NH_4^+ and NO_3^-).

Phospholipid fatty acid (PLFA) analysis was used to determine whether microbial community structure was affected by restoration of the native vegetation. With the exception of Archaea, all other living organisms contain PLFAs as a component of their cellular membranes (White et al. 1996; Hedrick et al. 2000). These compounds can be used as biomarkers to identify functional groups of microbes such as Gram-positive bacteria or arbuscular mycorrhizal (AM) fungi (Zelles and Bai 1994; White et al. 1996; Hedrick et al. 2000). Phospholipid fatty acids are preferable to the use of fatty acids alone as fatty acids can persist in soils for long periods of time representing a legacy of past microbial communities. Phospholipid fatty acid represent living organisms (White et al. 1996), thus ensuring capture of the current microbial community response to a disturbance such as exotic plant invasion or restoration activities. Samples were collected within 24 h of rainfall or wetting of soils to a 10-year average rainfall volume. Soil samples were passed through a 2-mm sieve and lyophilized prior to extraction. Phospholipid fatty acids were extracted from 6 g of soil following the modified Bligh–Dyer method (Frostegard et al. 1991). Quantification of fatty acids was obtained using a gas chromatograph (HP6980; Hewlett Packard, Palo Alto, CA, USA) with a flame ionization detector and HP3365 ChemStation Software. Phospholipid fatty acid peaks were converted to PLFA identities and abundances using MIDI Sherlock Microbial Identification System (MIDI, Inc., Newark, NJ, USA) followed by comparison of peak areas with a known internal standard 19:0 of known concentration. Bacterial biomarkers included: 14:0, 15:0 iso, 15:0 antiso, 16:0 iso, 16:0 iso G, 16:1 w9c, 16:1 w7c, 16:0, 16:1 2OH, 17:1 alcohol, 17:0 iso, 17:0 antiso, 17:0 cyclo, 17:1 w8c, 18:1 w5c, 18:0, 19:0 cyclo c11–12, 22:0 and 24:0 and fungi: 18:2 w6c, 18 1w9c and 17:0 and AM fungi: 16:1 w5c. Nomenclature for PLFAs followed Lechevalier and Lechevalier (1988), Vestal and White (1989), Zelles (1999), Myers et al. (2001) and Hebel et al. (2009).

Measurements of ecosystem function

Laboratory incubations for potential N mineralization were performed over a 30-day period in soil samples maintained at 25 °C and 60 % humidity. NH_4^+ and NO_3^- were extracted with a 2-M KCl 4:1 solution (Riley and Vitousek 1995) and shipped on dry ice for analysis at the University of California Analytical Laboratory at UC Davis (anlab.ucdavis.edu). Net mineralization was calculated as the change in NH_4^+ minus the change in NO_3^- over time, and net nitrification was calculated as the change of NO_3^- over time following Riley and Vitousek (1995). Potential soil respiration rates were determined using laboratory incubations. Soils were maintained at 20 % soil moisture and 25 °C in sealed glass jars for 10 days. Jar headspace concentrations of CO_2 (ppm) were determined using a LiCor 800 infrared gas analyser (Lincoln, NE, USA) and converted to a rate function of μmol CO_2-C/g soil × day (Chatterjee et al. 2008).

Plant species per cent cover and richness were analysed using repeated-measures multivariate analysis of variance (MANOVA) to assess how the vegetative community responded to restoration. Plant biomass and litter, soil chemistry, soil-extractable N, potential soil respiration and N mineralization data were analysed with ANOVA followed by Tukey's HSD to determine whether restoration altered soil chemical pools and cycling rates. Non-normal data were log(x + 1) or square root transformed when appropriate and a Kruskal-Wallis nonparametric test was performed in cases where the data could not be transformed to normality. Microbial biomass and F:B were analysed using ANOVA to determine coarse microbial community compositional shifts between treatments. Principal component analysis (PCA) was used to create ordination diagrams to compare microbial community compositions, which were then further analysed by ANOVA of PC1 and PC2 values to determine if community composition differed following restoration and across sampling dates. The analyses were conducted using JMP9 (SAS Institute 2009) with an alpha level of $P \leq 0.05$.

Results

Restoration shifted plant species dominance from exotic to native grassland plant species. More specifically, restoration reduced exotic forbs by 59 % at the inland site and 75 % at the coastal site and exotic grasses by 15 % at the inland and 39 % at the coastal site. There was also a 79 and 93 % increase in native grasses at inland and coastal sites, respectively (Table 2). A complete species list and individual cover values are reported in Dickens (2010). Restoration promoted a shift in the quality and quantity of aboveground litter inputs to soil. Litter cover was 25 % higher in restored plots ($P <$ 0.0001) than in unrestored plots at the inland site during germination and 38 % higher at senescence ($P = 0.0002$) but litter cover was unaffected by restoration at the coastal site. In 2007, the drought year, the coastal site accumulated 30 % greater litter than the inland site, but in 2008, an average rain year, 85 % less litter cover than the inland site ($P < 0.0001$ both years). Restoration at the inland grassland site led to a 300 % increase in native grass biomass ($P < 0.001$). Biomass data for the coastal site was not available because plots were unintentionally destroyed during management practices prior to biomass collection. Plant tissue C content varied across all species tested (Table 3). *Erodium brachycarpum* (decreased by 50 % inland), *Brassica nigra* (decreased by 112 % coastal)

and *Avena barbata* (increased by 48 % inland and 100 % coastal) had the lowest leaf tissue N concentrations, whereas the exotic grasses *Brachypodium distachyon* (decreased by 54 % coastal) and *Bromus rubens* (decreased from 3 to 0 %) had the highest. The native grass, *Stipa pulchra,* had an intermediate N concentration and increased by 39 % (inland) and 40 % (coastal) (Table 3). Overall changes in tissue chemistry appear small, but in fact species with the most different tissue chemistry from the native *S. pulchra* are the species that decreased the most with restoration leaving those more similar to *Stipa* as dominant exotic species.

Restoration led to shifts in microbial biomass, microbial community structure and fungal : bacterial (F : B) ratio, but shifts were variable across seasons. Microbial biomass was 29 times lower following restoration at the inland site during germination (Table 4). However, microbial biomass was approximately doubled with restoration during senescence at the coastal site. Fungal : bacterial ratios, while not different between unrestored and restored treatments at the inland site, increased at the coastal

Table 3. Plant leaf tissue chemical composition for some of the most common species encountered at the two project sites. Five samples of each plant species were analysed and averaged per species.

Functional group	Species	N	C	C/N
Native grass	S. pulchra	1	42.7	42.1
Exotic grass	A. barbata	0.7	41.2	62
	B. distachyon	1.3	42.3	31.3
	B. rubens	1.7	42.5	25.5
	Festuca myuros	0.9	42.6	49.7
Exotic forb	E. brachycarpum	0.6	42.7	67.4
	B. nigra	0.9	41.2	56.5

Table 2. Common species mean per cent cover of inland and coastal grassland plant functional groups during the peak of the 2007–08 season. Repeated-measures MANOVA were conducted to assess differences in plant composition between treatments of unrestored and restored grasslands over 3 years during the 2006–09 growing seasons.

Grassland type	Functional groups	Unrestored	Restored	P-values		
				Treatment	Time	Time × treatment
Inland grassland	Native grass	8.3 (1)	40.1 (1)	<0.0001	<0.0001	0.071
	Native forb	4.1 (10)	3.2 (9)	0.015	<0.0001	0.047
	Exotic forb	59.6 (8)	24.7 (5)	0.073	<0.0001	<0.0001
	Exotic grass	47.0 (5)	39.7 (4)	0.372	<0.0001	0.009
Coastal grassland	Native shrubs	2.4 (1)	0.0 (0)	0.362	0.670	0.670
	Native grasses	3.0 (1)	41.5 (1)	<0.001	0.074	0.048
	Exotic forbs	50.8 (5)	12.9 (4)	<0.001	<0.001	0.001
	Exotic grasses	67.8 (2)	41.5 (2)	0.066	<0.001	0.196

Table 4. The common PLFA biomarkers (μmol PLFA g^{-1} soil) and corresponding microbial taxa from the inland and coastal grasslands and between sites during the 2007–08 season. Means are shown for biomarkers making up >2 % of total PLFA abundance. Asterisks indicate the level of significance between treatments. *$P \leq 0.1$, **$P \leq 0.05$ and ***$P \leq 0.001$ determined with ANOVA.

Grassland type	Microbial functional group	Germination		Peak		Senescence	
		Unrestored	Restored	Unrestored	Restored	Unrestored	Restored
Inland grassland	General	240 048	53 142***	21 534	28 807	34 587	33 510
	General bacteria	1 470 719	124 444***	59 108	125 379	183 234	152 469
	Gram positive	897 933	77 569**	76 223	83 427	107 136	91 697
	Gram negative	414 437	34 637***	35 871	38 600	58 637	46 054
	Fungi	695 573	46 975*	63 432	55 442	82 956	72 111
	AM fungi	135 076	11 818***	12 502	15 280	19 198	16 899
	Microeukaryote	21 184	2086	0	2073	3065	3073
	Protozoa	5 216 391	0**	1188	1131	1112	1048
	Proteobacteria	0	2800***	2074	2019	3364	3314
	Pseudomonas	13 969	2187	1756	1495	3164	2396
	Microbial biomass	7 792 960	267 741**	252 280	267 362	362 341	314 489
	F : B	0.461	0.382	0.555	0.414	0.461	0.469
Coastal grassland	General	14 022	10 959	18 573	15 032	6007	16 676**
	General bacteria	3386	3422	4226	3940	708	4313**
	Gram positive	24 697	18 450	25 853	21 221	12 092	24 348*
	Gram negative	15 204	13 713	14 417	12 504	2082	15 775**
	Fungi	15 408	16 942	16 988	17 943	13 269	21 107*
	AM fungi	4893	3539	5093	4034*	850	5245**
	Protozoa	665	2346	691	293	0	505
	Proteobacteria	9574	9975	10 280	10 059	9771	14 262*
	Microbial biomass	90 565	82 187	97 554	86 315	45 193	105 756**
	F : B	0.362	0.465	0.382	0.476***	0.975	0.481**
Between site	Inland	Coastal	Inland	Coastal	Inland	Coastal	
	General	146 595	12 491**	25 171	16 802	34 049	11 341***
	General bacteria	797 581	3404**	130 267	4083***	167 852	2510***
	Gram positive	487 751	21 573**	79 825	23 537***	99 417	18 220***
	Gram negative	224 537	14 458**	37 236	13 461***	52 345	8929***
	Fungi	371 274	16 175**	59 437	17 466***	77 534	17 188***
	AM fungi	73 447	4216**	13 891	4563***	18 049	3047***
	Microeukaryote	11 635	0*	1037	0	3069	0***
	Protozoa	2 608 196	1506*	1160	492	1081	253
	Proteobacteria	1400	9775***	2047	10 169***	3339	12 016***
	Pseudomonas	8078	0*	1696	0**	2780	0***
	Microbial biomass	4 030 350	86 376*	259 821	91 935***	338 415	75 475***
	F : B	0.421	0.413	0.485	0.429	0.465	0.728*

site restored plots during season peak, but were lower during plant senescence (Table 4). The greatest numbers of PLFA biomarkers at both sites were from bacterial functional groups with markers for fungi, protozoa and proteobacteria in lower abundance (Table 4). The inland site also had biomarkers for microeukaryotes and pseudomonads in low abundances. Concentrations of biomarkers from all functional groups except microeukaryotes and *Pseudomonas* differed between unrestored and restored plots during plant germination. Soils sampled during plant senescence at the coastal site and at germination at the inland site had increased AM fungal marker 16 : 1 w5c **[see Supporting Information]**. The microbial community as a whole, as defined by PLFA biomarkers was differentiated by both restoration treatment and season at both sites (Fig. 2; **see Supporting Information**). There

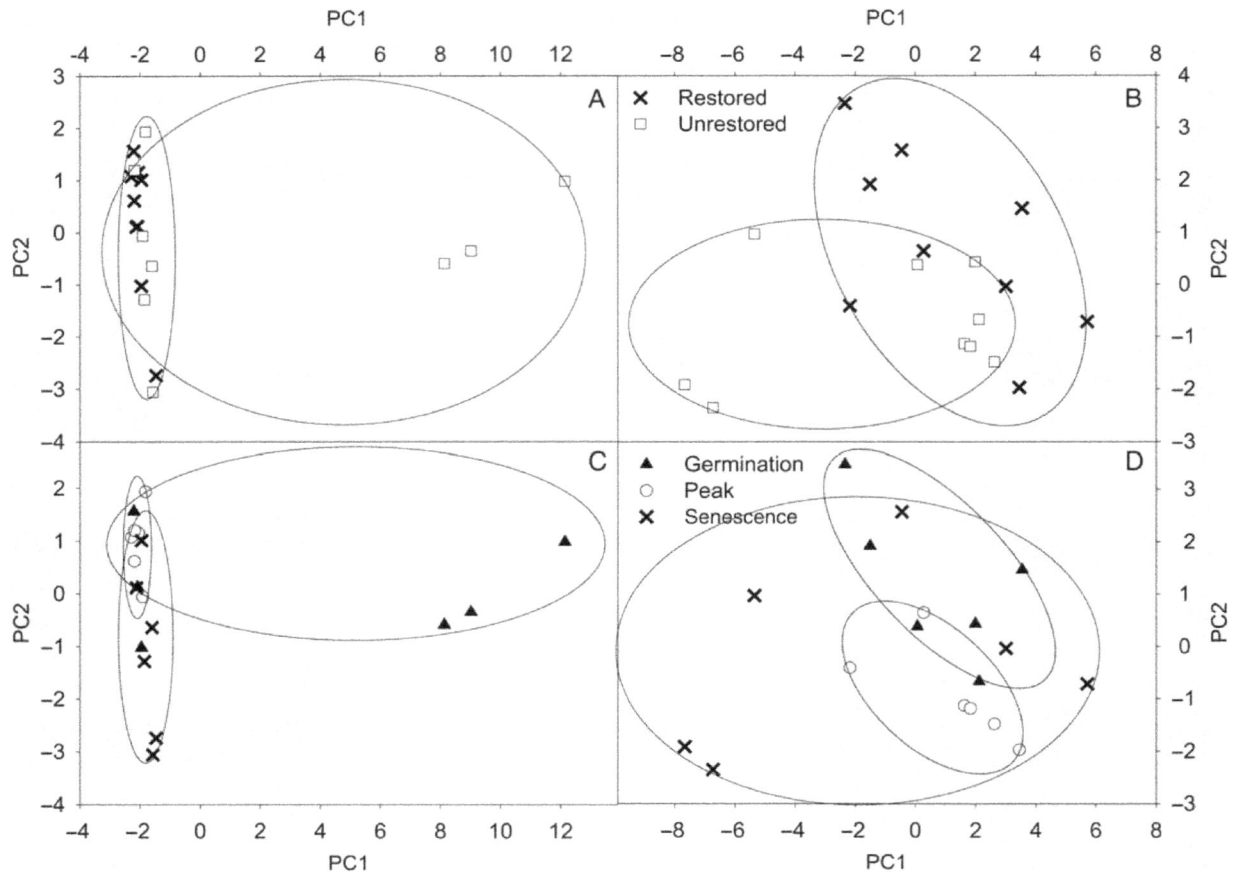

Figure 2. Principal component analysis results for PLFA microbial community analysis at the inland site (A and C) and the coastal site (B and D) during the 2007–08 growing season. Restored data points of graph (A) refer to restored-burned treatments of the inland site and those of graph (B) refer to restored-weeded treatments of the coastal site. Graphs (A) and (B) assess differences between treatment while (C) and (D) assess differences between sampling dates. PC1 explains 83 % variation and PC2 has a cumulative per cent variation of 91 % for the inland site, while the coastal site cumulative variance explained by PC1 is 59 % and PC2 is 70 %. Ellipses indicate statistically different microbial communities determined by ANOVA of PC values.

were several between-site differences in response to restoration microbial communities differed in both functional group and mass between sites and within the growing season. The inland site generally had greater microbial biomass and AM fungi than the coastal site, but sites had similar F : B ratios except at senescence when coastal F : B ratios nearly doubled. There was a similar decrease in average microbial biomass between the two sites over the season (inland = 16 %, coastal = 13 %).

Chemical properties of soil N shifted in the form of both increased and reduced NH_4-N according to the season and site, increased NO_3-N and total extractable N and altered N cycling rates. Restoration of the inland site reduced NH_4-N during germination in the 2007–08 season ($P = 0.004$), but increased NH_4-N during the peak of the growing season ($P = 0.009$; Fig. 3). In contrast, restoration of the coastal site only reduced NH_4-N during senescence ($P = 0.020$; Fig. 3B). Restoration leads to a greater availability of NO_3-N ($P = 0.058$) and total extractable N ($P = 0.0002$)

during the peak season at the inland site and of NO_3-N during senescence at both sites (inland $P = 0.005$; coastal $P = 0.005$; Fig. 3A and B). Restoration did not impact total extractable N at the coastal site. Peak season extractable N patterns were consistent across the 3 years of 2007–09 at the inland site where total extractable N (2007, $P = 0.040$, 2008, $P = 0.002$, 2009, $P = 0.001$) and NO_3-N in 2007 ($P = 0.041$) and 2009 ($P < 0.001$) increased following restoration. The coastal site had increased total extractable N but reduced NO_3-N with restoration in the drought year, 2007, with no differences in any form of extractable N in 2008.

Total soil N and C, pH and P were unaffected by restoration (Table 5). Soil potential respiration was unaffected by restoration at the inland grassland sites. Plots at the coastal site were unintentionally destroyed before soil respiration sampling was conducted so potential soil respiration data were not available for that site. Potential N mineralization was reduced by restoration only in August

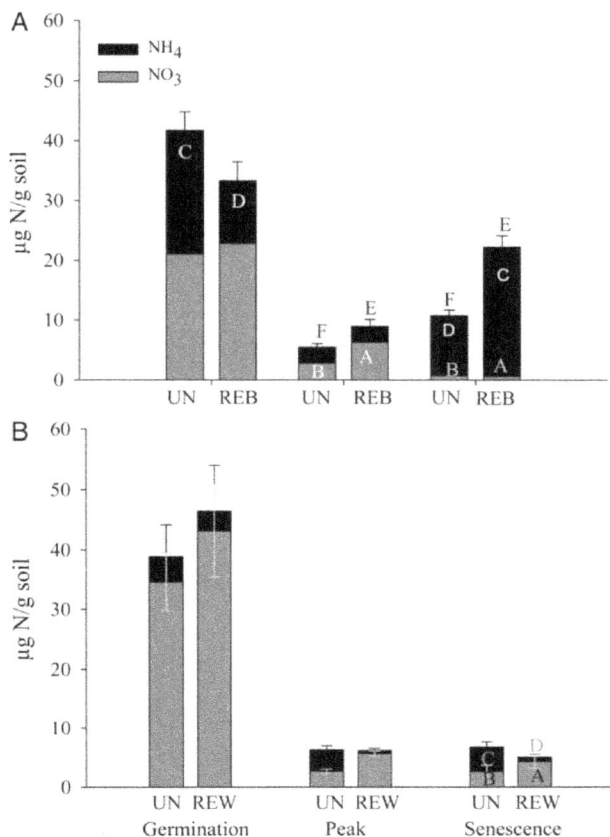

Figure 3. Soil-extractable N during the 2007–08 season at the inland site (A) and the coastal site (B). Treatments are: UN = unrestored, REB = restored by burning at the inland site and REW = restored by weeding and mowing at the coastal site. Letters indicate significant differences using ANOVA followed by Tukey–Kramer HSD test: NO_3 = A and B, NH_4 = C and D and total extractable N = E and F. Bars indicate standard error and letters significant differences ($P \leq 0.05$).

soils and only at the inland site ($P = 0.017$). Potential nitrification rates were increased with restoration for soils collected in August at the inland site ($P = 0.002$; Fig. 4C) and in March at the coastal site ($P = 0.011$; Fig. 4B).

Discussion

Microbial shifts in community structure and altered extractable N pools and N cycling indicate that exotic plant–soil feedbacks were decoupled following restoration at both sites regardless of differences in soil properties between sites. Decoupling is used here to mean that established interactions between exotic plants and soil organisms were disrupted and replaced to some degree by interactions between native plants and soil organisms (Bardgett *et al.* 2013). Native grasses increased at both sites and exotics were reduced leading to shifts in plant inputs, particularly by reducing the proportion of exotic

Table 5. Soil chemical data (means and standard errors) for the burned, inland site and the coastal site collected the summer of 2006. ANOVA was conducted to assess differences in soil chemical characteristics between treatments of unrestored and restored grasslands during the 2006–07 growing season.

	Inland unrestored	SE	Inland restored-burned	SE	Inland treatment P-value	Coastal unrestored	SE	Coastal restored-weeded	SE	Coastal treatment P-value	Inland site	SE	Coastal site	SE	Site P-value
Total N (%)	0.2	0.01	0.2	0.00	1.000	0.2	<0.10	0.2	<0.10	0.162	0.2	0.0	0.2	0.0	<0.0001
Total C (%)	2.2	0.08	2.2	0.06	0.885	3.2	0.20	3.1	0.20	0.727	2.2	0.1	3.2	0.1	<0.0001
Soil organic matter (%)	8.5	0.30	8.0	0.20	0.136	13.3	0.30	13.0	0.20	0.541	8.3	0.2	13.1	0.2	<0.0001
C/N	13.4	0.07	13.3	0.14	0.499	14.0	0.53	15.7	0.80	0.105	13.4	0.1	14.9	0.5	0.007
NH_4 (µg g^{-1})	6.5	0.25	9.2	2.43	0.284	7.4	0.28	8.3	0.60	0.308	7.8	1.2	7.8	0.4	0.997
NO_3 (µg g^{-1})	3.6	1.66	1.8	0.45	0.295	5.8	1.31	31.9	8.30	0.001	2.7	0.9	18.8	5.2	0.004
Total extractable N (µg g^{-1})	10.1	1.84	11.0	2.38	0.783	13.1	1.51	40.2	8.60	0.007	10.5	1.5	26.7	5.4	0.006
Olsen-P (µg g^{-1})	5.1	0.71	4.1	0.79	0.351	18.8	1.10	22.1	3.60	0.401	4.6	0.5	20.4	1.9	<0.0001
pH	5.9	0.07	5.9	0.04	0.980	8.0	<0.10	8.0	0.10	0.537	5.9	0.0	8.0	0.0	<0.0001

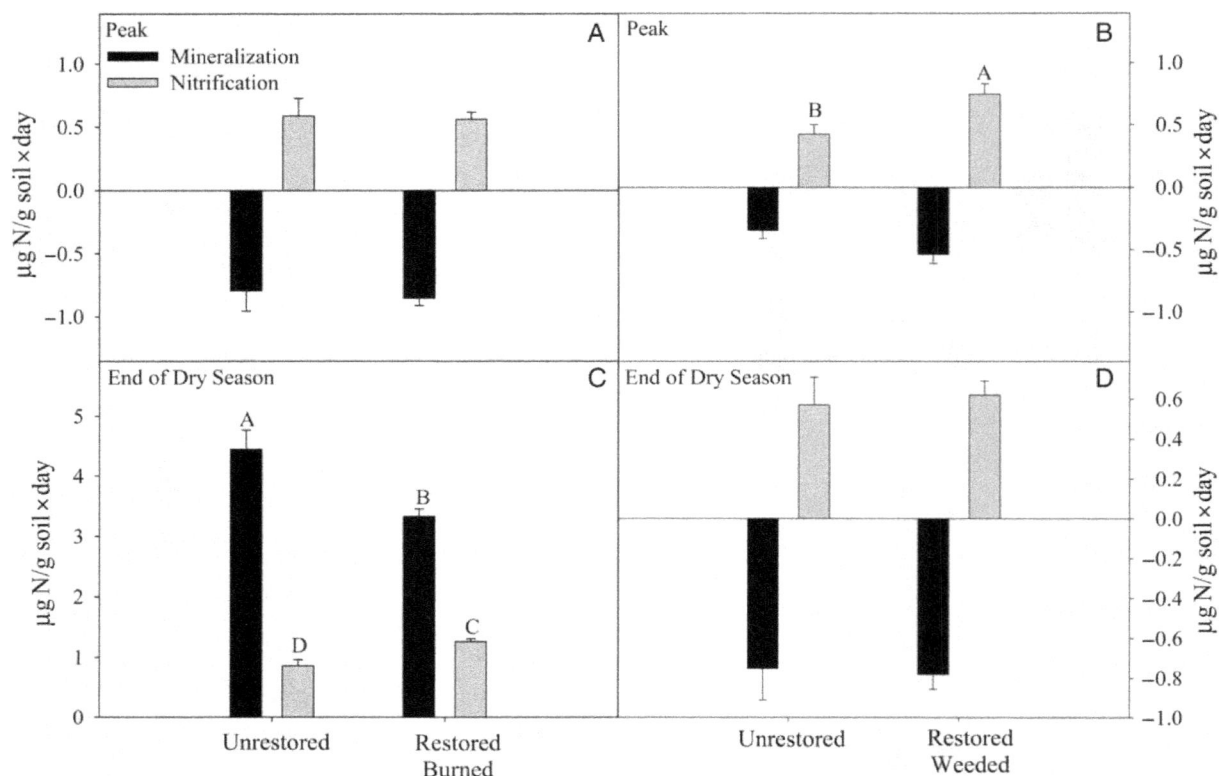

Figure 4. Potential N mineralization and nitrification from 30-day laboratory incubations from the inland site (A and C) and the coastal site (B and D) for soils collected in the March (peak) and August (end of the summer dry season) of 2008. Letters indicate significant differences using the Tukey–Kramer HSD test following ANOVA ($P \leq 0.05$). Bars indicate standard error.

species litter having higher C : N in both sites and increasing the proportion of *S. pulchra* litter with intermediate quality. The reduction of exotic plant inputs followed by replacement by native plant inputs altered the microbial community, increased NO_3-N availability and nitrification rates and decreased NH_4-N availability and N mineralization rates. Although no uninvaded grasslands are available as reference sites, these rapid changes indicate that these grasslands have some capacity for soil resilience.

Differences in microbial community biomass, F : B and individual markers (indicated by concentrations of PLFA biomarkers) between unrestored and restored soils support our hypothesis that restoration of the native plant community would decouple the previously existing exotic plant–soil feedbacks and allow for establishment of native plant–soil feedbacks. Potthoff *et al.* (2009) found similar shifts in PLFA profiles that they interpreted to indicate resilience of the soil microbial community to disturbance. Restoration reduced microbial biomass values during germination indicating a stronger response of soil microbes in unrestored soils following the first rains, likely due to higher root activity of germinating exotics, rapid decomposition of exotic annual grassland seedlings

due to self-thinning (Bartolome 1979; Savelle 1997; Eviner and Firestone 2007) or decomposition of the previous year's microbial biomass. Shifts in microbial community in response to seasonal changes in temperature and moisture, such as those found here, are expected due to species-specific growth requirements (Pommerville 2007). Mycorrhizal fungi may decline in perennial grassland soils invaded by exotic annual grasses, so we expected fungal biomarkers to be lower in unrestored plots (Hawkes *et al.* 2006). The grasslands of this experiment had reduced fungal biomarkers, 18 : 2 w6c, 18 : 1w9c and 16 : 1 w5cat germination, but increased fungal markers 18 : 2 w6c and 18 : 1w9c the remainder of the season. Similar to Hawkes' *et al.* (2006) findings, our coastal grassland site had lower AM fungi PLFA markers in unrestored plots dominated by exotic annuals, but this did not occur until late in the growing season (i.e. plant senescence). Native plant species tend to have a later phenology than exotics in the semi-arid grasslands and thus experience peak growth rates later into the season than exotic plants (Jackson and Roy 1986; Holmes and Rice 1996). AM fungi associated with exotics have reduced abundance earlier in the season when their exotic, annual host plants senesce (Nelson and Allen

1993). Other fungal groups were also in high concentration during senescence at both sites. Therefore, phenology of the dominant plant species either annual exotics or native annuals to native perennial grassland species was related to the activity of soil microorganisms.

The anticipated shift in N availability occurred and also showed seasonal patterns corresponding to plant phenologies. Total extractable N and NO_3-N increased with restoration during peak and senescence periods of the growing season as also observed by Jackson et al. (1988, 1989). One of the more striking results was that extractable N concentrations at the inland site were higher in unrestored soils at plant germination but quickly became significantly lower within 2–3 months, suggesting increased rates of N uptake by plants in the unrestored plots. Plants may take up as much NO_3^- as becomes available (Jackson et al. 1988, 1989). This effect of exotic plant removal on increased mineral N has been observed in both grassland and coastal sage scrub in other studies of semi-arid environments (Dickens 2010; Dickens and Allen 2014). In this study, plant uptake of N was still low at germination, but began to increase rapidly as plant growth reached its maximum rates. The peak and senescence sampling dates at these sites correspond to the periods of maximum annual growth and transition to reproduction phases during which their N use would be highest. California grassland natives tend to germinate and complete their life cycles later than exotic annuals (Jackson and Roy 1986; Holmes and Rice 1996) and likely have continued nutrient uptake closer to the senescence sampling date. Stipa pulchra recycles about half its annual N internally and thus may not take up N as soon or at rates as high as those observed for exotic annuals (Clark 1977; Jackson et al. 1988; Hooper and Vitousek 1998). Stipa pulchra is the dominant native so less rapid and total uptake rates would translate to greater overall extractable N left in the soil throughout the season. Restoration led to reduced NH_4-N and N mineralization but increased nitrification, indicating greater immobilization of N following restoration. In other grasslands, exotic grass invasion is associated with increased N mineralization, which is attributed to a greater abundance of ammonia-oxidizing bacteria (Hawkes et al. 2005). Here, the biomass of bacteria was higher in unrestored soils, although our PLFA assay could not identify whether these ammonia-oxidizing bacteria were reduced by restoration.

Long-term invasion and anthropogenic disturbance may be one mechanism explaining the resistance of soil total C and N pools and potential respiration to changes under restored plant community conditions. Total soil C and N pools and C cycling may show resistance or may only be slowly responsive to changes in vegetation. These soils have likely been invaded by exotic annual grasses for more than a century (Minnich 2008), so sufficient time has likely passed for total C and N pools to change in response to invasion. Similar resistance of C and N pools were observed in grasslands of northern California when community composition was altered to test legacy effects of plant–soil interactions (Potthoff et al. 2009). However, decreased soil C was found in invaded grasslands in central California compared with reference patches of native grassland (Koteen et al. 2011). Kindscher and Tieszen (1998) found that tall grass prairie soils may require >35 years to recover C from agricultural use following restoration. The shifts in litter input at our grasslands may not have been great enough to lead to altered C and N pools. Differences and variability in litter quality in grasslands are often subtle making responses to changes in litter difficult to detect (Eviner and Firestone 2007). This suggests that our grasslands could have been resistant to impacts of the initial exotic annual invasion or that restoration must occur for a longer time than 9 years to detect total soil N and C responses. Another contributing factor is the reinvasion of exotic grass following restoration. While native vegetation continued to dominate, exotic species had a continuous impact on the soil.

Between-site differences in restoration responses of extractable N availability and microbial community structure were primarily seasonal. This corresponded with our hypothesis that soil responses are sensitive to environment. In this case, the important environmental influences included soil nutrient and climatic differences. Greater immobilization at the coastal than the inland site may be the result of a 15 % lower soil C:N ratio and 5 % greater soil organic matter content than at the inland site. Higher organic matter in conjunction with lower C:N soil values would allow for higher rates of N mineralization while also leading to increased immobilization overall in restored plots of the coastal site (Knops et al. 2002; Berger and Jackson 2003). Instead, the inland site had a more stable microbial community (F:B was unchanged) and a steady use of nutrients over the season followed by a second peak of microbial activity at senescence. So while differences in soil nutrient conditions, timing of responses to restoration and specific PLFA concentrations occurred between sites, the patterns of increased NO_3-N and nitrification, reduced N mineralization and altered microbial community composition following restoration occurred at both sites. This indicates that grassland soils were responsive to changes in vegetation and may therefore be resilient to invasion.

Conclusions

Restoration of invaded grasslands decoupled exotic plant–soil feedbacks related to microbial community structure, extractable N and N cycling. This study indicates that the soils of these systems are dynamic and change in response to exotic or native vegetation type and seasonal variation in soil moisture. Semi-arid grasslands in general are known to be unstable in productivity and reliant on seasonal precipitation patterns (Talbot *et al.* 1939). Measured changes in extractable soil N and microbial characteristics in response to removal of exotic plants indicated that grassland soils are not resistant to the impacts of plant community shifts, but have the capacity for resilience regardless of the method of exotic plant control (i.e. prescribed burn, mowing and weeding). This indicates that the method of exotic plant removal is not important in these grasslands, but that removal of exotic plants and decoupling exotic plant–soil feedbacks are required for grassland soils to diverge from invaded conditions. In contrast, the lack of change in total soil C and N pools and potential soil respiration may be an indication that, for these soil characteristics, these grasslands are resistant to invasion. Stable pools of C and N may buffer these soils, enabling resilience of the more labile and rapidly responding mineral N and microbial characteristics. However, the absence of uninvaded grasslands does not allow us to rule out changes in C and N pools that may have occurred long ago or will require a more complete restoration and longer time frames for recovery. The differences between sites regarding the timing of microbial activity and N cycling highlight the importance of matching sampling efforts to seasonality of plant and microbial growth patterns. Overall NO_3-N use and net N cycling differences between restored and unrestored plots were similar between the coastal and inland sites, indicating that shifts in plant community composition from exotic to native-dominated communities produce the same impact on N regardless of site history, restoration methods and differences in soil type.

Sources of Funding

This work has been funded by the University of California-Riverside and specifically the Department of Botany and Plant Sciences and the Center for Conservation Biology, the Shipley-Skinner Riverside County Endowment, the UC Fisher Vegetation Scholarship, The Nature Conservancy and the National Science Foundation (DEB 04-21530).

Contributions by the Authors

S.J.M.D. and E.B.A. formed the research questions and all authors contributed to the development, analysis of data and manuscript drafting. S.J.M.D. implemented the project in the field and laboratory.

Acknowledgements

The authors extend their gratitude to Dr Jodie Holt, Dr Mike Allen and Dr Darrel Jenerette for thoughtful discussions. For assistance in data collection they thank Christopher True, Dr Robert Cox, Dr Leela Rao, Dr Robert (Chip) Steers, Dr Heather Schneider, Dr Mike Bell and Dr Kris Weathers, as well as Andrew Sanders for assistance with plant identification. They also thank their funding sources: The University of California-Riverside and specifically the Department of Botany and Plant Sciences and the Center for Conservation Biology, the Shipley-Skinner Riverside County Endowment, the UC Fisher Vegetation Scholarship, The Nature Conservancy and the National Science Foundation (DEB 04-21530).

Supporting Information

The following Supporting Information is available in the online version of this article –

Table S1. The common phospholipid fatty acid (PLFA) biomarkers (μmol PLFA g^{-1} soil) and the corresponding microbial functional groups from the inland site during the 2007–08 season. Asterisks indicate the level of significance between treatments. **$P \leq 0.05$.

Table S2. The common PLFA biomarkers (μmol g^{-1} soil) and the corresponding microbial functional groups from the coastal site during the 2007–08 season. Asterisks indicate the level of significance between treatments. *$P \leq 0.001$, **$P \leq 0.05$.

Table S3. Soil microbial PLFA principal component (PC) per cent weights at both locations. Positive and negative signs indicate the direction of the weighting along the corresponding PC. Cumulative per cent explained equals the variance within the PLFA data explained by successive PCs.

Literature Cited

Bardgett RD, Manning P, Morriën E, De Vries FT. 2013. Hierarchical responses of plant–soil interactions to climate change: consequences for the global carbon cycle. *Journal of Ecology* **101**: 334–343.

Bartolome JW. 1979. Germination and seedling establishment in California annual grassland. *Journal of Ecology* **67**:273–281.

Batten KM, Scow KM, Davies KF, Harrison SP. 2006. Two invasive plants alter soil microbial community composition in serpentine grasslands. *Biological Invasions* **8**:217–230.

Berg G, Smalla K. 2009. Plant species and soil type cooperatively shape the structure and function of microbial communities in the rhizosphere. *FEMS Microbial Ecology* **68**:1–13.

Berger M, Jackson LE. 2003. Microbial immobilization of ammonium and nitrate in relation to ammonification and nitrification rates in organic and conventional cropping systems. *Soil Biology and Biochemistry* **35**:29–36.

Bever JD, Westover KM, Antonovics J. 1997. Incorporating the soil community into plant population dynamics: the utility of the feedback approach. *Journal of Ecology* **85**:561–573.

Biswell HH. 1956. Ecology of California grasslands. *Journal of Range Management* **9**:19–24.

Brady NC, Weil RR. 1996. *The nature and property of soils*. Upper Saddle River, New Jersey: Prentice-Hall, Inc.

Chatterjee A, Vance GF, Pendall E, Stahl PD. 2008. Timber harvesting alters soil carbon mineralization and microbial community structure in coniferous forests. *Soil Biology and Biochemistry* **40**:1901–1907.

Christian JM, Wilson SD. 1999. Long-term ecosystem impacts of an introduced grass in the northern Great Plains. *Ecology* **80**: 2397–2407.

Cione NK, Padgett PE, Allen EB. 2002. Restoration of a native shrubland impacted by exotic grasses, frequent fire, and nitrogen deposition in Southern California. *Restoration Ecology* **10**:376–384.

Clark FE. 1977. Internal cycling of N-15 in shortgrass prairie. *Ecology* **58**:1322–1333.

D'Antonio CM, Malmstom C, Reynolds SA, Gerlach J. 2007. Chapter 6: ecology of non-native invasives. In: Stromberg M, Corbin J, D'Antonio C, eds. *California grasslands ecology and management*. Berkeley, CA: University of California Press, 67–86.

Dickens SJM. 2010. *Invasive plant–soil feedbacks and ecosystem resistance and resilience: a comparison of three vegetation types in California*. PhD Dissertation, University of California Riverside, USA.

Dickens SJM, Allen EB. 2014. Soil nitrogen cycling is resilient to invasive annuals following restoration of coastal sage scrub. *Journal of Arid Environments* **110**:12–18.

Ehrenfeld JG. 2003. Effects of exotic plant invasions on soil nutrient cycling processes. *Ecosystems* **6**:503–523.

Eviner VT. 2004. Plant traits that influence ecosystem processes vary independently among species. *Ecology* **85**:2215–2229.

Eviner VT, Chapin FS. 2003. Functional matrix: a conceptual framework for predicting multiple plant effects on ecosystem processes. *Annual Review of Ecology and Systematics* **34**:455–485.

Eviner VT, Firestone MK. 2007. Mechanisms determining patterns of nutrient dynamics. In: Stromberg M, Corbin J, D'Antonio C, eds. *California grasslands ecology and management*. Berkeley, CA: University of California Press, 94–106.

Frostegard A, Tunlid A, Baath E. 1991. Microbial biomass measured as total lipid phosphate in soils of different organic content. *Journal of Microbiological Methods* **14**:151–163.

Gillespie IG, Allen EB. 2004. Fire and competition in a southern California grassland: impacts on the rare forb *Erodium macrophyllum*. *Journal of Applied Ecology* **41**:643–652.

Grayston SJ, Wang SQ, Campbell CD, Edwards AC. 1998. Selective influence of plant species on microbial diversity in the rhizosphere. *Soil Biology and Biochemistry* **30**:369–378.

Hawkes CV, Wren IF, Herman DJ, Firestone MK. 2005. Plant invasion alters nitrogen cycling by modifying the soil nitrifying community. *Ecology Letters* **8**:976–985.

Hawkes CV, Belnap J, D'Antonio C, Firestone MK. 2006. Arbuscular mycorrhizal assemblages in native plant roots change in the presence of invasive exotic grasses. *Plant and Soil* **281**:369–380.

Hebel CL, Smith JE, Cromack K. 2009. Invasive plant species and soil microbial response to wildfire burn severity in the cascade range of Oregon. *Applied Soil Ecology* **42**:150–159.

Hedrick DB, Peacock A, Stephen JR, McNaughton SJ, Bruggemann J,

White DC. 2000. Measuring soil microbial community diversity using polar lipid fatty acid and denaturing gradient gel electrophoresis data. *Journal of Microbiological Methods* **41**:235–248.

Holmes TH, Rice KJ. 1996. Patterns of growth and soil-water utilization in some exotic annuals and native perennial bunchgrasses of California. *Annals of Botany* **78**:233–243.

Hooper DU, Vitousek PM. 1998. Effects of plant composition and diversity on nutrient cycling. *Ecological Monographs* **68**: 121–149.

Jackson LE, Roy J. 1986. Growth patterns of Mediterranean annual and perennial grasses under simulated rainfall regimes of southern France and California. *Acta Oecologica* **7**:191–212.

Jackson LE, Strauss RB, Firestone MK, Bartolome JW. 1988. Plant and soil-nitrogen dynamics in California annual grassland. *Plant and Soil* **110**:9–17.

Jackson LE, Schimel JP, Firestone MK. 1989. Short-term partitioning of ammonium and nitrate between plants and microbes in an annual grassland. *Soil Biology and Biochemistry* **21**:409–415.

Kindscher K, Tieszen LL. 1998. Floristic and soil organic matter changes after five and thirty-five years of native tallgrass prairie restoration. *Restoration Ecology* **6**:181–196.

Knops JMH, Bradley KL, Wedin DA. 2002. Mechanisms of plant species impacts on ecosystem nitrogen cycling. *Ecology Letters* **5**: 454–466.

Koteen L, Baldocchi DD, Harte J. 2011. Invasions of non-native grasses causes a drop in soil carbon storage in California grasslands. *Environmental Research Letters* **6**:1–10.

Kulmatiski A, Beard KH. 2008. Decoupling plant-growth from land-use legacies in soil microbial communities. *Soil Biology and Biochemistry* **40**:1059–1068.

Kulmatiski A, Beard KH. 2011. Long-term plant growth legacies overwhelm short-term plant growth effects on soil microbial community structure. *Soil Biology and Biochemistry* **43**:823–830.

Lechevalier H, Lechevalier MP. 1988. Chemotaxonomic use of lipids: an overview. In: Ratledge C, Wilkindon SC, eds. *Microbial lipids*. New York, NY: Academic Press, 869–902.

Minnich R. 2008. *California's fading wildflowers: lost legacy and biological invasion*. Berkeley, CA: University of California Press.

Myers RT, Zak DR, White DC, Peacock A. 2001. Landscape-level patterns of microbial community composition and substrate use in upland forest ecosystems. *Soil Science Society of America Journal* **65**:359–367.

Nelson JW, Zinn CJ, Strahorn AT, Watson EB, Dunn JE. 1919. *Soil survey of the Los Angeles area, California. Advanced sheets-field operations of the Bureau of soils, 1916*. Washington: United States Department of Agriculture.

Nelson LL, Allen EB. 1993. Restoration of *Stipa pulchra* grasslands: effects of mycorrhizae and competition from *Avena barbata*. *Restoration Ecology* **1**:40–50.

Pommerville JC. 2007. *Alcamo's fundamentals of microbiology*. Sudbury, Massachusetts: Jones and Bartlett Publishers.

Potthoff M, Steenwerth KL, Jackson LE, Drenovsky RE, Scow KM, Joergensen RG. 2006. Soil microbial community composition as affected by restoration practices in California grassland. *Soil Biology and Biochemistry* **38**:1851–1860.

Potthoff M, Jackson LE, Solow S, Jorgensen RG. 2009. Below and above ground responses to lupine and litter mulch in a California grassland restored with native bunchgrasses. *Applied Soil Ecology* **42**:124–133.

Riley RH, Vitousek PM. 1995. Nutrient dynamics and nitrogen trace gas flux during ecosystem development in montane rain-forest. *Ecology* **76**:292–304.

Santiago LS. 2007. Extending the leaf economics spectrum to decomposition: evidence from a tropical forest. *Ecology* **88**:1126–1131.

Santiago LS, Schuur EAG, Silvera K. 2005. Nutrient cycling and plant-soil feedbacks along a precipitation gradient in lowland Panama. *Journal of Tropical Ecology* **21**:461–470.

SAS Institute. 2009. *JMP*, Version 9. Cary, NC: SAS Institute Inc.

Savelle GD. 1977. *Comparative structure and function in a California annual and native bunchgrass community.* Dissertation. University of California, Berkeley, USA.

Talbot MW, Biswell HH, Hormay AL. 1939. Fluctuations in the annual vegetation of California. *Ecology* **20**:394–402.

Vestal JR, White DC. 1989. Lipid analysis in microbial ecology—quantitative approaches to the study of microbial communities. *Bioscience* **39**:535–541.

Wardle DA. 2002. *Communities and ecosystems: linking the aboveground and belowground components: monographs in population biology.* New Jersey, NY: Princeton University Press.

Wardle DA, Bardgett RD, Klironomos JN, Setala H, van der Putten WH, Wall DH. 2004. Ecological linkages between aboveground and belowground biota. *Science* **304**:1629–1633.

White DC, Stair JO, Ringelberg DB. 1996. Quantitative comparisons of *in situ* microbial biodiversity by signature biomarker analysis. *Journal of Industrial Microbiology* **17**:185–196.

Wolfe BE, Klironomos JN. 2005. Breaking new ground: soil communities and exotic plant invasion. *Bioscience* **55**:477–487.

Yoshida LC, Allen EB. 2004. N-15 uptake by mycorrhizal native and invasive plants from a N-eutrophied shrubland: a greenhouse experiment. *Biology and Fertility of Soils* **39**:243–248.

Zelles L. 1999. Fatty acid patterns of phospholipids and lipopolysaccharides in the characterization of microbial communities in soil: a review. *Biology and Fertility of Soils* **29**:111–129.

Zelles L, Bai QY. 1994. Fatty-acid patterns of phospholipids and lipopolysaccharides in environmental-samples. *Chemosphere* **28**:391–411.

6

Effects of disturbance on vegetation by sand accretion and erosion across coastal dune habitats on a barrier island

Thomas E. Miller*

Department of Biological Science, Florida State University, Tallahassee, FL 32306, USA

Guest Editor: Elise S. Gornish

Abstract. Coastal geomorphology and vegetation are expected to be particularly sensitive to climate change, because of disturbances caused by sea-level rise and increased storm frequency. Dunes have critical reciprocal interactions with vegetation; dunes create habitats for plants, while plants help to build dunes and promote geomorphological stability. These interactions are also greatly affected by disturbances associated with sand movement, either in accretion (dune building) or in erosion. The magnitude and intensity of disturbances are expected to vary with habitat, from the more exposed and less stable foredunes, to low-lying and flood-prone interdunes, to the protected and older backdunes. Permanent plots were established at three different spatial scales on St George Island, FL, USA, where the vegetation and dune elevation were quantified annually from 2011 to 2013. Change in elevation, either through accretion or erosion, was used as a measure of year-to-year disturbance over the 2 years of the study. At the scale of different dune habitats, foredunes were found to have the greatest disturbance, while interdunes had the least. Elevation and habitat (i.e. foredune, interdune, backdune) were significantly correlated with plant community composition. Generalized linear models conducted within each habitat show that the change in elevation (disturbance) is also significantly correlated with the plant community, but only within foredunes and interdunes. The importance of disturbance in exposed foredunes was expected and was found to be related to an increasing abundance of a dominant species (*Uniola paniculata*) in eroding areas. The significant effect of disturbance in the relatively stable interdunes was surprising, and may be due to the importance of flooding associated with small changes in elevation in these low-lying areas. Overall, this study documents changes in the plant community associated with elevation, and demonstrates that the foredune and interdune communities are also associated with the responses of specific species to local changes in elevation due to accretion or erosion.

Keywords: Climate change; coastal zones; disturbance; geomorphology; ordination; plant community.

* Corresponding author's e-mail address: miller@bio.fsu.edu

Introduction

Coastal sand dunes provide the first line of defense against storms and high water levels in many parts of the world (Sallenger 2000; Ruggiero et al. 2001; Feagin et al. 2005). As such, coastal ecosystems are particularly sensitive to sea-level rise and any changing frequencies of tropical storms and hurricanes, all of which are predicted to occur with global climate change (e.g. Duran and Moore 2013; Prisco et al. 2013). These climate effects may be especially important because dunes have a high ecosystem value as habitat for endemic plants and animals, while sheltering bay (e.g. seagrass, oyster beds and saltmarsh) habitats, as well as inland wetlands and marshes (Martinez and Psuty 2004; Gutierrez et al. 2011). Dunes can also be important for protecting coastal towns, as well as the economic activities they provide (e.g. fisheries, tourism).

The development and maintenance of coastal dune habitats requires a plentiful supply of sand, strong winds to move sand inland and an obstacle, usually plants, to stop the sand and create dunes. Thus the plants on dunes have long been recognized as key components of coastal habitats. The vegetation on dunes has served as a model system for influential studies of plant succession and ecology (Cowles 1899; Oosting and Billings 1942); because dunes form near shore, dunes that are progressively inland create chronosequences for vegetation studies.

The relationship between plants and dunes is both reciprocal and complex (e.g. Stallins and Parker 2003). The plants are thought to control sand movement and determine the shape and position of the dunes (e.g. Moreno-Casasola 1986), while the dune structure can determine the abiotic factors such as soil moisture and nutrients that control plant establishment, growth and reproduction (Ehrenfeld 1990). Many previous studies of coastal dune vegetation identified harsh physical factors such as salt spray, soil moisture and sand movement as the primary factors responsible for the zonation patterns parallel to the beach (e.g. Oosting and Billings 1942; Miller et al. 2010; Bitton and Hesp 2013).

Fewer studies have considered the feedbacks between plants and the dune geomorphology. It has been shown that certain species, frequently grasses, are correlated with dune formation on beach plateaus, generally promoting the development of foredunes (Gibson and Looney 1994; Stallins and Parker 2003; Bitton and Hesp 2013). Dune building plants have also been shown to slow moving sand particles (Zarnetske et al. 2012) and to be highly tolerant of burial and, for marine coasts, high salinity. But, less is known about how plants influence the continued growth and maintenance of foredunes, or how they affect more inland areas such as interdunes (also called overwash plateaus) and older backdune areas. For example, some plants have been associated with dune accretion (Bitton and Hesp 2013), while other plants may hinder dune formation, promoting lower and flatter areas of interdunes (Ehrenfeld 1990; Wolner et al. 2013). But, almost nothing is known of interactions between vegetation and geomorphology across different spatial scales (e.g. habitats) in coastal dunes.

This lack of knowledge about dune processes at various spatial scales makes it difficult to predict the effects of climate change on sandy coastal habitats. The short- and long-term effects of both sea-level rise and increased storm frequencies are relatively unknown, but are expected to be significant, given the low elevation and dynamic nature of the geomorphology of coastal dunes. Understanding the effects of the disturbance caused by sand accretion or erosion on dune plant communities will help to elucidate the mechanisms by which global climate change can affect areas such as barrier islands, sandy bars and spits. Ultimately, this should also help for predicting and preparing for the effects of climate change.

This study uses multi-year plant and elevation surveys of St George Island, FL, USA to (i) quantify the dynamic relationship between dune elevation and plant communities. The study also determines (ii) whether the nature of the relationships among elevation, change in elevation and the plant community changes with spatial location, as one moves from the newer and more disturbed foredunes to the low and wet interdunes, and finally to older, more protected backdunes.

Methods

Site description

St George Island, FL, USA (29°46′00″N, 84°41′30″) is typical of barrier islands that form with low tidal ranges on wave-dominated coasts. It is ~45 km long and 1 km wide, off the Florida panhandle in the northern Gulf of Mexico. Such islands worldwide share a number of habitats maintained by the wind and wave forces that created the islands themselves. At the ocean side of the island, foredunes are created and maintained by the sand blown from the beach plateau. They have high (3–5 m) and dynamic dune ridges subject to wind, spray and high tides. Foredunes are lower in overall plant diversity and are frequently quite dry (Miller et al. 2010). Behind the foredunes are interdunes, which are low, and relatively flat and homogeneous in elevation. The sand in the interdune contains more organic material than other dune areas and is often wet or flooded. Saltwater will inundate interdunes with major storms, but they more frequently fill with freshwater from rains as the lens under the island fills. Finally, further inland from the interdunes are backdunes, which consist of irregular but stable ridges (1–2 m height) separated by troughs and can extend to the bay side of the island. Backdunes are much more stable than foredunes and have the highest plant diversity (Miller et al. 2010). They contain some woody species, as well as species also common to both interdunes and fore-dunes. Older and wider islands may also include later stages of dune and vegetation succession, but these are not present at the relatively young field site used in this study.

Sampling design

This work was conducted in the St George Island State Park, which occupies the easternmost 14 km of the island. A study was established in 1999 to follow the vegetation on the actively growing eastern tip of St George Island and has been continued annually since (except for 2002; see Miller et al. 2010). Initially, two grids were set up in each of the three habitats (foredune, interdune and backdune); one more replicate grid was set up in each habitat in 2010. Each of the nine grids consists of 49 plots in a 7×7 array, with 10 m between plots. A wooden stake marks each plot and the vegetation in a 1 m^2 area to the northeast of each stake is censused in the late fall of every year. The per cent cover of each species per plot is recorded, along with any special environmental factors such as sand disturbance or flooding. Elevations for each of the 441 plots were determined each fall from 2011 to 2013 using a rotating laser level (Topcon RL-H3C). No elevation 'standard' is available in this area as past hurricanes have uprooted benchmarks. Because of the unstable nature of the landscape, an average elevation of 98 stakes in backdune areas thought to be stable were used as a standard elevation to compare across years. The topographic differences among the three habitats can be clearly seen by looking at these elevations (Fig. 1).

Data analysis

Data used in this paper are from the 3 years of 2011–13. To quantify the plant community in any given year, the per cent cover data were converted to the presence/absence of each species on each plot to minimize noise from the error associated with estimating per cent cover (i.e. Hirst and Jackson 2007). Then nonmetric multidimensional scaling (NMDS) was applied using the 'vegan' package of R (R Core Team 2014), with a maximum of 20 random starts in search of a stable solution. In general, stable solutions were not found, but repeated runs of metaMDS gave very similar results.

To minimize the effects of rare species, only species that were found on more than 5 % of the plots were included in the ordination. The effects of habitat and grid within habitat were evaluated using PERMANOVA through the 'adonis' function in the vegan package.

The elevations in 2011 and 2013 were compared to get the change in elevation over the 2-year span (see example in Fig. 2), showing either accretion or erosion (analyses by single-year spans found similar results). To show the effect of sand accretion or loss for the previous 2 years on plant community structure in 2013 while accounting for the effects of elevation, the vegetation in 2013 was characterized for each 1 m^2 plot using the NMDS scores determined above, and then the first and second axes scores were used as the dependent variables in generalized linear models (glm function in R), with the independent variables of (i) habitat, (ii) elevation in 2011 and (iii) change in elevation between 2011 and 2013. Replicate grid ID did not have a significant contribution from the model for any habitat and was not included in the results presented here. The change in elevation has a very awkward distribution, with both extreme high and low outliers that could not be easily transformed. To allow the data to be analysed, only the extreme outliers were transformed, by assuming that all values >0.2 m were equal to 0.2 m and all less than -0.2 m were equal to -0.2 m. Because there were strong interactions between the effects of some of the independent variables, separate GLM analysis were then conducted within each habitat, using just elevation and change in elevation as variables. The loadings of individual species in the NMDS were then used to determine which individual species were contributing the most to overall patterns in each habitat.

To determine which species were most correlated with disturbance, individual species per cent cover were correlated with elevation and with change in elevation per plot within each habitat using non-parametric Kendall τ values. Because this involves many comparisons, P values were corrected for a false discover rate (Benjamini and Hochberg 1995).

Results

The elevation of the dunes varied over almost 4 m in these relatively young dunes (Fig. 3), with the highest areas along foredune ridges (Fig. 1). Foredunes were on average the highest and most variable of the three habitats, while interdunes were not surprisingly the lowest and least variable. The change in elevation from 2011 to 2013 was generally very small (Fig. 3), with notable exception of dunes gaining over 0.6 m or losing over 1.5 m. These were generally due to the collapse of foredune

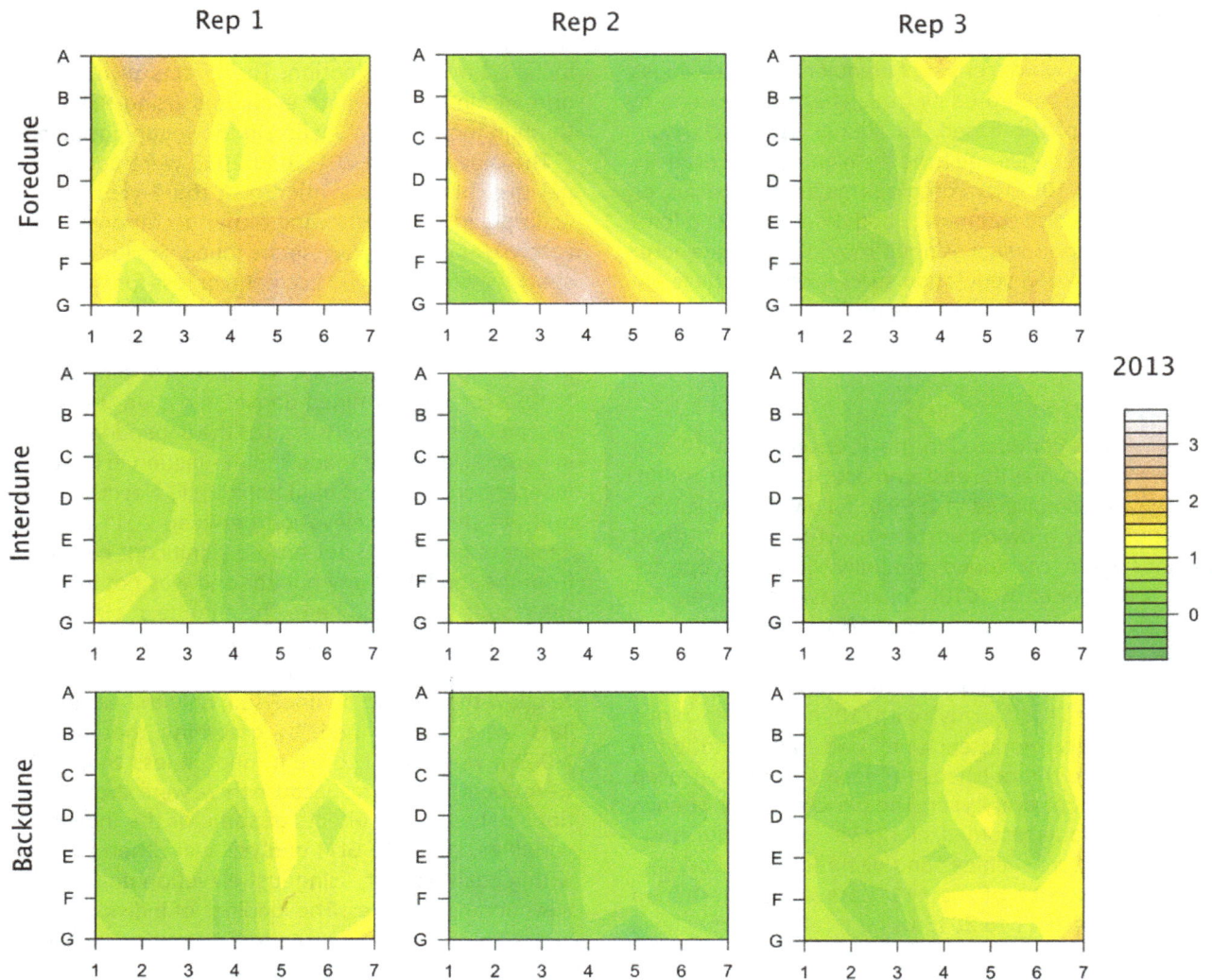

Figure 1. Topographic maps from 2013 of each of the nine grids used in this study. Grids are 60 × 60 m, with a 7 × 7 grid of points and each point 10 m from the next. The scale at right is in metres, and the 0 value is arbitrary. Extrapolated points were estimated using 'filled.contour' in R (R Core Team 2014).

ridges into leeward areas more inland. This caused a loss of dune elevation along the ridge but an increase in elevation in the lower areas were the sand ended up.

Over 60 species have been documented across the three dune habitats, with species richness increasing from foredunes to interdunes to backdunes (Miller *et al.* 2010). Vegetation on the dunes also varied among habitats, as shown by the NMDS (Fig. 4), with a significant effect of habitat ($F = 56.1$, $P < 0.001$) and replicate grid nested within habitat ($F = 10.1$, $P < 0.001$). As noted in Miller *et al.* (2010), interdunes are generally dominated by species associated with wetter areas, such as *Juncus* spp., *Phyla nodiflora* and *Paspalum distichum*, while foredunes and backdunes have species associated with drier areas, such as *Uniola paniculata*, *Schizachyrium maritima* and *Ipomoea imperata*.

The analysis of the full model predicting NMDS scores for 2013 based on habitat, elevation in 2011 and change in elevation from 2011 to 2013 suggested that there were significant effects of elevation, habitat and change in elevation, with elevation particularly loading on the first NMDS axis (Table 1). However, because of the significant interactions, especially between elevation and habitat, the data must be analysed separately for each habitat to determine the effects of elevation and change in elevation.

The NMDS analysis of the separate dune habitats also confirmed that elevation was a major contributor to plant community structure across foredunes, interdunes and backdunes (Table 2). However, the change in elevation was significant only in foredune and interdune habitats; only elevation significantly contributed to NMDS

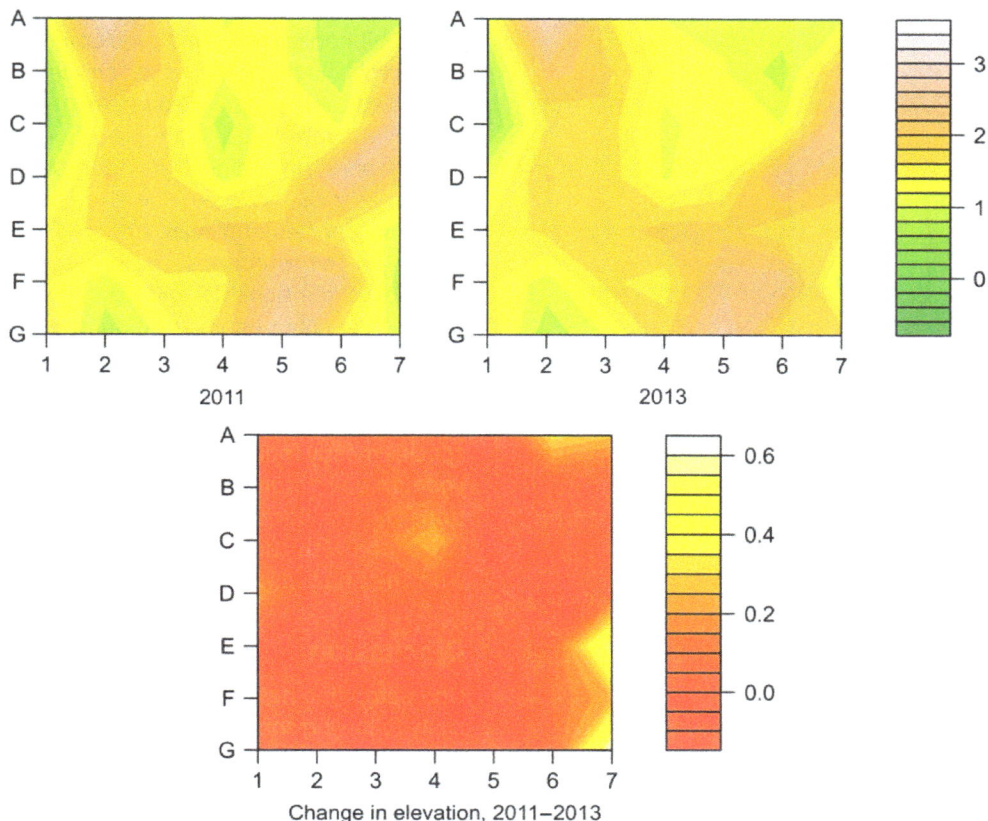

Figure 2. An example of the change in the topography of a foredune plot on St George Island from 2011 to 2013. The top two figures show the same plot in the first and last year of this study, with north towards the top. The lower figure shows the difference in elevation between the top two plots; note the much smaller scale. For this foredune plot, elevations are increasing on the eastern, more shoreward, side as winds carry sands up the beach plateau.

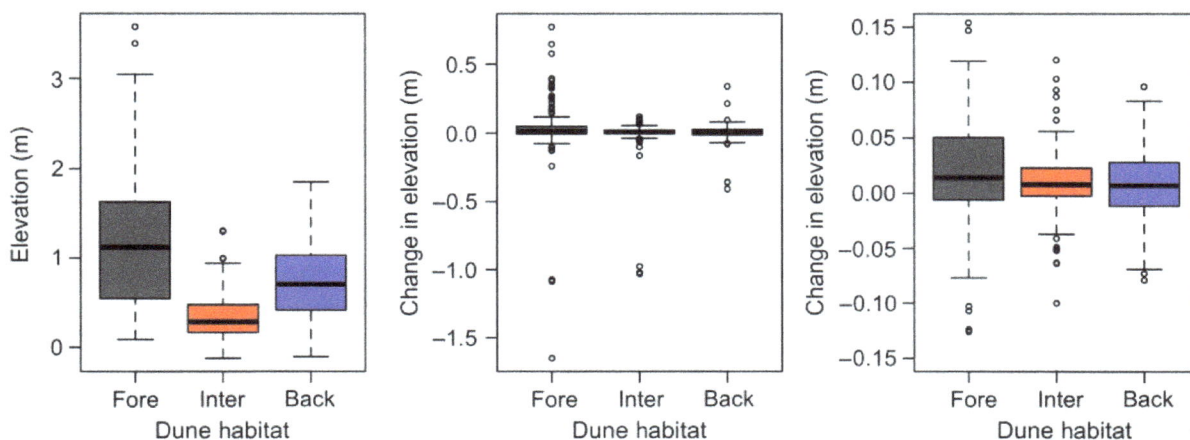

Figure 3. Boxplots showing the median (bar), 50 % confidence intervals (CI) (box) and 97 % CI (whiskers) for elevation (left) and change in elevation (centre and right) for the 147 plots in each habitat. The centre plot includes the outliers, while the right plot shows the same boxplots more closely, so the median and CI can be seen.

scores in backdunes. There were generally no significant interactions between elevation and change in elevation, except in interdunes for the first NMDS axis.

Finally, at the scale of the individual plots in each habitat, one can ask which species were significantly correlated with elevation or change in elevation and how.

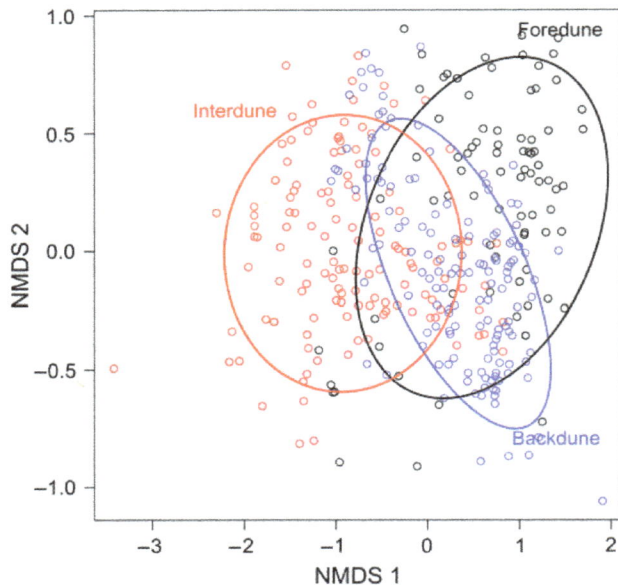

Figure 4. Ordination of the vegetation on 441 plots on St George Island, from 2013, using NMDS (stress = 0.112). Ovals denote ∼75 % CI for the 147 plots found in each of the three habitats, foredunes (black), interdunes (red) and backdunes (blue).

Table 1. F-values from generalized linear models of the effects of dune elevation and change in elevation from 2011 to 2013 on NMDS axis scores for vegetation on St George Island, from 2013 (**P < 0.01, ***P < 0.001). See text for details.

	NMDS axis 1	NMDS axis 2
Elevation	1177.70***	3.01
Habitat	136.05***	116.27***
Change in elevation	32.616***	9.89***
Elevation × habitat	55.149***	9.78***
Elevation × change in elevation	0.68	0.002
Habitat × change in elevation	16.09***	9.68***
Elevation × habitat × change in elevation	5.49**	0.08

Several species were positively correlated with elevation across all habitats (Table 3), including most notably the grasses *Uniola paniculata* and *Schizachyrium maritimum*, as well as the forb *Ipomoea imperati*. Other species were negatively correlated with elevation across all three dune habitats, including *Phyla nodiflora*, *Eragrostis lugens* and *Fimbristylis spadacea*. However, this study particularly concerns species that were affected by changes in elevation. In the foredunes, the only species exhibiting a significant relationship with the change in elevation was the very abundant *Uniola paniculata*, which appears to increase in abundance as dune elevation decreases. In interdunes, the low-lying *Polypremum procumbens* was positively correlated with increasing elevation, while a species mostly found associated with standing water, *Polygonum punctatum*, was correlated with lowering elevations. In the backdunes, no species abundances were correlated with changes in elevation, which is consistent with the analyses of the ordination scores.

Discussion

Understanding how dunes and vegetation interact, especially following sand accretion and loss, will be critical for determining the vulnerability of coastal areas to climate change (e.g. sea-level rise and changing storm frequencies). The static patterns of dunes and vegetation have been well documented in a variety of coastal systems (e.g. Gibson and Looney 1994; Stallins and Parker 2003; Martinez and Psuty 2004). However, few studies have followed both dunes and vegetation through time in order to quantify the dynamic relationship between geomorphology and vegetation and no prior studies have been carried out at different scales across different dune habitats. This study demonstrates that the relationships among geomorphology, vegetation and disturbance vary in important ways in different parts of the dune ecosystem.

At the scale of the entire coastal ecosystem, elevation was the dominant factor correlated with vegetation patterns in 2013 (Table 1), consistent with other studies (e.g. Moreno-Casasola 1986; Bitton and Hesp 2013). However, foredunes, interdunes and backdunes can have very

Table 2. F-values from generalized linear models of the effects of elevation and change in elevation from 2011 to 2013 by habitat on NMDS axis scores for vegetation on St George Island in 2013 (*P < 0.05, **P < 0.01, ***P < 0.001). See text for details.

	Foredune		Interdune		Backdune	
	NMDS1	NMDS2	NMDS1	NMDS2	NMDS1	NMDS2
Elevation in 2011	133.68***	31.98***	390.36***	5.37*	271.938***	3.48
Change in elevation from 2011 to 2013	8.82**	18.17***	85.13***	6.34*	1.032	2.51
Elevation × change in elevation	0.133	1.77	15.7***	0.01	0.611	0.66

Table 3. Correlations between species per cent cover and elevation and species per cent cover and change in elevation, within each habitat. Analyses were done for species occurring in >5 % of the plots in that habitat. Values are Kendall τ values, with significance corrected for multiple comparisons (*$P < 0.05$, **$P < 0.01$, ***$P < 0.001$). NA, species not available.

Species	Foredune		Interdune		Backdune	
	Elevation	Change	Elevation	Change	Elevation	Change
Bulbostylis ciliatifolia	NA	NA	−0.03	0.06	−0.21**	−0.04
Cenchrus incertus	−0.05	−0.06	0.34***	−0.12	0.03	−0.03
Centella asiatica	−0.15	−0.00	−0.35***	0.10	−0.47***	−0.03
Chamaesyce maculata	−0.03	−0.01	NA	NA	0.04	−0.07
Cnidoscolus stimulosus	NA	NA	NA	NA	0.24**	0.03
Cynanchum angustifolium	−0.47***	0.04	−0.11	0.03	−0.36***	−0.12
Cyperus croceus	−0.09	0.11	0.09	0.03	0.10	0.05
Rhynchospora colorata	NA	NA	−0.16*	−0.02	−0.27***	0.04
Eragrostis lugens	−0.40***	0.03	−0.08	0.11	−0.41***	0.01
Fimbristylis spadaceae	−0.45***	−0.01	−0.11	0.08	−0.23**	0.04
Fuirena scirpoidea	NA	NA	−0.02	0.05	−0.15*	−0.05
Heterotheca subaxillaris	−0.07	−0.01	0.25**	0.02	0.11	−0.05
Hydrocotyle bonariensis	0.07	−0.06	0.12	−0.10	−0.16*	−0.13
Ipomea imperati	0.23**	0.01	0.12	−0.06	0.36***	0.01
Iva imbricata	0.22**	0.07	NA	NA	NA	NA
Juncus megacephalus	NA	NA	−0.07	−0.03	−0.15	0.07
Muhlenbergia capillaris	−0.16	0.06	0.09	0.12	−0.43***	−0.02
Oenothera humifusa	0.03	−0.15	0.41***	0.14	0.03	0.10
Panicum aciculare	NA	NA	0.08	0.00	−0.41***	−0.07
Panicum amarum	−0.07	0.03	0.20**	0.14	0.04	0.10
Paronychia erecta	NA	NA	NA	NA	−0.06	−0.16
Paspalum vaginatum	−0.30**	0.00	−0.48***	−0.12	NA	NA
Physalis angustifolia	NA	NA	0.33***	−0.05	0.04	−0.09
Phyla nodiflora	−0.47***	0.06	−0.40***	0.06	−0.22**	0.09
Polypremum procumbens	NA	NA	0.34***	0.20*	−0.19**	−0.07
Polygonum punctatum	NA	NA	0.10	−0.22*	NA	NA
Schizachyrium maritimum	0.31***	−0.13	0.50***	−0.07	0.30***	−0.05
Sclerea verticillata	NA	NA	−0.06	−0.03	−0.24**	0.02
Setaria parvifolia	NA	NA	NA	NA	−0.29***	−0.06
Smilax auriculata	NA	NA	0.09	0.11	0.15*	−0.02
Spartina patens	−0.05	0.12	−0.09	0.11	−0.17*	−0.08
Sporobolus virginicus	−0.12	−0.03	−0.01	0.13	−0.09	0.01
Uniola paniculata	0.45***	−0.20*	0.51***	−0.01	0.39***	0.03

different types of vegetation (Fig. 4) and so habitat also explained a significant amount of variation in plant communities among plots. Despite these dominant effects of elevation and habitat, the amount of sand accretion or loss (disturbance) at each site also had a significant effect on the plant community structure, which has implications for the effects of climate change. Because it appears that the effect of elevation changes with dune habitat (Table 1), further analyses were conducted at the scale of the individual habitats.

At the scale of each dune habitat, the statistical models continue to show that elevation is the dominant factor correlated with vegetation patterns (Table 2). However, the change in elevation also has a significant correlation with vegetation patterns in the foredunes and interdunes, suggesting that vegetation in these habitats may be more affected by climate change than vegetation in the backdunes (Table 2). This is particularly interesting in light of the patterns of dune elevation and change in elevation across the three habitats (Figs 1 and 3). It is well known that foredunes tend to be taller and more subject to both dune building and erosion than other dune habitats, which is consistent with the patterns observed on St George from 2011 to 2013. So, it seems reasonable the higher sand movement, creating both increases and decreases in elevation, would affect the plant communities on foredunes. However, interdunes are the lowest and show the least variation in elevation over this time period. So, it is somewhat surprising that there is a correlation between vegetation pattern and change in elevation for interdunes. One possible explanation is that the lower-lying habitats are more affected by flooding; small changes in elevation can determine whether the plot is flooded or has saturated soil or is dry. Thus, small changes in elevation may have greater effects in interdunes than any other dune habitat, which also makes interdunes particularly susceptible to flooding associated with climate change.

At the scale of the individual species and plots, there are many significant correlations between the cover of particular species and elevation, as would be predicted from prior studies (e.g. Miller *et al.* 2010; Bitton and Hesp 2013; Wolner *et al.* 2013). These correlations are likely due to effects of soil moisture or soil nutrient levels, or both (see Miller *et al.* 2010). Some of the species showing the strongest associations with elevation are also among the most abundant species and characteristic of particular habitats and elevations. In particular, *U. paniculata* is considered important for stabilizing dune ridges; it is interesting that this species has the same association across all three dune habitats. *Schizachyrium maritima* has also been found to be an important species for restoration after hurricanes (Gornish and Miller 2013) and also favours higher elevation on dunes. *Iva imbricata* was also positively correlated with increasing dune elevation and is another species that may be important for stabilizing foredunes (see Colosi and McCormick 1978). Other species that increase in per cent with increasing elevation are *Oenothera humifusa* in interdunes (this species appears to be salt-intolerant; Miller *et al.* 2010), *Ipomoea imperata* and *Physalis angularis*, all of which can be relatively abundant on dune ridges in dune habitats.

Species that were negatively associated with elevation also include some relative dominants in each habitat. *Eragrostis lugens* and *Fimbristylis spadacea* are abundant in lower areas of foredunes and backdunes, but are generally not found in the wetter areas associated with interdunes. The grasses *Muhlenbergia capillaris* and *Panicum aciculare* can also dominate lower areas in backdunes.

Of greater interest to this study were species that were correlated with changes in elevation, as this might suggest species that either influence or respond to sand movement and may be more important for climate change responses. Despite the significant effects of change in elevation on ordination scores, relatively few species showed strong positive or negative correlations with change in elevation over 2011–13. *Uniola paniculata* did show an increase in abundance as dunes lost elevation on foredunes. This was somewhat surprising, as *U. paniculata* has been implicated as a dune builder. However, on St George Island, *Uniola* is also dominant on older foredunes that are starting to erode as they become blocked by newer dunes forming towards the beach, which may cause a net negative association with elevation. *Polypremum procumbens* was significantly associated with increases in elevation in interdunes, *Polygonum punctatum*, a wetland specialist in dunes, was associated with decreases. These species are good indicators of low and high dune areas respectively, but both are relatively infrequent and unlikely to have a large role in reciprocal interactions between vegetation and dune morphology.

This study demonstrates that changes in elevation in coastal areas are correlated with changes in the associated plant community, which is consistent with previous studies such as Moreno-Casasola (1986) and Wolner *et al.* (2013). Further, it demonstrates that these relationships change with dune habitat, which predicts how effects of climate change will vary spatially. Foredune vegetation is very much affected by the more variable and disturbed nature of the dynamic sand movement near the shore. Plants that help build or sustain dunes will be particularly important. Interdunes are low, wet and less variable in elevation and there are suggestions that plants may play a significant role in actually preventing dunes from building (Ehrenfeld 1990; Stallins 2005; Wolner *et al.* 2013). Backdunes retain dune and trough associated variation in elevation, but are much more stable than other areas. The influence of long-lived species, including woody species, suggests that these areas may also be the result of longer-term successional processes. Overall, the interactions between dune geomorphology and their associated communities are complex and vary at different spatial scales.

Just as relationships between dune geomorphology and vegetation vary with dune habitat, they are also likely to change across different spans of years. The years used in this study were particularly quiescent; the closest storm, Tropical storm Debby, passed more than 100 miles south of St George Island in 2012. It will be important to follow these same patterns during extremes of storms and droughts that are known to have major effects on dune geomorphology and vegetation (see Hesse and Simpson 2006; Miller *et al.* 2010).

Conclusions

Previous studies have documented correlations between environmental traits and vegetation, primarily in fore-dunes. However, few studies have correlated disturbance in dunes with plant community composition and no prior studies have investigated how the effect of disturbance varies at different spatial scales. This study used a unique long-term data base from a barrier island in the northern Gulf of Mexico to correlate the vegetation patterns with dune habitat, elevation and the change in elevation from 2011 to 2013. Generalized linear models suggest that both elevation and change in elevation (disturbance) do affect the vegetation, but that these effects differ among foredunes, interdunes and backdunes. A particularly notable result is that, while interdunes are subject to only minor disturbances, these can have significant effects on the plant community, perhaps because of changes in the hydrology of this lower-lying habitat. Overall, this study provides a better view of the links between dynamic dune geomorphology and the plant community, which may be important for predicting future effects of climate change on coastal ecosystems.

Sources of Funding

This work was funded in part by an award from the National Science Foundation (DEB 1050469) to T.E.M. and by funds from Florida State University.

Contributions by the Authors

T.E.M. is responsible for all the organization of the field-work, data management, analyses and writing for this study.

Acknowledgements

The author thanks Elise Gornish for helping with much of this study and for encouraging the completion of this analysis. Abigal Pastore, Will Ryan and Marina Lauck provided ideas and comments on the writing. The graduate students in the Field Quantitative Method class have been cheerful and entertaining fieldwork for this study for the past 12 years.

Literature Cited

Benjamini Y, Hochberg Y. 1995. Controlling the false discovery rate: a practical and powerful approach to multiple testing. *Journal of the Royal Statistical Society Series B* **57**:289–300.

Bitton MC, Hesp PA. 2013. Vegetation dynamics on eroding to accreting beach-foredune systems, Florida panhandle. *Earth Surface Processes and Landforms* **38**:1472–1480.

Colosi JC, McCormick JF. 1978. Population structure of *Iva imbricata* in five coastal dune habitats. *Bulletin of the Torrey Botanical Club* **105**:175–186.

Cowles HC. 1899. The ecological relations of the vegetation on the sand dunes of Lake Michigan (concluded). *Botanical Gazette* **27**: 361–391.

Duran O, Moore LJ. 2013. Vegetation controls on the maximum size of coastal dunes. *Proceedings of the National Academy of Sciences of the USA* **110**:17217–17222.

Ehrenfeld JG. 1990. Dynamics and processes of barrier island vegetation. *Reviews in Aquatic Sciences* **2**:437–480.

Feagin RA, Sherman DJ, Grant WE. 2005. Coastal erosion, global sea-level rise, and the loss of sand dune plant habitats. *Frontiers in Ecology and the Environment* **3**:359–364.

Gibson DJ, Looney PB. 1994. Vegetation colonization of dredge spoil on Perdido Key, Florida. *Journal of Coastal Research* **10**:133–143.

Gornish ES, Miller TE. 2013. Using long-term census data to inform restoration methods for coastal dune vegetation. *Estuaries and Coasts* **36**:1014–1023.

Gutierrez JL, Jones CG, Byers JE, Arkema KK, Berkenbusch K, Commito JA, Duarte CM, Hacker SD, Lambrinos JG, Hendriks IE, Palomo MG, Wild C. 2011. Physical ecosystem engineers and the functioning of estuaries and coasts. In: Heip CH, Middelburg JJ, Philippart CJ, eds. *Treatise on estuarine and coastal science, vol. 7: functioning of ecosystems at the land-ocean interface*. Amsterdam: Elsevier, 53–81.

Hesse PP, Simpson RL. 2006. Variable vegetation cover and episodic sand movement on longitudinal desert sand dunes. *Geomorphology* **81**:276–291.

Hirst CN, Jackson DA. 2007. Reconstructing community relationships: the impact of sampling error, ordination approach, and gradient length. *Diversity and Distributions* **13**:361–371.

Martinez ML, Psuty NP. 2004. *Coastal dunes: ecology and conservation*. Berlin: Springer.

Miller TE, Gornish ES, Buckley HL. 2010. Climate and coastal dune vegetation: disturbance, recovery, and succession. *Plant Ecology* **206**:97–104.

Moreno-Casasola P. 1986. Sand movement as a factor in the distribution of plant communities in a coastal dune system. *Vegetatio* **65**: 67–76.

Oosting HJ, Billings WD. 1942. Factors affecting vegetational zonation on coastal dunes. *Ecology* **23**:131–142.

Prisco I, Carboni M, Acosta ATR. 2013. The fate of threatened coastal dune habitats in Italy under climate change scenarios. *PLoS ONE* **8**:e68850.

R Core Team. 2014. *R: a language and environment for statistical computing.* Vienna, Australia: R Foundation for Statistical Computing.

Ruggiero P, Komar PD, McDougal WG, Marra JJ, Beach RA. 2001. Wave runup, extreme water levels, and the erosion of properties backing beaches. *Journal of Coastal Research* **17**:407–419.

Sallenger AH. 2000. Storm impact scale for barrier islands. *Journal of Coastal Research* **16**:890–895.

Stallins JA. 2005. Stability domains in barrier island dune systems. *Ecological Complexity* **2**:410–430.

Stallins JA, Parker AJ. 2003. The influence of complex systems interactions on barrier island dune vegetation pattern and process. *Annals of the Association of American Geographers* **93**:13–29.

Wolner CW, Moore LJ, Young DR, Brantley ST, Bissett SN, McBride RA. 2013. Ecomorphodynamic feedbacks and barrier island response to disturbance: insights from the Virginia Barrier Islands, Mid-Atlantic Bight, USA. *Geomorphology* **199**:115–128.

Zarnetske PL, Hacker SD, Seabloom EW, Ruggiero P, Killian JR, Maddux TB, Cox D. 2012. Biophysical feedback mediates effects of invasive grasses on coastal dune shape. *Ecology* **93**:1439–1450.

7

Suppression of annual *Bromus tectorum* by perennial *Agropyron cristatum*: roles of soil nitrogen availability and biological soil space

Robert R. Blank*, Tye Morgan and Fay Allen

Great Basin Rangelands Research Unit, USDA-Agricultural Research Service, Reno, NV, USA

Associate Editor: Inderjit

Abstract. Worldwide, exotic invasive grasses have caused numerous ecosystem perturbations. Rangelands of the western USA have experienced increases in the size and frequency of wildfires largely due to invasion by the annual grass *Bromus tectorum*. Rehabilitation of invaded rangelands is difficult; but long-term success is predicated on establishing healthy and dense perennial grass communities, which suppress *B. tectorum*. This paper reports on two experiments to increase our understanding of soil factors involved in suppression. Water was not limiting in this study. Growth of *B. tectorum* in soil conditioned by and competing with the exotic perennial *Agropyron cristatum* was far less relative to its growth without competition. When competing with *A. cristatum*, replacing a portion of conditioned soil with fresh soil before sowing of *B. tectorum* did not significantly increase its growth. The ability of conditioned soil to suppress *B. tectorum* was lost when it was separated from growing *A. cristatum*. Soil that suppressed *B. tectorum* growth was characterized by low mineral nitrogen (N) availability and a high molar ratio of NO_2^- in the solution-phase pool of $NO_2^- + NO_3^-$. Moreover, resin availability of $NO_2^- + NO_3^-$ explained 66 % of the variability in *B. tectorum* aboveground mass, attesting to the importance of *A. cristatum* growth in reducing N availability to *B. tectorum*. Trials in which *B. tectorum* was suppressed the most were characterized by very high shoot/root mass ratios and roots that have less root hair growth relative to non-suppressed counterparts, suggesting co-opting of biological soil space by the perennial grass as another suppressive mechanism. Greater understanding of the role of biological soil space could be used to breed and select plant materials with traits that are more suppressive to invasive annual grasses.

Keywords: Plant–soil relationships; root competition.

Introduction

Invasive exotic grasses are causing ecosystem perturbations with lasting consequences worldwide (Lenz *et al.* 2003; Ogle *et al.* 2003; Milton 2004; Dogra *et al.* 2010; Speziale *et al.* 2014). Especially pernicious invaders are the exotic annual grasses (DiTomaso 2000; Blumler 2006). Rehabilitation of annual grass-degraded lands can be exceedingly difficult, expensive and prone to failure (Young 1992; Jacobs *et al.* 1998; Cox and Allen 2008).

* Corresponding author's e-mail address: bob.blank@ars.usda.gov

In the intermountain region of the western USA, the Eurasian annual grass *Bromus tectorum* is responsible for landscape-level conversion of native *Artemisia* spp. ecosystems to annual grass dominance (Mack 1981; D'Antonio and Vitousek 1992; Billings 1994; Knapp 1996). The major pathway by which *B. tectorum* assumes dominance is by first occupying safesites within the community (often facilitated by disturbance) and expanding from those sites to a critical density and biomass, whereby conditions for large-scale wildfires are promulgated. Following the wildfire, *B. tectorum* readily dominates the site due to lack of competition, its inherently high growth rate, prolific seed production and ability to rapidly utilize post-fire elevated available nutrients (Fig. 1A; Mack 1981; Knapp 1996). Native species recruiting post-wildfire, including perennial grasses, find it difficult to compete against *B. tectorum* from the seedling stage (Francis and Pyke 1996; Arredondo *et al.* 1998; Brooks 2003; Humphrey and Schupp 2004; Blank 2010).

Fortunately, some plant communities resist invasion by *B. tectorum*; they are able to suppress its growth (James *et al.* 2008; Blank and Morgan 2012a; Chambers *et al.* 2014). Common threads to this resistance/suppression are well-established, healthy, and properly-spaced populations of perennial grasses (Fig. 1B–D, Humphrey and Schupp 2004; McGlone *et al.* 2011). Understanding how soil biochemical attributes affect suppression and how these attributes interact with a perennial grass offers hope of greater success in rehabilitating exotic annual grass-degraded ecosystems (D'Antonio and Thomsen 2004). The nature of suppression is complex and involves biotic and abiotic processes that temporally interact with soil type, the array of plant communities and characteristics of the invasive species (Huenneke *et al.* 1990; Tilman 1997; Naeem *et al.* 2000; Gundale *et al.* 2008). Established perennial plants can simply reduce soil resources to levels below which annual grasses are no longer as competitive (Wedin and Tilman 1990; Claassen and Marler 1998; Prober and Lunt 2009). Suppression of annual grasses may also involve root competition other than nutrient depletion whence perennial roots simply occupy biological soil space and, through chemical signalling, allelopathy included, may forestall competing roots from entering their space (Monk and Gabrielson 1985; McConnaughay and Bazzaz 1992; Schenk 2006).

Figure 1. (A) A far too typical landscape scene in northern Nevada, USA, several years after a wildfire. This landscape, once occupied by *Artemisia wyomingesis* and perennial grasses, is now dominated by *B. tectorum* and represents an environment exceedingly difficult to rehabilitate. Photographic examples showing perennial grass suppression of *B. tectorum*. (B) A high-elevation community in the Virginia Range, Nevada, USA. In the foreground is the native perennial *Pseudoroegneria spicata* with no presence of *B. tectorum*. (C) *Agropyron cristatum* sown after a wildfire in the early 1990s near Midas, Nevada. Although individual plants suppress *B. tectorum*, the density of perennial grasses is insufficient to prevent re-invasion by the exotic annual. Surface soil litter is mainly from *B. tectorum*. (D) A dense and robust community of *A. cristatum* planted after a 1985 wildfire near the Peterson Range, Nevada, which should resist re-invasion by *B. tectorum* if managed properly. Characteristic of all these suppressed areas is a ring around the perennial grasses that contain no plants of *B. tectorum* even though seedbank analyses indicate the presence of germinable seeds.

In a previous study, we explored the mechanistic underpinnings of perennial grass suppression of *B. tectorum* (Blank and Morgan 2012b). The data suggest that perennial grass roots reduced soil nitrogen (N) and phosphorous (P) availability and occupied biological soil space, thereby reducing *B. tectorum* growth. This paper reports on additional experiments to more definitively elucidate soil factors involved in suppression of *B. tectorum*. The perennial grass used was *Agropyron cristatum*, a native to Russia and central Asia. This grass is often used to rehabilitate degraded rangelands in the western USA, and well-established stands effectively suppress *B. tectorum* (Evans and Young 1978; Wicks 1997). Two hypotheses were tested: (i) soil conditioning brought about by established *A. cristatum* will reduce availability of soil mineral N and P to levels low enough to significantly reduce growth *of B. tectorum*, and (ii) occupation of biological soil space by roots of *A. cristatum* will cause roots of *B. tectorum* to alter their architecture, morphology and activity resulting in reduced growth, i.e. suppression.

Methods

Two experiments were conducted in a greenhouse at Reno, NV, USA (39°32′17.20″N; 119°48′22.89″W). Prior to each experiment, soil substrate was freshly collected from a *Krascheninnikovia lanata* (winterfat) site, invaded by *B. tectorum* for about 12 years, ~80 km northwest of Reno, NV, USA (40°7′59.43″N; 120°6′56.18″W). This soil, conditioned by *B. tectorum*, has elevated soil N availability relative to nearby soil conditioned by native vegetation (Blank and Morgan 2013). Surface soils (0–25 cm, corresponding to the A horizon) were composited from an area of ~10 m². Soils, loamy sand in texture, were sieved to <4 mm to remove coarse fragments and medium-to-large roots and homogenized by hand mixing on a greenhouse bench. Four replicates of this soil were analysed for various attributes (see below). This original soil—referred to as fresh soil—is taken from a soil classified as a coarse-loamy, mixed, superactive, calcareous, mesic Typic Torriorthent.

Experiment 1 quantified the suppression of *B. tectorum* (cheatgrass) by established *A. cristatum* (crested wheatgrass). Twelve replicate clear plastic rhizotrons, 5 × 30 × 100 cm depth, were filled with equal volumes of soil. The outsides of the rhizotrons were covered with insulation that could be removed from the back to observe rooting patterns. Prior to seed planting, rhizotrons were paired in adjoining plastic containers to maintain a slight angle so that roots would readily intercept the clear rhizotron backing for observation, and deionized water was added to reach field capacity—~6 % by weight for the soils used—over the entire rhizotron. Two

seeds of *A. cristatum* were sown in the rhizotrons 6 cm from each edge to leave an 18 cm space in between for later planting of *B. tectorum*, and allowed to establish for 68 days, the conditioning phase. We define conditioning as the plant species-dependent engendering specific traits such as carbon flow, root exudation, nutrient uptake, root occupancy of soil space, alteration of the soil microbial community etc. that might affect competitive interactions. During establishment, *A. cristatum* was supplemented with 500–1000 mL of deionized water per week depending on depletion in the rhizotron as gauged by visual inspection when opaque backs were removed. After establishment, four treatments were randomly imposed to three replicate rhizotrons. In one treatment, *B. tectorum* was sown directly between *A. cristatum* in the conditioned soil as a test for maximal suppression. For the next treatment, 500 g of soil were removed from between the established *A. cristatum* plants, replaced with 500 g of fresh soil and *B. tectorum* sown in the new soil. The purpose of this treatment was to test how fresh soil mitigated against suppression. In another treatment, 500 g of conditioned soil were removed from between the established *A. cristatum* plants; then a nylon mesh (2-mm opening) was placed in the excavated area, 500 g of fresh soil were placed over the mesh and *B. tectorum* sown. Our purpose was to examine how reduced root movement into the fresh soil from *A. cristatum*, but still allowing diffusion of gases, solutes and microbes from the adjacent conditioned soil, affects suppression. For the last treatment, 500 g of soil were removed from between the established *A. cristatum* plants; then a plastic barrier was placed in the hole and filled with 500 g of fresh soil, and sown to *B. tectorum*. This treatment tested *B. tectorum* suppression upon total blocking of *A. cristatum* roots, which would affect biological soil space and exposure to potential pathogenic organisms and allelochemicals in soil conditioned by *A. cristatum*. Deionized water was immediately applied to soil above the newly sown seeds of *B. tectorum*. A small subsample of the homogenized 500 g conditioned soil was analysed for several soil attributes (see below). We also grew *B. tectorum* without competition in small containers filled with 500 g of either fresh soil or conditioned soil (from the soil excavated from rhizotrons). During growth of *B. tectorum*, rhizotrons and containers were watered twice weekly; water was not limiting to *B. tectorum* in this study. Supplemental lighting, using four high-pressure sodium lamps each producing 124 000 lumens at 2100 K temperature, was used to assure at least 12 h of daylight. After 70 days of growth, *B. tectorum* was clipped at the soil surface, dried for 48 h at 70 °C and weight recorded. Soil within the rooting zone of *B. tectorum* was excavated and roots reserved. Subsamples of roots from each treatment and replicates were washed, immediately observed with a light microscope and photographs taken. These subsamples were then added to the original sample, dried for 48 h at 70 °C and weight recorded. After harvest, soil within the rooting zone of B. tectorum of each treatment was homogenized and analysed for solution-phase anions (Cl⁻, NO_2^-, NO_3^-, SO_4^{2-} and ortho-P) using immiscible displacement (Mubarak and Olsen 1976) with quantification by ion chromatography (Dionex® AS11-HC column with gradient elution) and mineral N, defined as $NH_4^+ + NO_2^- + NO_3^-$, by 1.5 M KCl extraction (Keeney and Nelson 1982).

Experiment 2 explored the role of soil nutrient availability of N and P in the suppression process and tested to a greater extent if and how conditioned soil affects suppression. Twelve rhizotrons were filled with freshly collected soil and planted to A. cristatum (see Experiment 1). Soil was conditioned by A. cristatum for 64 days with lighting and watering as described for Experiment 1. Four treatments were imposed following conditioning by A. cristatum. In four randomly chosen rhizotrons, B. tectorum was sown directly between established A. cristatum to test for maximal suppression (Treatment 1). The remaining eight rhizotrons had their backs removed and soil was separated by depths (0–30 cm (Treatment 2), 30–60 cm (Treatment 3) and 60–90 cm (Treatment 4)), and homogenized along with any roots present. For each soil depth separate, 2500 g was placed in containers and B. tectorum immediately sown. These treatments tested the suppressive ability of conditioned soil, by depth, without live plants of A. cristatum, but with different amounts of now inactive roots present depending on soil depth, greatest in the 0–30 cm depth increment and least in the 60–90 cm depth increment. In similar-sized containers, B. tectorum was sown in fresh soil to serve as unsuppressed controls (six replicates). To gauge the influence of B. tectorum growth on post-harvest soil attributes, five replicates of unplanted controls in fresh soil were prepared in similar-sized containers. For all experimental units, one anion and cation exchange resin capsule (Unibest®) was placed at 15 cm directly beneath where B. tectorum was sown to gauge nutrient availability. After 64 days of growth, above-ground and root biomass of B. tectorum were harvested, dried and weighed. Resin capsules were removed, washed extensively with deionized water, dried and treated with 40 mL of 1 N HCl and shaken for 30 min. Resin availability of NH_4^+, $NO_2^- + NO_3^-$ and ortho-P were quantified using a Lachat® autoanalyser. Soil in the rooting zone of B. tectorum was homogenized and analysed for mineral N and soil-solution anions as stated in Experiment 1. Availability of micronutrients was determined using the DTPA method (Lindsay and Norvell 1978).

Nitrogen mineralization potential was quantified using a moist 30-day incubation procedure (Bundy and Meisinger 1994). Total soil C and N were quantified using a LECO Truspec® analyser.

The data structure for Experiment 1 includes eight treatments with replication for a total of 38 experimental units. Experiment 2 had 10 treatments with replication for a total of 43 experimental units. For each experiment, a separate ANOVA was performed and means separated using Tukey's honest significant difference test. A backward selective regression was used to identify variables possibly related to above-ground B. tectorum biomass. The procedure was applied separately to Experiments 1 and 2 and to the combined data set.

Results

Experiment 1

Competition against established A. cristatum, in either conditioned or fresh soil, significantly reduced above-ground biomass of B. tectorum, relative to its growth in fresh soil without competition (Fig. 2). Growth of B. tectorum improved using fresh soil above a mesh, but not significantly so, relative to its growth competing with A. cristatum in either fresh or conditioned soil. Above-ground biomass of B. tectorum was far greater when fresh soil was placed in a plastic barrier between established A. cristatum. When competing with A. cristatum, B. tectorum was marked by very high shoot/root mass ratios relative to its ratios when not in competition (Fig. 2). The most suppressed plants of B. tectorum were characterized by minimal root branching and some consisted of one very long root. In the most suppressed B. tectorum trials, roots had fewer and shorter root hairs based upon microscopic inspection.

Nutrient attributes quantified for Experiment 1 differed significantly among treatments (Table 1). Soil mineral N content was greatest in the fresh soil (0.450 mmol kg⁻¹) and did not decline significantly after conditioning by A. cristatum (0.332 mmol kg⁻¹). Following harvest of B. tectorum, soil in its rooting zone of all experimental units had significantly less mineral N than the fresh soil; notable was the far lower mineral N remaining after plant growth in its rooting zone when competed with A. cristatum. The molar proportion of NO_2^- in the solution-phase $NO_2^- + NO_3^-$ pool varied widely among treatments. Notable are the very high values in trials where A. cristatum competed against B. tectorum, with exception of the plastic barrier treatment. Solution-phase ortho-P was less variable among treatments than mineral N, and plant growth facilitated elevated P values relative to fresh soil. Using a backwards regression variable-selection

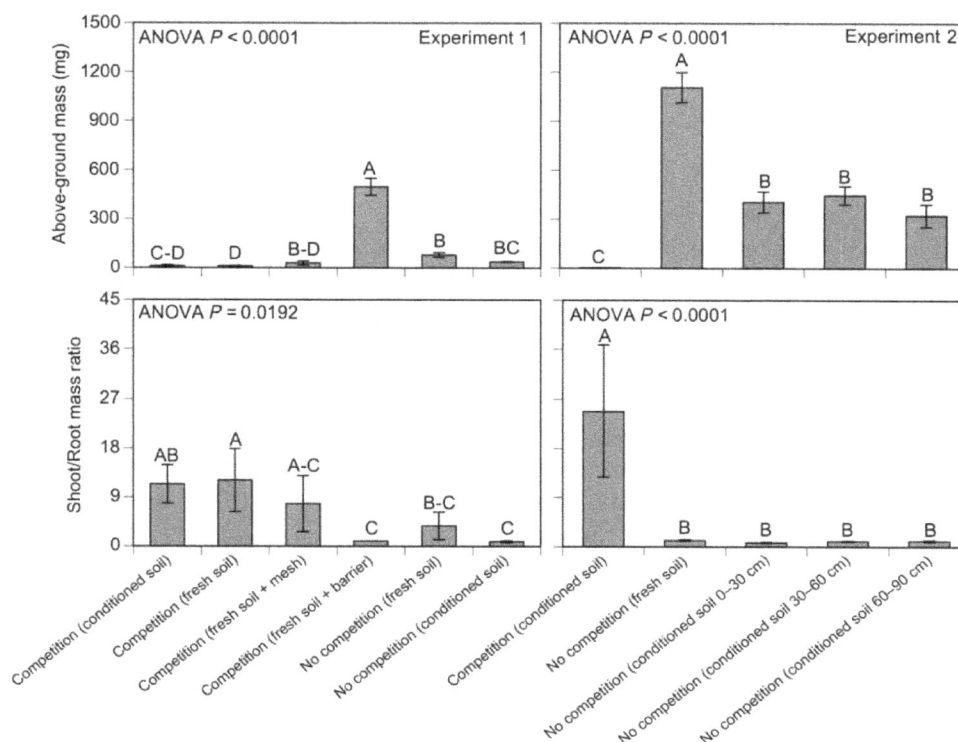

Figure 2. Above-ground biomass and shoot/root mass ratios of *B. tectorum* plants following harvest of Experiments 1 and 2. For each panel, ANOVA results are provided and bars with non-overlapping letters are significantly different at the $\leq$0.05 level.

Table 1. Selected soil attributes for Experiment 1.[1]

Treatment	Mineral N (mmol kg^{-1})	Mole NO$_2^-$ (%)	Ortho-P (μmol L^{-1})
Freshly collected field soil	0.450^A	49^{BC}	20.6^B
Conditioned soil prior to sowing *B. tectorum*[2]	0.332^{AB}	16^C	30.8^A
Conditioned soil post-harvest *B. tectorum*[3]	0.057^C	81^{AB}	20.1^B
Fresh soil post-harvest *B. tectorum*[4]	0.048^C	72^{AB}	29.1^{AB}
Fresh soil above mesh post-harvest *B. tectorum*[4]	0.059^C	95^A	27.3^{AB}
Fresh soil above plastic barrier post-harvest *B. tectorum*[4]	0.064^C	46^{BC}	29.5^{AB}
Fresh soil post-harvest *B. tectorum* without competition[5]	0.228^{BC}	10^C	34.6^A
Conditioned soil post-harvest *B. tectorum* without competition[5]	0.154^C	7^C	34.9^A
ANOVA	<0.0001	<0.0001	0.0004

[1]For each column, means with different superscripted letters are significantly different at the <0.05 level; mineral N is total NH$_4^+$ + NO$_2^-$ + NO$_3^-$ extractable by KCl; mole NO$_2^-$ is the molar proportion of NO$_2^-$ in the solution-phase pool of NO$_2^-$ + NO$_3^-$. Attributes unaffected by treatment included solution-phase NO$_2^-$, NO$_3^-$ and SO$_4^{2-}$.

[2]Soil from a homogenized subsample taken between *A. cristatum* that established for 60 days in rhizotrons.

[3]Soil from a homogenized subsample taken from the rooting zone of *B. tectorum* in competition with *A. cristatum* in rhizotrons.

[4]Soil from homogenized subsamples of the fresh soil and the fresh soil placed above the mesh or plastic barrier in rhizotrons.

[5]Soil from a homogenized subsample of entire container.

procedure, root mass explained 88 % of above-ground biomass; but no measured soil nutrient attributes significantly predicted above-ground biomass of *B. tectorum*.

Experiment 2

Similar to Experiment 1, established *A. cristatum* suppressed the growth of *B. tectorum* relative to its non-competed growth in fresh or conditioned soil (Fig. 1). Suppressed *B. tectorum*, akin to Experiment 1, had very high shoot to root mass ratios. Growth of *B. tectorum* was not nearly as suppressed when grown in conditioned soil taken from three depths in rhizotrons planted to *A. cristatum* (Fig. 1). Moreover, shoot to root mass ratios of *B. tectorum* grown non-competed in this conditioned soil were similar to those of non-competed *B. tectorum* grown in fresh soil (Experiment 1).

Following harvest of *B. tectorum*, mineral N and resin availability of $NO_2^- + NO_3^-$ were greatest in unplanted controls (Table 2). Relative to all other treatments, mineral N was by far lowest (0.026 mmol kg^{-1}) in the 0–30 cm depth increment when competed with *A. cristatum* in the rhizotrons. After conditioning of soils by *A. cristatum* growth, mineral N was not significantly reduced relative to fresh soil. Moreover, following the harvest of *B. tectorum* not competing with *A. cristatum*, soil mineral N was not significantly reduced relative to fresh soil. Resin availability of $NO_2^- + NO_3^-$ mirrored mineral N data with the unplanted controls having the greatest resin availability and the competed rhizotron values (placed at 15 cm) the least. The molar proportion of NO_2^- in the solution-phase $NO_2^- + NO_3^-$ pool was by far greatest in the conditioned soil sown to *B. tectorum* treatment. In general, plant growth, be it *A. cristatum* or *B. tectorum*, facilitated an increase in soil-solution *ortho*-P relative to fresh soil. Soil-solution *ortho*-P values were quite similar among the treatments with plant growth and only solution *ortho*-P of the fresh soil was significantly less (Table 2). There were no significant differences among the samples measured in resin availability of P (Table 2). Micronutrient availability of Zn did not vary much among treatments with the only significant difference between the fresh soil and the non-competed conditioned soil from the 30–60 cm depth increment. Manganese availability differed considerably among treatments with the highest values occurring in the rhizotron soils at depths of 30–60 and 60–90 cm and the lowest values in the fresh soil, the unplanted control soils and the non-competed soils. Using a backwards regression variable-selection procedure, applied to only data set 2, a combination of root biomass, resin-available $NO_2^- + NO_3^-$, and solution-phase NO_3^- explained 94 % of above-ground biomass variability. With combined data sets, root biomass and solution-phase NO_3^- explained 87 % of the variability in *B. tectorum* above-ground biomass (Fig. 2).

Discussion

We partially accept hypothesis 1 that established *A. cristatum* will reduce availability of soil mineral N

and P to levels low enough to suppress growth *of B. tectorum*. In regard to availability of soil P, there simply is no evidence from our data that established *A. cristatum* has reduced its availability sufficiently to suppress *B. tectorum* (Tables 1 and 2). We do accept the hypothesis that established *A. cristatum* has reduced the availability of soil N and thereby suppressed *B. tectorum*. Firstly, in both experiments, *B. tectorum* competing against established *A. cristatum* was significantly suppressed relative to its growth un-competed (Fig. 2). Secondly, following harvest of *B. tectorum*, mineral N was far less in soils with established *A. cristatum* relative to soil in non-competed trials (Tables 1 and 2). Thirdly, for Experiment 2, 66 % of the variability in above-ground biomass of *B. tectorum* is explained by resin availability of $NO_2^- + NO_3^-$ (Fig. 3). Finally, solution-phase NO_3^- was a significant variable in predicting above-ground mass of *B. tectorum* in the combined data set (Fig. 3).

It is not surprising that lowered soil N availability, due to established *A. cristatum*, would suppress the growth of *B. tectorum*. Many annual grasses, including *B. tectorum*, are nitrophiles and their growth is stimulated by additions of mineral N (Huenneke *et al.* 1990; Brooks 2003; Vasquez *et al.* 2008). Conversely, growth of annual grasses are often suppressed when mineral N is lowered by manipulating solution culture (Muller and Garnier 1990) or by addition of labile C sources that immobilize soil N (McLendon and Redente 1992; Young *et al.* 1998; Blank and Young 2009).

Besides availability of N, other aspects of the soil N cycle may be involved, at least tangentially, in suppression of *B. tectorum*. Soil conditioned by *A. cristatum* and competing with *B. tectorum* had very high molar proportions of NO_2^- in the solution-phase $NO_2^- + NO_3^-$ pool (Tables 1 and 2). Moreover, that soils only conditioned by *A. cristatum* prior to sowing *B. tectorum* had far lower molar NO_2^- levels suggests that the grasses interact with the soil differently when combined than they do individually. Perennial grasses differentially affect soil N cycling (Wedin and Tilman 1990; Vinton and Burke 1995), but we are unaware of any literature that tested the combined effect of a perennial grass and an annual grass on the soil N cycle. Roots of grasses can inhibit nitrite-oxidizers (Munro 1966); but why then did the greatest molar content occur only upon growth of *B. tectorum*. *Bromus tectorum* has high affinity to uptake N in the $NO_3^- - N$ form relative the $NH_4^+ - N$ form (MacKown *et al.* 2009), but we are unaware of any data on its ability to uptake the $NO_2^- - N$ form. If *B. tectorum* does not have a high affinity to uptake the $NO_2^- - N$ form, then perennial grasses that inhibit nitrate-oxidizers would likely elevated their suppressive ability.

Table 2. Selected soil attributes for Experiment 2.[1]

Treatment	Mineral N (mmol kg⁻¹)	Resin N (μmol)	Mole NO₂⁻ (%)	Solution P (μmol L⁻¹)	Resin P (μmol)	DTPA Zn (μmol kg⁻¹)	DTPA Mn (μmol kg⁻¹)
Freshly collected field soil	0.430[BC]	nd	5.2[BC]	17.8[C]	nd	5.91[AB]	53.1[A]
Conditioned soil, 0–30 cm, prior to placing in containers and sowing B. tectorum[2]	0.160[CD]	nd	0.3[C]	35.2[AB]	nd	4.76[AB]	12.4[D]
Conditioned soil, 30–60 cm, prior to placing in containers and sowing B. tectorum[2]	0.260[CD]	nd	0.8[C]	30.7[B]	nd	3.94[B]	16.1[CD]
Conditioned soil, 60–90 cm, prior to placing in containers and sowing B. tectorum[2]	0.250[CD]	nd	0.5[C]	32.9[AB]	nd	3.59[B]	16.4[CD]
Conditioned soil, post-harvest B. tectorum[3]	0.026[D]	0.6[C]	50.3[A]	36.6[A]	1.18	4.54[AB]	25.2[B]
Container soil following B. tectorum harvest growing in fresh soil[4]	0.550[B]	35.0[A]	0.6[C]	36.2[A]	0.92	6.38[A]	20.3[BC]
Container soil following B. tectorum harvest growing in conditioned soil, 0–30 cm[4]	0.315[C]	10.5[BC]	11.1[BC]	35.1[AB]	1.21	4.81[AB]	17.0[CD]
Container soil following B. tectorum harvest growing in conditioned soil, 30–60 cm[4]	0.364[BC]	18.7[B]	9.6[BC]	37.7[A]	1.21	4.46[B]	18.3[B-D]
Container soil following B. tectorum harvest growing non-competed in conditioned soil, 60–90 cm[4]	0.414[BC]	11.6[BC]	11.6[BC]	34.8[AB]	1.26	4.53[AB]	19.8[BC]
Fresh soil unplanted control[4]	1.280[A]	45.6[A]	20.6[B]	34.5[AB]	0.75	5.56[AB]	21.1[BC]
ANOVA	<0.0001	<0.0001	<0.0001	<0.0001	0.6754	0.0023	<0.0001

[1]For each column, means with different superscripted letters are significantly different at the <0.05 level; mineral N is total $NH_4^+ + NO_2^- + NO_3^-$ extractable by KCl; resin N includes $NO_2^- + NO_3^-$; mole NO_2^- is the molar proportion of $NO_2^- + NO_3^-$ in the solution-phase pool of $NO_2^- + NO_3^-$; nd, not determined. Attributes unaffected by treatment included 30-day aerobic incubated NH_4^+ and NO_3^-, net N mineralization potentials, total C and N, and DTPA extractable Fe and Cu.

[2]Soils taken after 64 days conditioning by A. cristatum from rhizotrons and homogenized by depth.

[3]Soils collected from within rooting zone of B. tectorum in rhizotrons that were conditioned by A. cristatum for 64 days.

[4]Soils from homogenized sample of entire container.

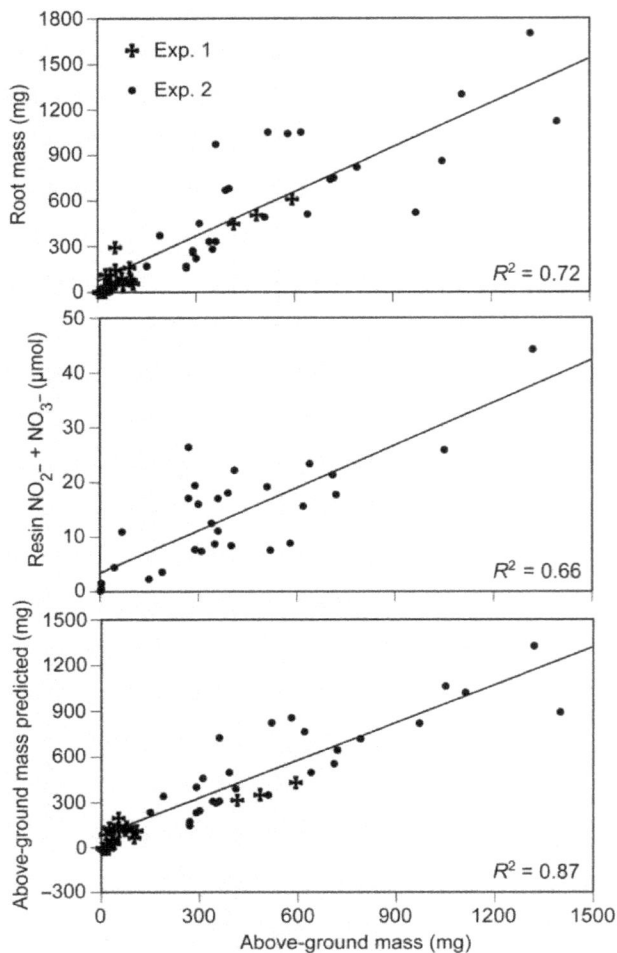

Figure 3. Graphs showing variables strongly related to above-ground biomass as determined by backward selection regression. Top graph, combined Experiments 1 and 2, relates root biomass with above-ground biomass. Middle graph relates resin availability of $NO_2^- + NO_3^-$ with above-ground biomass; resin data were only collected for Experiment 2. Bottom graph, combined Experiments 1 and 2, relates predicted above-ground biomass using the combination of root biomass and solution-phase NO_3^-.

We accept hypothesis 2 that occupation of biological soil space by established roots of *A. cristatum* will suppress growth of *B. tectorum*. Compelling aspects of our data include high shoot to root ratios of *B. tectorum* when competing with established *A. cristatum* (Fig. 2), response of *B. tectorum* to varying soil treatments (Fig. 2), and distinct elongated root architectures with far fewer root hairs in the most suppressed trials. The concept of biological soil space implies that physical space is a resource in itself, beyond that of access to nutrients and water (McConnaughay and Bazzaz 1991, 1992). In this construct, occupation of physical space by roots of established *A. cristatum* will constrain root growth of *B. tectorum*. The mechanistic underpinnings of

suppression via biological soil space may involve root signalling or root toxicity (Schenk 2006). It is possible that the elevated shoot to root ratios in *B. tectorum* competing against established *A. cristatum* in this study is likely less due to reduced availability of N than interactions with pre-existing roots of *A. cristatum*. Low soil N availability should stimulate rather than decrease root growth (Hill et al. 2006). If reduced biological soil space due to established *A. cristatum* roots is partly responsible for suppression of *B. tectorum*; then replacement of conditioned soil between established *A. cristatum* plants should have increased biological space for *B. tectorum* root growth and also increased nutrient availability resulting in less suppression—yet suppression still occurred. One possibility is that roots of *A. cristatum* may have proliferated in the fresh soil and simply occupied soil space faster than roots of the newly sown *B. tectorum*. Visual inspection upon harvesting *B. tectorum* did reveal the presence of *A. cristatum* roots. Moreover, very low post-harvest mineral N levels in the fresh soil (Table 1) lend support to the re-occupation of biological soil space by *A. cristatum* as such small plants of *B. tectorum* simply could not have depleted that much mineral N. We also expected a mesh would limit new root encroachment by *A. cristatum* and the fresh soil placed above the mesh would have much un-occupied biological soil space for roots of *B. tectorum* to proliferate; yet, *B. tectorum* planted in this soil was still suppressed and had high shoot to root ratios. Indeed, as the mesh was removed from the rhizotrons following harvest of *B. tectorum*, visible inspection indicated very few roots of *A. cristatum* had penetrated the mesh. Nonetheless, the fresh soil added above the mesh had very low mineral N content after harvest, in fact the lowest among all the treatments. Clearly, enough *A. cristatum* roots had penetrated the mesh to reduce mineral N content and thereby partially suppressed *B. tectorum* via lowered N availability. In all individual trials of *B. tectorum* competing with established *A. cristatum*, no matter the treatment, post-harvest *B. tectorum* had high shoot to root ratios (Fig. 2) and decreased root hair formation. Recent research has demonstrated that root competition is far more complex than simple resource depletion (see review by Schenk 2006). In this new construct of root to root interactions, it is possible that newly establishing roots of *B. tectorum* sense the presence of *A. cristatum* roots and do not grow appreciably into the fresh soil provided. Alternatively, established roots of *A. cristatum* may exude toxic substances that affect *B. tectorum* root architecture; unfortunately our experimental protocols are not able to rigorously test this conjecture.

Soil conditioned by *A. cristatum*, then homogenized (roots of *A. cristatum* were also homogenized), potted and sown to *B. tectorum* produced far more above-ground biomass than it did when competing with *A. cristatum*. Our expectation was that the conditioned soil separated from *A. cristatum* would retain its ability to suppress *B. tectorum* because the soil would have depleted N availability, at least initially. Moreover, we expected conditioned soil from the 0–30 cm depth increment would have greater filling of biological soil space with established roots of *A. cristatum* and therefore be more suppressive to *B. tectorum* than conditioned soil from lower depths. In fact, conditioned soil removed from rhizotrons did not suppress *B. tectorum* and shoot to root mass ratios were not elevated as in competed trials. Lack of suppression in this situation may be explained by the following. Firstly, the now dead roots of *A. cristatum* have mineralized and contributed N to enhance *B. tectorum* growth. The relatively high mineral N and resin available N levels post-harvest for these trials lend credence to this possibility. Secondly, the lack of an established root system of *A. cristatum* due to homogenization prior to sowing *B. tectorum* in containers may free up biological soil space for *B. tectorum* resulting in lower root to root signalling and exudation of toxins (Schenk 2006).

We expected that *B. tectorum* growing in the fresh soil above the plastic barrier in the rhizotrons would have above-ground biomass similar to its growth, non-competed, in fresh soil. Why then did the plastic barrier facilitate even greater growth of *B. tectorum*? Speculating, given the equal watering regimes used in all experimental units, the plastic barrier could have reduced water flow beyond the rooting zone of *B. tectorum* and essentially provided greater water availability.

Conclusions

The non-native perennial grass *A. cristatum*, when established, suppresses the growth of the exotic annual grass, *B. tectorum*. Reduced soil N availability and co-opting of soil space by perennial grass roots are potential soil factors involved in suppression. If only it were as easy to establish perennial grasses on *B. tectorum*-invaded rangelands as it is in the greenhouse, rehabilitation of *B. tectorum* degraded rangelands would be easier and far less expensive. The use of non-native plant materials to facilitate rehabilitation of exotic annual grass-invaded rangelands is controversial (D'Antonio and Meyerson 2002). Some researchers, however, make the case that particular non-natives possess attributes that allow faster and more effective rehabilitation (Asay et al. 2001; Ewel and Putz 2004). The reality that *A. cristatum* suppresses *B. tectorum* so effectively offers opportunities to use this species and other non-native competitive grasses

as a successional bridge to encourage subsequent native plant recruitment (Cox and Anderson 2004; Brown et al. 2008; Davies et al. 2013). Perennial grasses differ markedly in their ability to suppress annual grasses (Borman et al. 1990, 1991). A portion of the suppressive ability of *A. cristatum* is via utilization of soil N resources such that it is less available to *B. tectorum*; however, the annual is also an effective competitor for soil N (Monaco et al. 2003). If perennial grasses do not strongly couple root uptake of N with the timing of its availability in soil, pulses of availability can occur leading to less suppression. We believe greater understanding of aspects of suppression via biological soil space can be a fruitful area of research. What specific properties do established perennial grasses engender to biological soil space to resist subsequent growth of alien annual grasses? Is allelopathy involved? Is alteration of the soil microbial community involved? Understanding specific mechanisms could direct plant breeding strategies to develop perennial grasses more suppressive to exotic annual grasses.

Sources of Funding

This research is funded through USDA-Agricultural Research Service Project No: 5370-13610-001-00D.

Contributions by the Authors

R.R.B. designed the experiment, analysed the data and was the principle writer. T.M. designed and built rhizotrons. T.M. and F.A. collected soils, monitored the experiments and conducted soil analyses.

Acknowledgement

We thank the associate editor and anonymous reviewers for constructive criticisms leading to a more impactful paper.

Literature Cited

Arredondo JT, Jones TA, Johnson DA. 1998. Seedling growth of intermountain perennial and weedy annual grasses. *Journal of Range Management* **51**:584–589.

Asay KH, Horton WH, Jensen KB, Palazzo AJ. 2001. Merits of native and introduced Triticeae grasses on semiarid rangelands. *Canadian Journal of Plant Science* **81**:45–52.

Billings WD. 1994. Ecological impacts of cheatgrass and resultant fire on ecosystems in the Western Great Basin. In: Monson SB, Kitchen SG, eds. *Proceedings—ecology and management of annual rangelands*. Ogden, UT: US Forest Service, Intermountain Res Stat Gen Tech Rep INT-GTR-313, 22–30.

Blank RR. 2010. Intraspecific and interspecific pair-wise seedling competition between exotic annual grasses and native perennials: plant–soil relationships. *Plant and Soil* **326**:331–343.

Blank RR, Morgan T. 2012*a*. Mineral nitrogen in a crested wheatgrass stand: implications for suppression of cheatgrass. *Rangeland Ecology and Management* **65**:101–104.

Blank RR, Morgan T. 2012*b*. Suppression of *Bromus tectorum* L. by established perennial grasses: I. Potential mechanisms. *Applied and Environmental Soil Science* **2012**:Article ID 632172; doi: 10.1155/2012/632172.

Blank RR, Morgan T. 2013. Soil engineering facilitates downy brome (*Bromus tectorum*) growth—a case study. *Invasive Plant Science and Management* **6**:391–400.

Blank RR, Young JA. 2009. Plant–soil relationships of *Bromus tectorum* L: interactions among labile carbon additions, soil invasion status, and fertilizer. *Applied and Environmental Soil Science* **2009**:Article ID 929120; doi:10.1155/2009/929120.

Blumler MA. 2006. Geographical aspects of invasion: the annual bromes. *Middle States Geographer* **39**:1–7.

Borman MM, Krueger WC, Johnson DE. 1990. Growth patterns of perennial grasses in the annual grassland type of southwest Oregon. *Agronomy Journal* **82**:1093–1098.

Borman MM, Krueger WC, Johnson DE. 1991. Effects of established perennial grasses on yields of associated annual weeds. *Journal of Range Management* **44**:318–322.

Brooks ML. 2003. Effects of increased soil nitrogen on the dominance of alien annual plants in the Mojave Desert. *Journal of Applied Ecology* **40**:344–353.

Brown CS, Anderson VJ, Claassen VP, Stannard ME, Wilson LM, Atkinson SY, Bromberg JE, Grant TA III, Munis MD. 2008. Restoration ecology and invasive plants in the semiarid West. *Invasive Plant Science and Management* **1**:399–413.

Bundy LG, Meisinger JJ. 1994. Nitrogen availability indices. In: Weaver RW *et al.*, eds. *Methods of soil analysis, part 2: microbiological and biochemical properties*. Madison, WI: Soil Science Society of America, 951–984.

Chambers JC, Bradley BA, Brown CS, D'Antonio C, Germino MJ, Grace JB, Hardegree SP, Miller RF, Pyke DA. 2014. Resilience to stress and disturbance, and resistance to *Bromus tectorum* L. invasion in cold desert shrublands of western North America. *Ecosystems* **17**:360–375.

Claassen VP, Marler M. 1998. Annual and perennial grass growth on nitrogen-depleted decomposed granite. *Restoration Ecology* **6**: 175–180.

Cox RD, Allen EB. 2008. Stability of exotic annual grasses following restoration efforts in southern California coastal sage scrub. *Journal of Applied Ecology* **45**:495–504.

Cox RD, Anderson VJ. 2004. Increasing native diversity of cheatgrass-dominated rangeland through assisted succession. *Journal of Range Management* **57**:203–210.

D'Antonio C, Meyerson LA. 2002. Exotic plant species as problems and solutions in ecological restoration: a synthesis. *Restoration Ecology* **10**:703–713.

D'Antonio CM, Thomsen M. 2004. Ecological resistance in theory and practice. *Weed Technology* **18**:1572–1577.

D'Antonio CM, Vitousek PM. 1992. Biological invasions by exotic grasses, the grass/fire cycle, and global change. *Annual Review of Ecology and Systematics* **23**:63–87.

Davies KW, Boyd CS, Nafus AM. 2013. Restoring the sagebrush component in crested wheatgrass–dominated communities. *Rangeland Ecology and Management* **66**:472–478.

DiTomaso JM. 2000. Invasive weeds in rangelands: species, impacts, and management. *Weed Science* **48**:255–265.

Dogra KS, Sood SK, Dobhal PK, Sharma S. 2010. Alien plant invasion and their impact on indigenous species diversity at global scale: a review. *Journal of Ecology and the Natural Environment* **2**: 175–186.

Evans RA, Young JA. 1978. Effectiveness of rehabilitation practices following wildfire in a degraded big sagebrush-downy brome community. *Journal of Range Management* **31**:185–188.

Ewel JJ, Putz FE. 2004. A place for alien species in ecosystem restoration. *Frontiers in Ecology and the Environment* **2**:354–360.

Francis MG, Pyke DA. 1996. Crested wheatgrass-cheatgrass seedling competition in a mixed-density design. *Journal of Range Management* **49**:432–438.

Gundale MJ, Sutherland S, DeLuca TH. 2008. Fire, native species, and soil resource interactions influence the spatio-temporal invasion pattern of *Bromus tectorum*. *Ecography* **31**:201–210.

Hill JO, Simpson RJ, Moore AD, Chapman DF. 2006. Morphology and response of roots of pasture species to phosphorus and nitrogen nutrition. *Plant and Soil* **286**:7–19.

Huenneke LF, Hamburg SP, Koide R, Mooney HA, Vitousek PM. 1990. Effects of soil resources on plant invasion and community structure in Californian serpentine grassland. *Ecology* **71**:478–491.

Humphrey LD, Schupp EW. 2004. Competition as a barrier to establishment of a native perennial grass (*Elymus elymoides*) in alien annual grass (*Bromus tectorum*) communities. *Journal of Arid Environments* **58**:405–422.

Jacobs JS, Carpinelli MR, Sheley RL. 1998. Revegetating weed-infested rangeland: what we've learned. *Rangelands* **20**:10–15.

James JJ, Davies KW, Sheley RL, Aanderud ZT. 2008. Linking nitrogen partitioning and species abundance to invasion resistance in the Great Basin. *Oecologia* **156**:637–648.

Keeney DR, Nelson DW. 1982. Nitrogen-inorganic forms. In: Page AL, ed. *Methods of soil analysis, part 2. Chemical and microbiological properties*. Madison, WI: Soil Science Society of America, 643–698.

Knapp PA. 1996. Cheatgrass (*Bromus tectorum* L) dominance in the Great Basin desert: history, persistence, and influences to human activities. *Global Environmental Change* **6**:37–52.

Lenz TI, Moyle-Croft JL, Facelli JM. 2003. Direct and indirect effects of exotic annual grasses on species composition of a South Australian grassland. *Austral Ecology* **28**:23–32.

Lindsay WL, Norvell WA. 1978. Development of a DTPA soil test for zinc, iron, manganese, and copper. *Soil Science Society of America Journal* **42**:421–428.

Mack RN. 1981. Invasion of *Bromus tectorum* L. into western North America: an ecological chronicle. *Agro-Ecosystems* **7**:145–165.

MacKown CT, Jones TA, Johnson DA, Monaco TA, Redinbaugh MG. 2009. Nitrogen uptake by perennial and invasive annual grass seedlings: nitrogen form effects. *Soil Science Society of America Journal* **73**:1864–1870.

McConnaughay KDM, Bazzaz FA. 1991. Is physical space a soil resource? *Ecology* **72**:94–103.

McConnaughay KDM, Bazzaz FA. 1992. The occupation and fragmentation of space: consequences of neighbouring roots. *Functional Ecology* **6**:704–710.

McGlone CM, Sieg CH, Kolb TE. 2011. Invasion resistance and persistence: established plants win, even with disturbance and high propagule pressure. *Biological Invasions* **13**:291–304.

McLendon T, Redente EF. 1992. Effects of nitrogen limitation on species replacement dynamics during early secondary succession on a semiarid sagebrush site. *Oecologia* **91**:312–317.

Milton SJ. 2004. Grasses as invasive alien plants in South Africa. *South African Journal of Science* **100**:69–75.

Monaco TA, Johnson DA, Norton JM, Jones TA, Connors KJ, Norton JB, Redinbaugh MB. 2003. Contrasting responses of Intermountain West grasses to soil nitrogen. *Journal of Range Management* **56**:282–290.

Monk CD, Gabrielson FC Jr. 1985. Effects of shade, litter and root competition on old-field vegetation in South Carolina. *Bulletin of the Torrey Botanical Club* **112**:383–392.

Mubarak A, Olsen RA. 1976. Immiscible displacement of the soil solution by centrifugation. *Soil Science Society of America Journal* **40**:329–331.

Muller B, Garnier E. 1990. Components of relative growth rate and sensitivity to nitrogen availability in annual and perennial species of *Bromus*. *Oecologia* **84**:513–518.

Munro PE. 1966. Inhibition of nitrite-oxidizers by roots of grass. *The Journal of Applied Ecology* **3**:227–229.

Naeem S, Knops JM, Tilman D, Howe KM, Kennedy T, Gale S. 2000. Plant diversity increases resistance to invasion in the absence of covarying extrinsic factors. *Oikos* **91**:97–108.

Ogle SM, Reiners WA, Gerow KG. 2003. Impacts of exotic annual brome grasses (*Bromus* spp.) on ecosystem properties of northern mixed grass prairie. *The American Midland Naturalist* **149**:46–58.

Prober SM, Lunt ID. 2009. Restoration of *Themeda australis* swards suppresses soil nitrate and enhances ecological resistance to invasion by exotic annuals. *Biological Invasions* **11**:171–181.

Schenk HJ. 2006. Root competition: beyond resource depletion. *Journal of Ecology* **94**:725–739.

Speziale KL, Lambertucci SA, Ezcurra C. 2014. *Bromus tectorum* invasion in South America: Patagonia under threat? *Weed Research* **54**:70–77.

Tilman D. 1997. Community invasibility, recruitment limitation, and grassland biodiversity. *Ecology* **78**:81–92.

Vasquez E, Sheley R, Svejcar T. 2008. Nitrogen enhances the competitive ability of cheatgrass (*Bromus tectorum*) relative to native grasses. *Invasive Plant Science and Management* **1**:287–295.

Vinton MA, Burke IC. 1995. Interactions between individual plant species and soil nutrient status in shortgrass steppe. *Ecology* **76**:1116–1133.

Wedin DA, Tilman D. 1990. Species effects on nitrogen cycling: a test with perennial grasses. *Oecologia* **84**:433–441.

Wicks GA. 1997. Survival of downy brome (*Bromus tectorum*) seed in four environments. *Weed Science* **45**:225–228.

Young JA. 1992. Ecology and management of medusahead (*Taeniatherum caput-medusae* ssp. *asperum* [SIMK.] MELDERIS). *Great Basin Naturalist* **52**:245–252.

Young JA, Trent JD, Blank RR, Palmquist DE. 1998. Nitrogen interactions with medusahead (*Taeniatherum caput-medusae* ssp. *asperum*) seedbanks. *Weed Science* **46**:191–195.

8

Mutualism-disrupting allelopathic invader drives carbon stress and vital rate decline in a forest perennial herb

Nathan L. Brouwer*, Alison N. Hale and Susan Kalisz

Department of Biological Sciences, University of Pittsburgh, Pittsburgh, PA 15260, USA

Associate Editor: Inderjit

Abstract. Invasive plants can negatively affect belowground processes and alter soil microbial communities. For native plants that depend on soil resources from root fungal symbionts (RFS), invasion could compromise their resource status and subsequent ability to manufacture and store carbohydrates. Herbaceous perennials that depend on RFS-derived resources dominate eastern North American forest understories. Therefore, we predict that forest invasion by *Alliaria petiolata*, an allelopathic species that produces chemicals that are toxic to RFS, will diminish plant carbon storage and fitness. Over a single growing season, the loss of RFS could reduce a plant's photosynthetic physiology and carbon storage. If maintained over multiple growing seasons, this could create a condition of carbon stress and declines in plant vital rates. Here we characterize the signals of carbon stress over a short timeframe and explore the long-term consequence of *Alliaria* invasion using *Maianthemum racemosum*, an RFS-dependent forest understory perennial. First, in a greenhouse experiment, we treated the soil of potted *Maianthemum* with fresh leaf tissue from either *Alliaria* or *Hesperis matronalis* (control) for a single growing season. *Alliaria*-treated plants exhibit significant overall reductions in total non-structural carbohydrates and have 17 % less storage carbohydrates relative to controls. Second, we monitored *Maianthemum* vital rates in paired experimental plots where we either removed emerging *Alliaria* seedlings each spring or left *Alliaria* at ambient levels for 7 years. Where *Alliaria* is removed, *Maianthemum* size and vital rates improve significantly: flowering probability increases, while the probability of plants regressing to non-flowering stages or entering prolonged dormancy are reduced. Together, our results are consistent with the hypothesis that disruption of a ubiquitous mutualism following species invasion creates symptoms of carbon stress for species dependent on RFS. Disruption of plant–fungal mutualisms may generally contribute to the common, large-scale declines in forest biodiversity observed in the wake of allelopathic invaders.

Keywords: Allelochemicals; *Alliaria petiolata*; carbon stress/carbon starvation; *Maianthemum racemosum*; mutualism disruption; root fungal symbiont; species invasion; vital rates.

* Corresponding author's e-mail address: brouwern@gmail.com

Introduction

The majority of flowering plant species form mutualisms with root fungal symbionts (RFS) such as arbuscular mycorrhizal fungi (AMF; 74 % of angiosperms; Brundrett 2009) and dark septate endophytes (DSE; ≥600 species; Jumpponen and Trappe 1998). Arbuscular mycorrhizal fungi and DSE live inside plant roots and deploy hyphae outside the root that increase water, nitrogen, phosphorus and other soil nutrients' availability to their plant partner (Smith and Read 2008; Newsham 2011). The RFS receive a substantial fraction of the plant partner's fixed carbon (for AMF up to 20 %; Smith and Read 2008).

Recent work highlights how anthropogenic changes in the environment, such as invasion, can negatively affect mutualisms (Tylianakis et al. 2008; Kiers et al. 2010). Invasive species can impact belowground processes and directly or indirectly alter soil microbial communities, including RFS. Mechanisms through which belowground impacts can occur (summarized in part by Wolfe and Klironomos 2005) include alterations in the quality, quantity and timing of litter inputs and subsequent changes in soil nutrient status (reviewed by Ehrenfeld 2003), direct changes to soil nutrient status through novel nutrient fixation strategies by the invader (e.g. Vitousek and Walker 1989), mutualist degradation (Vogelsang and Bever 2009) and allelopathy (e.g. Callaway et al. 2008; Grove et al. 2012). Specifically, allelochemicals can act as novel weapons that are directly toxic to plants or act indirectly on their associated microbes (Callaway and Ridenour 2004; Weir et al. 2004).

The invasion of North American forests by *Alliaria petiolata* (Brassicaceae, garlic mustard) is an emerging model system for investigations of allelopathic effects on belowground processes (Rodgers et al. 2008a). This species produces a suite of allelochemicals (Vaughn and Berhow 1999; Cipollini and Gruner 2007) that are toxic to RFS (Roberts and Anderson 2001; Stinson et al. 2006; Koch et al. 2011) even at low concentrations (Callaway et al. 2008; Cantor et al. 2011). Field studies document that areas infested with *Alliaria* exhibit shifts in soil fungal community composition with frequent reductions in AMF species richness (Burke et al. 2011; Lankau 2011a; Lankau et al. 2014), declines in total soil hyphal abundances (Cantor et al. 2011; Koch et al. 2011) and changes in the within-root community of AMF-dependent plants (Burke 2008; Bongard et al. 2013). Together, these studies suggest that within *Alliaria*-invaded ecosystems the function of the mutualistic fungal community can be compromised and that these changes contribute to *Alliaria*'s invasive success.

Herbaceous perennials dominate the temperate forest understories that *Alliaria* invades and these species as a group are typically highly- to obligately-dependent on RFS (Brundrett and Kendrick 1988; Whigham 2004). The fact that temperate forest soils are strongly resource limited (Whigham 2004; Gilliam 2014) likely drives the obligate nature of the relationship for many understory herbaceous perennials. Typically these species are slow growing (Gilliam 2014), exhibit high rates of RFS colonization (e.g. Brundrett and Kendrick 1988; Boerner 1990; Burke 2008) and have long-lived arbuscules (Brundrett and Kendrick 1990). Many also lack fine roots or root hairs (e.g. LaFrankie 1985) perhaps because their associated RFS hyphae fulfil this soil resource-gathering role. Since resources supplied by RFS are intimately tied to many plant metabolic functions (Schweiger et al. 2014), disruption of soil mutualisms is expected to severely limit the physiological rates of forest species (Hale et al. 2011). In the absence of RFS, plants generally exhibit reduced photosynthetic rates (Allen et al. 1981; Wright et al. 1998; Zhu et al. 2011) and subsequent carbon stress can curb their ability to carry out carbon-demanding functions such as growth (Lu and Koide 1994) and flowering (Koide et al. 1994).

Carbon stress is the reduction of a plant's pool of total non-structural carbohydrates (NSCs) (*sensu* Anderegg et al. 2012). In herbaceous perennials, chronic carbon stress can alter key vital rates including survival (Gremer and Sala 2013), flowering (Crone et al. 2009) and prolonged dormancy (Gremer et al. 2010). Invaders like *Alliaria* that alter the soil environment and essential RFS functions could induce carbon stress or 'carbon starvation' (*sensu* McDowell et al. 2008), ultimately diminishing the stability of populations of RFS-dependent native species.

Our prior experiments on the RFS-dependent understory perennial, *Maianthemum racemosum* (Ruscaceae, false Solomon's seal) confirm the dramatic physiological consequences of short-term RFS disruption by *Alliaria*'s allelochemicals. Key physiological traits including stomatal conductance, which is known to be highly dependent on RFS colonization (Augé et al. 2014), and photosynthetic rate both significantly declined in plants exposed to fresh *Alliaria* leaf litter (Hale et al. 2011). Soil respiration, to which fungi are the primary contributors (Anderson and Domsch 1975), was also reduced with *Alliaria* treatment. Importantly, in field plots invaded by *Alliaria* and in pot experiments with an *Alliaria* litter treatment, we demonstrated significant declines in the abundance of soil fungal hyphae relative to controls (37 % decline, Cantor et al. 2011; 29–38 % decline, A. N. Hale et al., submitted for publication). Together these data strongly support the idea that the observed physiological declines are driven by the inhibition of the RFS hyphal network in the soil (Hale et al. 2011).

Here we explore how the physiological stress of RFS-mutualism disruption in *Alliaria*-invaded forests could result in performance declines in an RFS-dependent forest perennial across two time scales. First, we ask: Given that *Alliaria*'s allelochemicals cause detectable shifts in the soil fungal community and alternative plant physiological rates, do they also cause declines in carbon storage in plants within a single growing season? In a greenhouse experiment we show that *Alliaria*-treated *Maianthemum* store significantly less carbon in their rhizome over one growing season relative to controls. Second, to determine the potential for short-term effects to scale up over time and affect population processes, we conducted a 7-year field experiment in an *Alliaria*-invaded forest in which *Alliaria* was weeded or left at ambient levels. We test whether *Maianthemum* exhibit lower growth rates consistent with carbon stress in the *Alliaria*-ambient plots. We also ask if *Alliaria* reduces size-based vital rates of *Maianthemum* and if so, how quickly these changes occur. We show that where *Alliaria* is present, *Maianthemum* have suppressed growth and vital rates relative to adjacent plot where *Alliaria* is removed.

Methods

Greenhouse study: assessing potential for carbon stress

The greenhouse study was conducted during the summer of 2010 in the greenhouse facilities at the University of Pittsburgh. In May, we obtained bare-root adult *Maianthemum* plants ($N = 42$) from a native plant nursery (Prairie Moon Nursery, Winona, MN, USA). Rhizomes ranged in size from 6.7 to 39.7 g fresh weight. We potted each rhizome in a 3 : 1 mixture of autoclaved Fafard potting soil and Turface. We inoculated plants with RFS by adding 150 g of field soil collected from areas adjacent to *Maianthemum* plants at our experimental field site (see details below). Pots were then placed in the greenhouse and watered every 2–3 days for 1 month, allowing the plants to complete stem elongation and establish the RFS mutualism.

In June, we assigned each plant to either an *Alliaria* treatment or a control treatment. To control for potential differences in initial carbohydrate status due to differences in plant age and/or size (e.g. Olano et al. 2006), we stratified the randomized assignment of rhizomes into the treatments to ensure that mean rhizome mass was the same in the *Alliaria* and control treatments. Plants in the *Alliaria* treatment were then exposed to *Alliaria* allelochemicals by placing 25 g of fresh *Alliaria* leaf tissue collected from a population with a recent history of invasion (<20 years) on top of the soil. When these plants were watered, the glucosinolates leached out of

the *Alliaria* leaves and into the soil (A. N. Hale et al., submitted for publication). As in previous experiments (Hale et al. 2011), plants in the control treatment received 25 g of fresh *Hesperis matronalis* (dame's rocket; Brassicaceae) leaf tissue. Like *Alliaria*, *Hesperis* is an invasive mustard in eastern North America (Leicht-Young et al. 2012). While *Hesperis* produces some glucosinolates (Larsen et al. 1992), RFS hyphae and vesicles have been observed within its root system (DeMars and Boerner 1995), indicating that *Hesperis* chemicals are less toxic to RFS than *Alliaria*. In the field, the high mortality rates of *Alliaria* seedlings and rosettes throughout the year (Davis et al. 2006) and the mortality of adults in the summer (Anderson et al. 1996) likely result in a sustained supply of allelochemicals into the soil. Thus, we re-applied fresh leaf tissue in both treatments every 2 weeks until the end of August to simulate a season-long supply of *Alliaria* allelochemicals.

We destructively harvested plants three times during the growing season (9 July, 6 August and at senescence) to assess the effect of the treatments on the carbohydrate status. For the last time point, we classified plants as being senesced when 40 % of the leaf tissue had yellowed and photosynthetic rates were <1.0 μmol m^{-2} s^{-1}. Details of the leaf gas exchange protocol for *Maianthemum* can be found in Hale et al. (2011). To harvest the plants, we carefully clipped the shoot and roots away from the rhizome. We also stained the roots of a subset of plants per treatment following Brundrett et al. (1984) to confirm RFS colonization. We then weighed the rhizome and immediately flash-froze it in liquid nitrogen. We stored samples at -80 °C until they could be lyophilized and ground. We followed the protocol of Zuleta and Sambucetti (2001) to analyse rhizome inulin (storage carbohydrate) and sucrose (mobile carbohydrate) content via high-performance liquid chromatography (HPLC). [Note: Starch is not present in the rhizome of *Maianthemum* (A. N. Hale et al., submitted for publication).] In brief, a 0.03 g dried sample for each plant is boiled while stirring with a magnetic stir bar. Once samples cool to room temperature, they are filtered through a 0.20 μm filter, and run on HPLC (Aminex HPX-87C anion-exchange column, deionized water at 85 °C was set as the mobile phase with a flux rate of 0.6 mL min^{-1}). Standards are used (inulin from dahlia tubers, Sigma-Aldrich; sucrose, Sigma-Aldrich) to confirm the identity of the sample peaks and to create standard curves to determine inulin and sucrose concentrations. Here, we express inulin and sucrose concentrations as a percentage of the HPLC dry sample mass. We also sum each plant's inulin and sucrose content to determine total NSC concentration (%).

To explore the effect of our treatments on rhizome carbohydrate status, we use a multivariate analysis of covariance (MANCOVA). Following a significant MANCOVA, individual ANCOVA tests are conducted for inulin, sucrose and total NSC. For all models, we include harvest date as a main effect because rhizome carbohydrate concentration varies over the growing season in perennial herbs (e.g. Lapointe 1998; Wyka 1999; Kleijn et al. 2005). We also include initial plant mass as a covariate to account for differences in carbohydrate storage that are related to plant size/age (MANCOVA model: total NSC + inulin + sucrose = treatment + harvest date + initial plant mass; ANCOVA models: carbohydrate = treatment + harvest date + initial plant mass). We calculate least squares means and standard errors for all ANCOVA models with a significant ($P < 0.05$) treatment effect. All analyses were conducted in SAS (v. 9.3, SAS Institute, Cary, NC, USA).

Field study: measuring impacts on vital rates of native plant populations

Study site. Our experimental plots are located in a beech-maple forest in southwest Pennsylvania [Trillium Trail Nature Reserve (hereafter TT), Allegheny County, PA, USA: 40°52′01.40″N; 79°90′10.75″W] with a rich herbaceous perennial understory flora (Knight et al. 2009). Based on previous work at TT (Burke 2008) and other temperate deciduous forests (e.g. Brundrett and Kendrick 1988), we estimate that 73 % of TT herbaceous perennials are AMF-dependent (Hale et al. 2011). We detected *Alliaria* allelochemicals in the soil of TT in concentrations that are toxic to AMF spores in lab assays (Cantor et al. 2011). Additionally, we showed that in soils where *Alliaria* occurs at TT, the density of fungal hyphae is lower (Cantor et al. 2011) and the fungal community composition shifts (Burke et al. 2011) relative to paired, non-invaded areas. *Maianthemum* plants collected at TT are heavily colonized by RFS, but their intra-root AMF community is significantly altered where *Alliaria* is present (Burke 2008). These results motivate further investigation of mutualism disruption by *Alliaria* in understanding mechanisms driving native plant performance declines.

Field experiment. We collected data on naturally occurring individuals of *M. racemosum* within six 14 × 14 m plots in TT from 2003 through 2013. Our six plots are split in half longitudinally so that each contains two experimental treatments: *Alliaria* removal (= low or no allelochemicals) or *Alliaria* present at ambient levels (= allelochemicals present). Annual removal of *Alliaria* from half of each plot (i.e. a 14 × 7 m area) began in spring 2006, ~15 years after *Alliaria* became established at this site (L. Smith, pers. comm.) This time frame for TT invasion coincides with the estimated *Alliaria* invasion history in the region that indicates that this invader has been present locally for <25 years (Lankau et al. 2009). We remove *Alliaria* concurrent with the onset of emergence of the perennial herb community. *Alliaria* individuals are removed as tiny seedlings, minimizing disturbance to the soil and other plants. Removed plants are discarded off site. In June of each year prior to *Alliaria* seed dispersal we erect a barrier at the border of the two treatments to block seed dispersal from the ambient into the *Alliaria* removal treatment. All *Maianthemum* plants emerging in the plots are permanently tagged and have annually been scored for individual size, stage (i.e. seedling, non-flowering, flowering and dormant) and deer browse status. Prior to initiation of the *Alliaria* removal treatment in 2006 there was no difference in *Alliaria* per cent cover between the plots ($\chi^2 = 0.11$, $P = 0.74$) or total per cent cover of all species ($\chi^2 = 0.038$, $P = 0.85$).

Plant vital rates. We assess the effect of *Alliaria* removal on *Maianthemum* growth and three vital rates: annual flowering frequency, retrogression of flowering plants to non-flowering the following year and the frequency of prolonged vegetative dormancy (Shefferson 2009). We test for differences using data collected prior to the implementation of the removal treatment (2003–06) and after the removal treatment began (2007–13). All models have the general form: response variable = treatment + year + treatment × year. To estimate differences in growth rate, we investigate the differences in average size between treatments for the initial cohort of plants first observed when the experiment began in 2003. The mean size of this cohort is estimated with a linear mixed model for each year since 2006 (Zuur et al. 2009). We model log(plant size) to improve normality of the residuals.

Annual flowering frequencies are modelled using a logistic mixed model. Retrogression frequencies were modelled without random effects for the years 2008–13 because of limited sample size. Our retrogression model, stated in terms of probability, is

$$\Pr(\text{Not Flowering}_{\text{time } t} | \text{Flowered}_{\text{time } t-1} \text{ and Not dormant}_{\text{time } t}).$$

Our sample for retrogression was therefore set by the number of plants that flowered the previous year (time $t - 1$) that emerged as either flowering or non-flowering the next year (time t).

Growth and vital rate analyses are conducted in R 3.1.0 (R Development Core Team 2014) using the *lme4* package (Bates et al. 2014). To account for repeated measures and blocking effects, we include random intercepts for individual plants and pairs of treatments within a plot. For each response variable we test for significant differences between annual means using the *multcomp* package in R (Bretz et al. 2010). We test for the presence of a long-term trend since 2006 in each treatment mean by specifying a trend contrast (Rosenthal and Rosnow 1985; Gurevitch and Chester 1986). All tests are planned contrasts so we do not correct for multiple comparisons. To further investigate trends in flowering frequencies, we also analyse these data using a two-level hierarchical model with time as a continuously varying main effect and year as a random effect.

Results of flowering and retrogression analyses are reported as effect sizes using odds ratios (OR) (Rita and Komonen 2008). Odds ratios have a lower bound of zero and no upper bound. Odds ratios of 1 indicate no difference between two treatments in the odds of an event happening. Statistical tests for OR therefore test whether they are different from 1. Odds ratios and their 95 % confidence intervals (CIs) are given in the text on their normal scale but graphed on a log scale to improve interpretation (*sensu* Galbraith 1988).

Mark-recapture models. We use mark-recapture models, a modified logistic regression approach (Kéry *et al.* 2005), to estimate the probability of prolonged vegetative dormancy. To test for pre-existing differences in dormancy rates, we conduct separate mark-recapture analyses of the 3 years prior to implementation of the removal treatment (2003–05) and the 7 years after the treatment began (2007–13). Mark-recapture results are assessed using the small sample size corrected information criteria AICc (AICc = AIC + $2k(k+1)/(n-k-1)$, where k = the number of parameters and n = sample size) to rank the explanatory ability of different models (Anderson 2010). To summarize the data we also analyse the entire data set (2003–13) and calculate the mean difference in dormancy rates between treatments. We first calculate dormancy rates for each treatment in each year, calculate the difference between these means and average the differences for the pre- and post-treatment time periods. We use the delta method (Powell 2007) in the R package *msm* to combine multiple standard errors and construct 95 % CIs around our final effect size estimates. Mark-recapture models are run in the R package *marked* (Laake *et al.* 2013).

Missing data due to herbivory. Deer browse compromised our ability to gain information on some individuals. Deer preferentially browse flowering *Maianthemum* and flowering individuals are of larger size than non-flowering individuals (N. L. Brouwer and S. Kalisz, unpubl. data). Accordingly, in the cases where an individual was browsed before its reproductive status was determined during the 10 annual censuses ($n = 103$ instances across 10 years), we assumed the browsed individual was flowering. Further, if browse occurred before an individual's size data was collected or size was otherwise unavailable, we used linear imputation (Gelman and Hill 2006) to estimate its

size (412 instances of size imputation out of 1481 total size records). Including imputed size data for the browsed plants prevents biasing our results against detecting a treatment effect (Hadfield 2008; Nakagawa and Freckleton 2008).

We imputed missing size data using estimates generated from multiple rounds of linear regression based on observed size data from the years prior to and after the missing data. We averaged these multiple estimates to arrive at a final imputed size estimate for each browsed individual. Linear regression models included all available covariates, including previous size, current status, treatment and reproductive output for flowering plants. We validated our imputations by comparing mean plant size and the overall size distribution in the population with and without imputed data **[see Supporting Information—Table S1]**.

Results

Greenhouse study: assessing potential for carbon stress

All *M. racemosum* plants examined exhibit colonization by internal RFS structures. However, *Maianthemum*'s rhizome carbohydrates were significantly affected by the *Alliaria* treatment (MANCOVA; Roy's greatest root = 7.57, $P = 0.002$), with plants in the *Alliaria* treatment experiencing a significant reduction in total NSC (Fig. 1; ANCOVA $F_{1,36} = 7.31$, $P = 0.01$). Specifically, plants treated with *Alliaria* stored, on average, 17 % less inulin relative

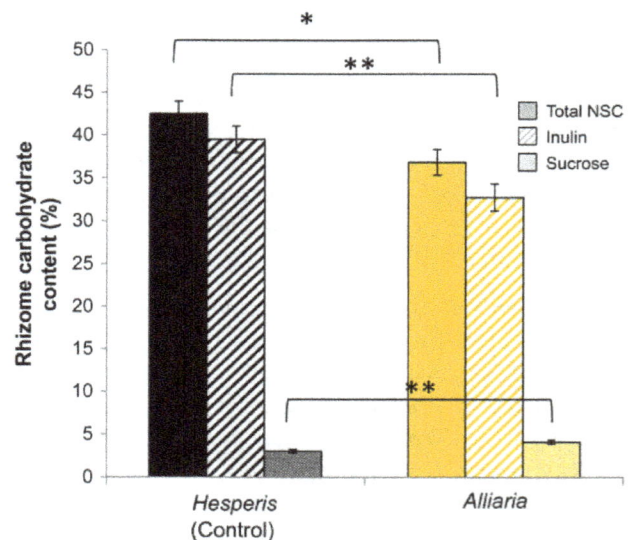

Figure 1. *Maianthemum racemosum* rhizome carbohydrate content (%) from *Alliaria* (yellow) and *Hesperis* (control; black) treatments in the greenhouse experiment. Total NSC content is shown in solid-coloured bars. Total NSC is a composite measure of stored sugars (inulin; bars with diagonal shading) and mobile sugars (sucrose; stippled bars). Values are least squares means from ANCOVAs ± 1 standard error. *$P < 0.05$; **$P < 0.005$.

to plants in the *Hesperis* treatment (Fig. 1; ANCOVA $F_{1,36} = 9.28$, $P = 0.004$). While plants in the *Alliaria* treatment had fewer stored sugars, they had higher sucrose concentrations in their rhizomes compared with plants in the *Hesperis* treatment (Fig. 1; ANCOVA $F_{1,36} = 12.88$, $P = 0.001$). The increase in mobile sugars did not compensate for the dramatic difference in stored sugars between treatments as total NSC in the *Alliaria*-treated plants was 13 % lower than that of *Hesperis*-treated plants. Harvest date was not a significant predictor of total NSC, inulin or sucrose.

Field study: impact on vital rates

Growth. Prior to implementation of the removal treatment, there was no difference in the mean size of plants in the initial 2003 cohort (Fig. 2; $P = 0.55$). By 2013 plants in the removal treatment are significantly larger than those in the ambient *Alliaria* treatment (mean difference $= 6.70$ cm, SE $= 2.96$; $P = 0.02$). There is a significant positive linear trend in size from 2006 to 2013 (trend contrast $P = 0.0056$) in the *Alliaria* removal plots but no trend in the ambient plots ($P = 0.91$).

Flowering. There is no significant difference in flowering probability across treatments for the first 6 years of the *Alliaria* removal (e.g. Fig. 3; $P_{2006} = 0.65$, $P_{2007} = 0.29$, $P_{2008} = 0.42$). However, by 2012 the flowering probability is 'leaning' (*sensu* Tukey 1991) in the predicted direction (OR $= 1.72$, CI$_{95\%} = 0.84$–3.52, $P = 0.14$) and by 2013 is significantly higher (OR $= 1.96$ CI$_{95\%} = 1.0$–3.87, $P = 0.051$) in the removal treatment. Across all years

(2006–13) there is an increasing trend in flowering probability in the removal treatment (trend contrast $P = 0.00008$) but no increase in the ambient treatment ($P_{\text{trend}} = 0.57$).

Analyses using time as a continuous variable and year as a random effect confirmed that flowering frequencies diverged between the treatments (treatment $\times$ time $\chi^2 = 6.81$, $P = 0.009$) with a significant positive linear trend in the removal treatment ($\beta_{\text{removal}\times \text{time}} = 0.18$, SE $= 0.069$) contrasted with evidence of a decrease in flowering probability in *Alliaria*-ambient plots ($\beta_{\text{time}} = -0.10$, SE $= 0.072$).

Retrogression. The number of flowering individuals was too low in 2005 and 2006 to accurately estimate retrogression of flowering plants in 2006 and 2007. By 2011, there was evidence that removal-treatment plants were less likely to retrogress (OR $= 0.28$ CI$_{95\%} = 0.052$–1.57, $P = 0.15$) and in 2012 they were significantly less likely to retrogress (OR $= 0.14$ CI$_{95\%} = 0.021$–0.96, $P = 0.045$). There was a significant decreasing trend in retrogression in the removal treatment from 2008 until 2013 ($P_{\text{trend}} = 0.011$) but no trend in the ambient treatment ($P_{\text{trend}} = 0.90$).

Dormancy. Dormancy rates were highly variable between years, ranging from <10 to >30 %, but estimated to be lower in the *Alliaria* removal treatment in six out of 7 years **[see Supporting Information—Table S2]**. For years prior to the implementation of the *Alliaria* removal treatment (2003–06) the best-ranked model contains only a year effect (Table 1) while for models of post-treatment years

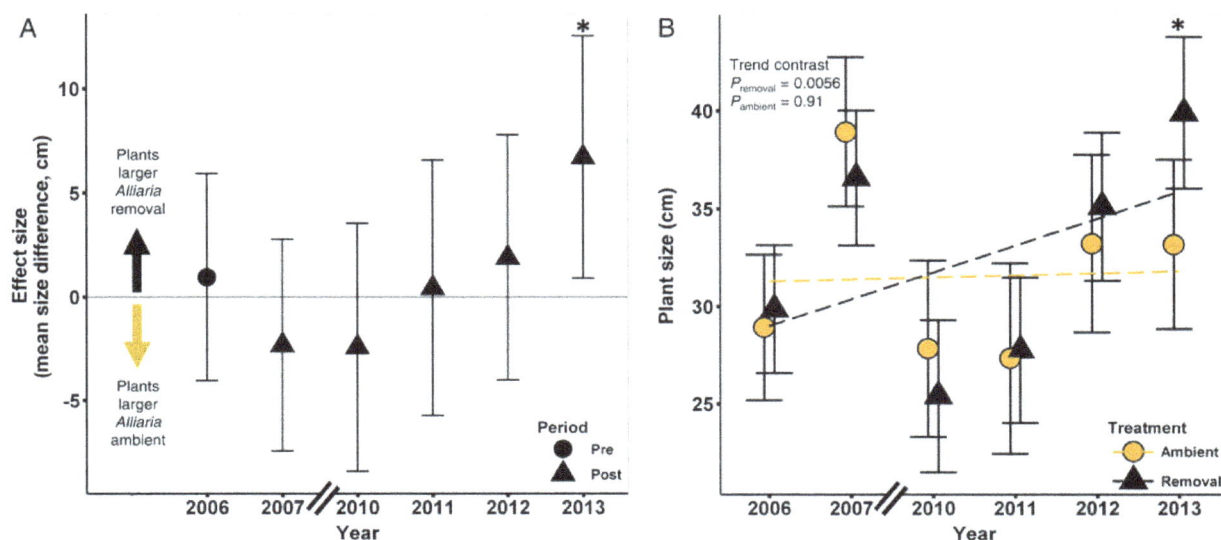

Figure 2. Effect of *Alliaria* on plant size of *Maianthemum* marked in the initial 2003 survey of the field experiment. (A) Mean difference (effect size) in plant size between *Alliaria* in ambient and removal treatments. (B) Annual mean plant sizes in both treatments and ANOVA trend contrasts. Error bars represent ± 95 % CIs. Asterisk indicates a significant difference in plant size between the two treatments ($P < 0.05$). Size data were not available for 2008 and 2009.

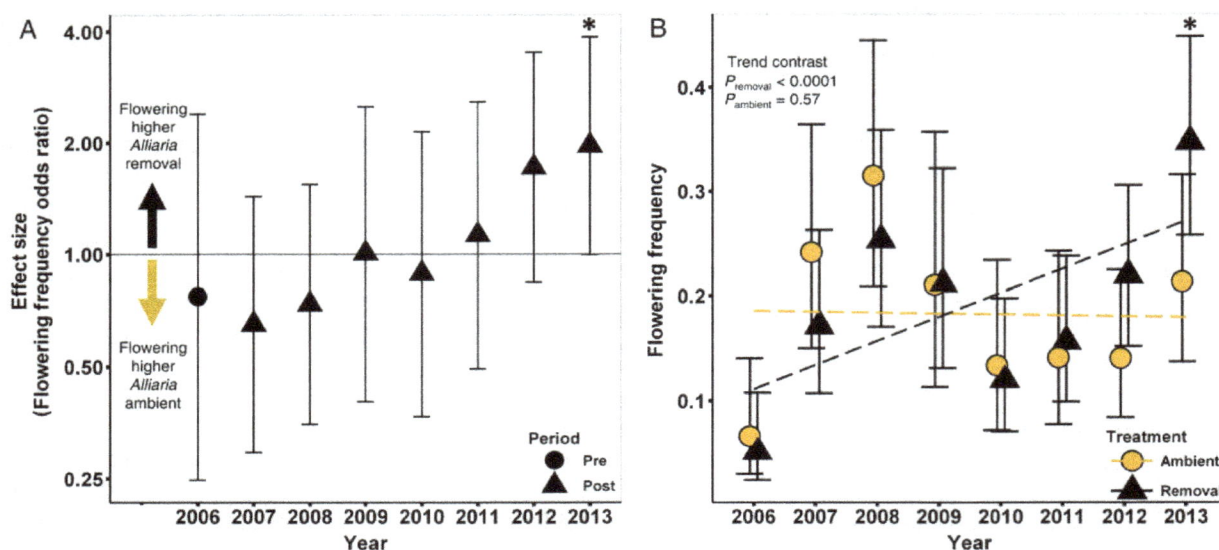

Figure 3. Effect of *Alliaria* on *Maianthemum* flowering frequency. (A) Mean difference (effect size, ES) in flowering frequency in *Alliaria*-ambient and removal plots. Effect size is expressed as an OR and plotted on the log scale. (B) Annual mean flowering frequencies for both treatments and ANOVA trend contrasts. Error bars represent ± 95 % CIs. Asterisk indicates a significant effect of *Alliaria* removal ($P < 0.05$).

Table 1. Ranking of mark-recapture models testing the effects of *Alliaria* removal on prolonged vegetative dormancy. Three sets of models were run over different time periods during the study: Set 1: years before *Alliaria* removal began (Pre-treatment); Set 2: years after the annual weeding treatment was initiated (post-treatment) and Set 3: all years. N, number of plants tracked over each time period; K, number of parameters in a model; Ln(lik), log likelihood. To calculate the mean pre-treatment and post-treatment effect size (Fig. 5) we used the parameters from the 'Removal × Year' model in the 'All years' model Set 3.

Set	Period	Model	N	K	AICc	ΔAICc	Ln(lik)
1	Pre-*Alliaria* removal (2003–06)	Year	158	5	452.8	0.00	−216.21
		Removal + Year		6	454.6	1.74	−215.00
		Removal × Year		9	466.2	11.59	−214.47
2	Post-*Alliaria* removal (2007–13)	Removal + Year	210	9	1166.4	0.00	−564.73
		Year		8	1172.4	6.03	−569.84
		Removal × Year		15	1187.2	14.76	−562.34
3	All years (2003–13)	Removal + Year	236	12	1646.3	0.00	−798.46
		Year		11	1652.5	6.23	−803.68
		Removal × Year		21	1680.3	27.74	−795.98

(2007–13) and the entire dataset (2003–13) the best models contain an effect of *Alliaria* removal, indicating that dormancy rates were typically lower in this treatment. There was an initially large difference in dormancy rates between plots that would be allocated to the two treatments in the first year of the study [**see Supporting Information—Table S2**], potentially resulting in the model of the pre-treatment years containing an *Alliaria* removal effect (AICc = 454.6) ranked almost as high as a year-only model (AICc = 452.8). However, since the year-only model has a lower AICc and fewer parameters, the larger model is not considered competitive (Arnold 2010). Moreover, in the other two pre-treatment years (2004 and 2005), there is no difference between dormancy estimates

[**see Supporting Information—Table S2**]. The results of model selection are reinforced by the calculation of average effect sizes for the period prior to *Alliaria* removal and after removal (Fig. 5). Prior to removal there is no significant difference between dormancy rates (ES = −0.05, $CI_{95\%}$ = −0.13–0.03) but after removal dormancy rates are ~7 % lower than in the *Alliaria*-ambient treatments (ES = −0.069, $CI_{95\%}$ = −0.12 to −0.2).

Discussion

To our knowledge this is the first study to explore the connections between an allelopathic invasive species' impacts on the soil biotic environment and changes in

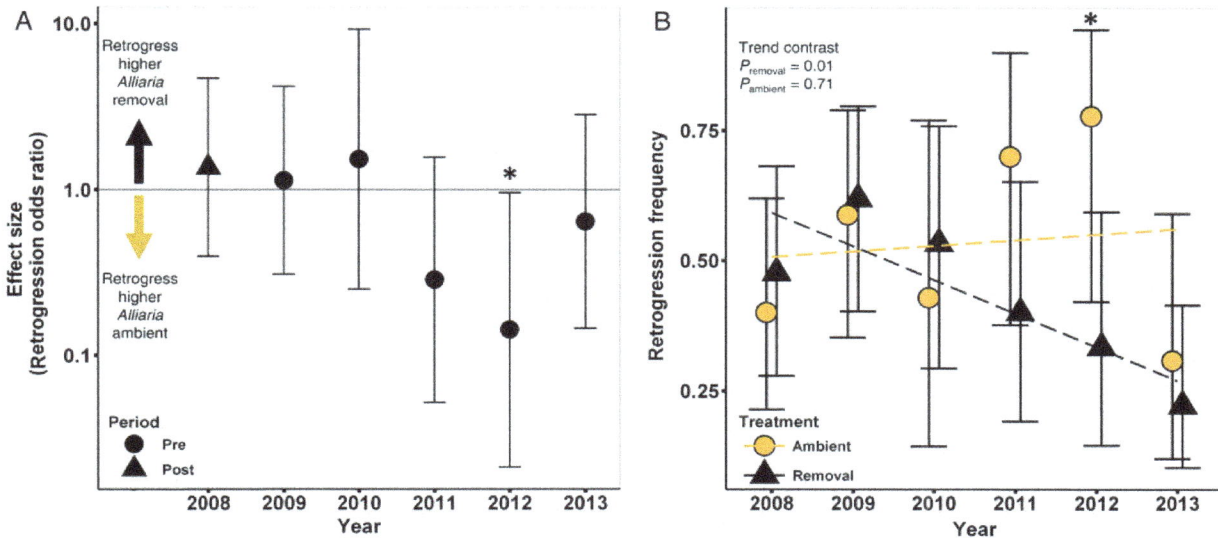

Figure 4. Effect of *Alliaria* on *Maianthemum* retrogression from flowering to non-flowering. (A) Annual mean difference in retrogression frequency (ES) in *Alliaria*-ambient and removal plot. Effect size is expressed as an OR and plotted on the log scale. (B) Mean retrogression frequencies in both treatments and ANOVA trend contrasts. Error bars represent $\pm$ 95 % CIs. Asterisk indicates a significant effect of *Alliaria* removal ($P < 0.05$). Retrogression is calculated conditional on a plant being observed above-ground and not dormant. Sample sizes for 2006 and 2007 were insufficient for vital rate calculation.

individual plants' carbon status and vital rates. The results presented here in conjunction with prior studies substantiate multiple steps in a physiologically based causal pathway between invasion and population-level impacts on native plants. Our prior work demonstrates that *Alliaria* treatment of soil around *Maianthemum* reduces the density of soil fungal hyphae (A. N. Hale *et al.*, submitted for publication) and plant photosynthetic rates (Hale *et al.* 2011). Here, our results demonstrate that treatment with *Alliaria* across the entire growing season results in negative effects on season-long carbon storage (Fig. 1). Relative to control plants, *Maianthemum* exposed to *Alliaria* stored 17 % less inulin in their rhizomes and experienced an overall reduction in total NSCs at the end of the season. Stomatal conductance modulates carbon fixation and is a key physiological rate affected by *Alliaria* exposure (Hale *et al.* 2011). Interestingly, a recent meta-analysis (Augé *et al.* 2014) comparing the effects of AMF inoculation on stomatal conductance (g_s) in field vs. greenhouse studies indicates that greenhouse experiments have smaller effect sizes than field studies. Thus, our carbon storage results are likely conservative estimates of the carbon impacts of mutualism disruption in the field.

Over time, chronic exposure to *Alliaria* was predicted to compound this carbon deficit and affect plant growth and vital rates. Results from our long-term field study of *Alliaria* removal are consistent with this prediction. Individual aboveground plant size (Fig. 2) and multiple carbon-intensive and size-dependent vital rates (Figs 3–5)

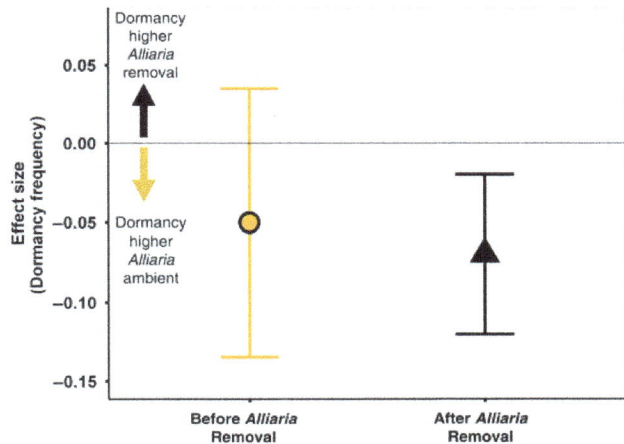

Figure 5. Effect size of *Alliaria* removal on the frequency of prolonged vegetative dormancy in *Maianthemum* before (2003–06; yellow) and after the treatment began (2007–13; black). Calculated with mark-recapture models; error bars represent $\pm$ 95 % CIs. Asterisk indicates a significant effect of *Alliaria* removal ($P < 0.05$).

are positively affected in *Alliaria* removal relative to *Alliaria*-ambient plots.

Other experimental studies where *Alliaria* and native plants are grown together in pots (Meekins and McCarthy 1999; Wixted and McGraw 2010; Lankau 2012; Smith and Reynolds 2014) or in the field (McCarthy 1997; Carlson and Gorchov 2004; Cipollini *et al.* 2008; Lankau 2011*b*) also find negative effects of *Alliaria* on native species. Competition, direct allelopathic phytotoxicity and allelopathic RFS-mutualism disruption are all mechanisms that could contribute to these results. Our greenhouse experiment adds support to the idea that it is *Alliaria*'s disruption of key belowground mutualists (RFS) rather than competition or direct phytotoxicity that accounts for its success as an invader. Below we discuss the general support or lack thereof for the likelihood of all three mechanisms.

Competition

We are aware of only two studies that have attempted to quantify reciprocal competition between *Alliaria* and focal plants. These pot studies found that *Alliaria* was equal to or weaker in competitive ability than three of four species tested (Meekins and McCarthy 1999; Leicht-Young *et al.* 2012). However, these studies are problematic in that they cannot separate competition from phytotoxicity or mutualism disruption. Bossdorf *et al.* (2004) found that *Alliaria* individuals from the native range outcompete *Alliaria* plants from the invaded range, supporting the hypothesis that invasive *Alliaria* express a different trade-off relative to their source populations. Invasive *Alliaria* are armed with novel allelochemical weapons but have evolved to be less competitive (Bossdorf *et al.* 2004). Further, field experiments demonstrate that native competitors can suppress *Alliaria* performance and abundance when the natives are not experiencing overabundant herbivore pressure (Eschtruth and Battles 2009*a*), as deer preferentially consume native plants and facilitate the high population growth and spread of *Alliaria* (Kalisz *et al.* 2014). In experimental studies that exclude deer from invaded sites, *Alliaria* abundance rapidly declines (Eschtruth and Battles 2009*b*; Knight *et al.* 2009; Kalisz *et al.* 2014). In total, these results underscore the widely held view that *Alliaria* is a relatively poor competitor (Rodgers *et al.* 2008*a*).

Direct phytotoxicity

Glucosinolates are known antimicrobial chemicals produced by members of the mustard family as defences against pathogens (Tierens *et al.* 2001). While *Alliaria*'s allelochemicals can be inhibitory to germinating seeds and inhibit new seedling root growth (lettuce and radish seed experiments: Vaughn and Berhow 1999; Roberts and Anderson 2001; Pisula and Meiners 2010; *Impatiens* and *Viola* seed experiments: Prati and Bossdorf 2004; Barto *et al.* 2010; Cipollini and Flint 2013), to our knowledge direct toxicity of *Alliaria* on mature plant tissues has never been demonstrated. *Alliaria* invades forest understories dominated by adult perennial plants dependent on RFS. The direct effect of allelochemicals is inversely proportional to target plant density or biomass (Weidenhamer 2006). Single-celled fungal spores and thin fungal hyphae

should be much more susceptible to *Alliaria* allelochemicals than mature plant tissues. Thus, while we cannot rule out direct phytotoxic effects of *Alliaria* on adult *Maianthemum* performance in our field or greenhouse experiments, a direct allelochemical effect is likely of small magnitude relative to indirect effects on RFS.

RFS-mutualism disruption

Mounting evidence shows that *Alliaria* can exert potent indirect effects on plants by suppressing RFS. Glucosinolates, like those produced by *Alliaria*, have a short half-life in the soil (<15 h; Gimsing *et al.* 2006). Yet, native plants grown in soils conditioned by *Alliaria*, treated with *Alliaria* tissue extracts, or collected from *Alliaria*-invaded sites all express reduced growth (Stinson *et al.* 2006; Callaway *et al.* 2008; Wolfe *et al.* 2008) despite the fact that the volatile allelochemicals were likely no longer present. Importantly, these studies demonstrate that *Alliaria* impacts are similar in magnitude to soil sterilization and that experimental soils result in lower colonization of roots by mycorrhizae (Stinson *et al.* 2006; Callaway *et al.* 2008; Wolfe *et al.* 2008). Finally, *Maianthemum* plants treated with *Alliaria* retain RFS structures internal to their roots, while exhibiting significant declines in soil hyphae (A. N. Hale *et al.*, submitted for publication). Together these experiments provide strong support for RFS-mutualism disruption and that its effects are of large magnitude relative to competition or direct phytotoxicity.

Mechanistically, our working model linking RFS-mutualism disruption to carbon stress is based on the following premises: If *Alliaria*'s allelochemicals destroy the hyphal network, yet the normally long-lived internal structures (Brundrett and Kendrick 1990) remain intact, then we would predict that the plant would increase carbon allocation to its RFS to provision the regrowth of the soil hyphal network, resulting in significant carbon stress for the plant. Loss of the hyphal network severely limits available soil nutrients and water to the plant (Newsham 2011; Augé *et al.* 2014). As a result, the plants photosynthesize less (Hale *et al.* 2011) and fix less carbon (NSC; Fig. 1). With this limited carbon pool, we suggest that plants may maintain concentrations of mobile sugars in the rhizome and roots to re-establish a functional RFS hyphal network that is repeatedly destroyed by our application of fresh *Alliaria* tissue. While our results are consistent with this working model (e.g. we observe greater sucrose concentrations in the rhizome of *Alliaria* vs. *Hesperis*-treated plants (Fig. 1)), additional experiments are needed to fully explore this hypothesis.

We note that the effects of allelopathic mutualism disruption by *Alliaria* could be amplified by additional factors. Like other invasive species of deciduous forests (Ehrenfeld *et al.* 2001; Poulette and Arthur 2012; Smith and Reynolds 2012; Kuebbing *et al.* 2014; Schuster and

Dukes 2014), *Alliaria* can affect multiple components of the soil environment. *Alliaria* increases soil nutrient availability (Rodgers *et al.* 2008*b*), litter decomposition rates and nitrogen loss (Ashton *et al.* 2005). Since the RFS community in general (Van Diepen *et al.* 2011) and specific RFS–plant interactions (e.g. Klironomos 2002) are sensitive to soil conditions, multiple invader-mediated changes to the soil environment could magnify the impacts of allelopathic RFS-mutualism disruption. These diverse and widespread consequences of invasive species for soil environments and RFS communities are alarming given the potentially central role RFS and other microbes play in the diversity, productivity and functioning of plant communities (van der Heijden *et al.* 2008).

Our greenhouse study indicates that *Maianthemum* carbon storage declines significantly in response to *Alliaria* treatment in just one growing season. In contrast, we observe a relatively slow recovery of individual size, growth and vital rates following *Alliaria* removal in our field study. The predicted significant trends indicative of recovery (Figs 2–4) emerged after a few years of *Alliaria* removal while significant differences within the single-year comparisons were not seen until ~6–7 years post removal (2012 or 2013). Two, non-mutually exclusive mechanisms could underlie this lag. First, the lag could be due to *Maianthemum*'s habit (LaFrankie 1985). In general, forest understory herbaceous perennials are light-limited, slow-growing, long-lived species (Whigham 2004) with slow responses to perturbation (Morris *et al.* 2008). Our data are consistent with the idea that following *Alliaria* removal, *Maianthemum* may take multiple years to re-gain sufficient carbon stores to allow size growth, sustain flowering and maintain low dormancy rates. Second, the observed lag in *Maianthemum* vital rate responses may be due to slow recovery of the RFS soil community following *Alliaria* removal, a phenomenon observed by Anderson *et al.* (2010) and Lankau *et al.* (2014). If populations of beneficial RFS have gone locally extinct and low dispersal distance limits RFS re-colonization (Rout and Callaway 2012), then the observed time lag of *Maianthemum* could be due to the low abundance of effective fungal partners. Given the reciprocal obligate dependence of AMF and forest herbaceous perennial plants, declines in the native understory community may drive reciprocal declines in the RFS soil community (Lankau *et al.* 2014).

Conclusions

Increases in invasive species are generally correlated with declines in native biodiversity (e.g. Butchart *et al.* 2010). However, the mechanistic underpinnings leading to native population collapse are rarely understood yet are the subject of numerous studies and invasion hypotheses

(Levine *et al.* 2003; Hulme *et al.* 2013). The disruption of plant soil feedbacks and root fungal symbioses are common aspects of plant invasions (i.e. Grove *et al.* 2012; Meinhardt and Gehring 2012; Ruckli *et al.* 2014; Shannon *et al.* 2014). As suggested by Hale and Kalisz (2012), chronic RFS-mutualism disruption could act as the first step in native plant biodiversity loss. In our system, the disruption of RFS by an allelopathic invader appears to begin a downward spiral in the physiological function (Hale *et al.* 2011), carbon status (Fig. 1) and ultimately vital rates (Figs 2–5) of a common native forest plant. Loss of these critical belowground mutualisms may be the proximate cause of plant mortality that is instead attributed to second-order effects (e.g. drought or herbivory) that are easier to observe (*sensu* McDowell 2011). Additional studies in invaded communities that explore the links between plant physiology, carbon allocation and population demographic performance are needed to determine the generality of these results. Mutualism disruption may be a widespread mechanism that helps explain how invasive species can cause large-scale changes to forest biodiversity observed in the wake of invasion (e.g. Rodgers *et al.* 2008*a*).

Sources of Funding

Funding was supplied by a United States National Science Foundation award DEB-0958676 to S.K., a NSF predoctoral fellowship to N.L.B. and a Phipps Conservatory Botany-in-Action award and an Andrew K. Mellon predoctoral fellowship to A.N.H.

Contributions by the Authors

S.K. conceived, designed, implemented and led data collection of the field experiment and assisted with the conception, design and implementation of the greenhouse experiment. A.N.H. designed, implemented and analysed data from the greenhouse experiment and assisted in data collected for the field experiment from 2008 to 2011. N.L.B. managed and analysed data from the field experiment and assisted with data collection for the field experiment since 2010. All three authors collaborated on the conception and writing of this article.

Acknowledgements

We thank the members of the Kalisz lab for field, lab and greenhouse assistance, and two anonymous reviewers whose detailed suggestions improved the manuscript.

Supporting Information

The following additional information is available in the online version of this article –

Table S1. Validation of imputed *Maianthemum* size data from field experiment. Original and imputed size data are compared using *t*-tests and Kolmogorov–Smirnov tests.

Table S2. Estimated frequency of prolonged vegetative dormancy of *Maianthemum* from field experiment using a Mark-Recapture model.

Literature Cited

Allen MF, Smith WK, Moore TS Jr, Christensen M. 1981. Comparative water relations and photosynthesis of mycorrhizal and non-mycorrhizal *Bouteloua gracilis* H.B.K. Lag ex steud. *New Phytologist* **88**:683–693.

Anderegg WRL, Berry JA, Smith DD, Sperry JS, Anderegg LDL, Field CB. 2012. The roles of hydraulic and carbon stress in a widespread climate-induced forest die-off. *Proceedings of the National Academy of Sciences of the USA* **109**:233–237.

Anderson DR. 2010. *Model based inference in the life sciences: a primer on evidence.* New York: Springer.

Anderson JPE, Domsch KH. 1975. Measurement of bacterial and fungal contributions to respiration of selected agricultural and forest soils. *Canadian Journal of Microbiology* **21**:314–322.

Anderson RC, Dhillion SS, Kelley TM. 1996. Aspects of the ecology of an invasive plant, garlic mustard (*Alliaria petiolata*), in central Illinois. *Restoration Ecology* **4**:181–191.

Anderson RC, Anderson MR, Bauer JT, Slater M, Herold J, Baumhardt P, Borowicz V. 2010. Effect of removal of garlic mustard (*Alliaria petiolata*, Brassicaeae) on arbuscular mycorrhizal fungi inoculum potential in forest soils. *The Open Ecology Journal* **3**:41–47.

Arnold TW. 2010. Uninformative parameters and model selection using Akaike's Information Criterion. *The Journal of Wildlife Management* **74**:1175–1178.

Ashton IW, Hyatt LA, Howe KM, Gurevitch J, Lerdau MT. 2005. Invasive species accelerate decomposition and litter nitrogen loss in a mixed deciduous forest. *Ecological Applications* **15**:1263–1272.

Augé R, Toler HD, Saxton AM. 2014. Arbuscular mycorrhizal symbiosis alters stomatal conductance of host plants more under drought than under amply watered conditions: a meta-analysis. *Mycorrhiza* **25**:13–24.

Barto EK, Friese C, Cipollini D. 2010. Arbuscular mycorrhizal fungi protect a native plant from allelopathic effects of an invader. *Journal of Chemical Ecology* **36**:351–360.

Bates D, Maechler M, Bolker B, Walker S. 2014. lme4: linear mixed-effect models using Eigen and S4. R package version 1.1-7. http://CRAN.R-project.org/package=lme4.

Boerner REJ. 1990. Role of mycorrhizal fungus origin in growth and nutrient uptake by *Geranium robertianum*. *American Journal of Botany* **77**:483–489.

Bongard CL, Fulthorpe RR, Bongard CL, Fulthorpe RR. 2013. Invasion by two plant species affects fungal root colonizers. *Ecological Restoration* **31**:253–263.

Bossdorf O, Prati D, Auge H, Schmid B. 2004. Reduced competitive ability in an invasive plant. *Ecology Letters* **7**:346–353.

Bretz F, Hothorn T, Westfall P. 2010. *Multiple comparisons using R.* London: CRC Press.

Brundrett MC. 2009. Mycorrhizal associations and other means of nutrition of vascular plants: understanding the global diversity of host plants by resolving conflicting information and developing reliable means of diagnosis. *Plant and Soil* **320**:37–77.

Brundrett MC, Kendrick B. 1988. The mycorrhizal status, root anatomy, and phenology of plants in a sugar maple forest. *Canadian Journal of Botany* **66**:1153–1173.

Brundrett MC, Kendrick B. 1990. The roots and mycorrhizas of herbaceous woodland plants. I. Quantitative aspects of morphology. *New Phytologist* **114**:457–468.

Brundrett MC, Piché Y, Peterson RL. 1984. A new method for observing the morphology of vesicular-arbuscular mycorrhizae. *Canadian Journal of Botany* **62**:2128–2134.

Burke DJ. 2008. Effects of *Alliaria petiolata* (garlic mustard; Brassicaceae) on mycorrhizal colonization and community structure in three herbaceous plants in a mixed deciduous forest. *American Journal of Botany* **95**:1416–1425.

Burke DJ, Weintraub MN, Hewins CR, Kalisz S. 2011. Relationship between soil enzyme activities, nutrient cycling and soil fungal communities in a northern hardwood forest. *Soil Biology and Biochemistry* **43**:795–803.

Butchart SHM, Walpole M, Collen B, van Strien A, Scharlemann JPW, Almond REA, Baillie JEM, Bomhard B, Brown C, Bruno J, Carpenter KE, Carr GM, Chanson J, Chenery AM, Csirke J, Davidson NC, Dentener F, Foster M, Galli A, Galloway JN, Genovesi P, Gregory RD, Hockings M, Kapos V, Lamarque JF, Leverington F, Loh J, McGeoch MA, McRae L, Minasyan A, Morcillo MH, Oldfield TEE, Pauly D, Quader S, Revenga C, Sauer JR, Skolnik B, Spear D, Stanwell-Smith D, Stuart SN, Symes A, Tierney M, Tyrrell TD, Vié JC, Watson R. 2010. Global biodiversity: indicators of recent declines. *Science* **328**:1164–1168.

Callaway RM, Ridenour WM. 2004. Novel weapons: invasive success and the evolution of increased competitive ability. *Frontiers in Ecology and the Environment* **2**:436–443.

Callaway RM, Cipollini D, Barto K, Thelen GC, Hallett SG, Prati D, Stinson K, Klironomos J. 2008. Novel weapons: invasive plant suppresses fungal mutualists in America but not in its native Europe. *Ecology* **89**:1043–1055.

Cantor A, Hale A, Aaron J, Traw MB, Kalisz S. 2011. Low allelochemical concentrations detected in garlic mustard-invaded forest soils inhibit fungal growth and AMF spore germination. *Biological Invasions* **13**:3015–3025.

Carlson AM, Gorchov DL. 2004. Effects of herbicide on the invasive biennial *Alliaria petiolata* (Garlic Mustard) and initial responses of native plants in a southwestern Ohio forest. *Restoration Ecology* **12**:559–567.

Cipollini D, Gruner B. 2007. Cyanide in the chemical arsenal of garlic mustard, *Alliaria petiolata*. *Journal of Chemical Ecology* **33**:85–94.

Cipollini KA, Flint WN. 2013. Comparing allelopathic effects of root and leaf extracts of invasive *Alliaria petiolata*, *Lonicera maackii*, and *Ranunculus ficaria* on germination of three native woodland plants. *Ohio Journal of Science* **112**:37–43.

Cipollini KA, McClain GY, Cipollini D. 2008. Separating above- and belowground effects of *Alliaria petiolata* and *Lonicera maackii* on the performance of *Impatiens capensis*. *The American Midland Naturalist* **160**:117–128.

Crone EE, Miller E, Sala A. 2009. How do plants know when other plants are flowering? Resource depletion, pollen limitation and mast-seeding in a perennial wildflower. *Ecology Letters* **12**:1119–1126.

Davis AS, Landis DA, Nuzzo V, Blossey B, Gerber E, Hinz HL. 2006. Demographic models inform selection of biocontrol agents for garlic mustard (*Alliaria petiolata*). *Ecological Applications* **16**:2399–2410.

DeMars BG, Boerner REJ. 1995. Arbuscular mycorrhizal development in three crucifers. *Mycorrhiza* **5**:405–408.

Ehrenfeld JG. 2003. Effects of exotic plant invasions on soil nutrient cycling processes. *Ecosystems* **6**:503–523.

Ehrenfeld JG, Kourtev P, Huang W. 2001. Changes in soil functions following invasions of exotic understory plants in deciduous forests. *Ecological Applications* **11**:1287–1300.

Eschtruth AK, Battles JJ. 2009a. Assessing the relative importance of disturbance, herbivory, diversity, and propagule pressure in exotic plant invasion. *Ecological Monographs* **79**:265–280.

Eschtruth AK, Battles JJ. 2009b. Acceleration of exotic plant invasion in a forested ecosystem by a generalist herbivore. *Conservation Biology* **23**:388–399.

Galbraith RF. 1988. A note on graphical presentation of estimated odds ratios from several clinical trials. *Statistics in Medicine* **7**:889–894.

Gelman A, Hill J. 2006. *Data analysis using regression and multilevel/hierarchical models.* Cambridge: Cambridge University Press.

Gilliam F. 2014. *The herbaceous layer in forests of eastern North America.* Oxford: Oxford University Press.

Gimsing AL, Sorensen JC, Tovgaard L, Jorgensen AMF, Hansen HCB. 2006. Degradation kinetics of glucosinolates in soil. *Environmental Toxicology and Chemistry* **25**:2038–2044.

Gremer JR, Sala A. 2013. It is risky out there: the costs of emergence and the benefits of prolonged dormancy. *Oecologia* **172**:937–947.

Gremer JR, Sala A, Crone EE. 2010. Disappearing plants: why they hide and how they return. *Ecology* **91**:3407–3413.

Grove S, Haubensak KA, Parker IM. 2012. Direct and indirect effects of allelopathy in the soil legacy of an exotic plant invasion. *Plant Ecology* **213**:1869–1882.

Gurevitch J, Chester ST Jr. 1986. Analysis of repeated measures experiments. *Ecology* **67**:251–255.

Hadfield JD. 2008. Estimating evolutionary parameters when viability selection is operating. *Proceedings of the Royal Society B: Biological Sciences* **275**:723–734.

Hale AN, Kalisz S. 2012. Perspectives on allelopathic disruption of plant mutualisms: a framework for individual- and population-level fitness consequences. *Plant Ecology* **213**:1991–2006.

Hale AN, Tonsor SJ, Kalisz S. 2011. Testing the mutualism disruption hypothesis: physiological mechanisms for invasion of intact perennial plant communities. *Ecosphere* **2**:art110.

Hulme PE, Pyšek P, Jarošík V, Pergl J, Schaffner U, Vilà M. 2013. Bias and error in understanding plant invasion impacts. *Trends in Ecology and Evolution* **28**:212–218.

Jumpponen A, Trappe JM. 1998. Dark septate endophytes: a review of facultative biotrophic root-colonizing fungi. *New Phytologist* **140**:295–310.

Kalisz S, Spigler RB, Horvitz CC. 2014. In a long-term experimental demography study, excluding ungulates reversed invader's explosive population growth rate and restored natives. *Proceedings of the National Academy of Sciences of the USA* **111**:4501–4506.

Kéry M, Gregg KB, Schaub M. 2005. Demographic estimation methods for plants with unobservable life-states. *Oikos* **108**:307–320.

Kiers ET, Palmer TM, Ives AR, Bruno JF, Bronstein JL. 2010. Mutualisms in a changing world: an evolutionary perspective. *Ecology Letters* **13**:1459–1474.

Kleijn D, Treier UA, Müller-Schärer H. 2005. The importance of nitrogen and carbohydrate storage for plant growth of the alpine herb *Veratrum album*. *New Phytologist* **166**:565–575.

Klironomos JN. 2002. Feedback with soil biota contributes to plant rarity and invasiveness in communities. *Nature* **417**:67–70.

Knight TM, Dunn JL, Smith LA, Davis J, Kalisz S. 2009. Deer facilitate invasive plant success in a Pennsylvania forest understory. *Natural Areas Journal* **29**:110–116.

Koch AM, Antunes PM, Barto EK, Cipollini D, Mummey DL, Klironomos JN. 2011. The effects of arbuscular mycorrhizal (AM) fungal and garlic mustard introductions on native AM fungal diversity. *Biological Invasions* **13**:1627–1639.

Koide RT, Shumway DL, Mabon SA. 1994. Mycorrhizal fungi and reproduction of field populations of *Abutilon theophrasti* Medic. (Malvaceae). *New Phytologist* **126**:123–130.

Kuebbing SE, Classen AT, Simberloff D. 2014. Two co-occurring invasive woody shrubs alter soil properties and promote subdominant invasive species. *Journal of Applied Ecology* **51**:124–133.

Laake JL, Johnson DS, Conn PB. 2013. marked: an R package for maximum likelihood and Markov Chain Monte Carlo analysis of capture–recapture data. *Methods in Ecology and Evolution* **4**:885–890.

LaFrankie JV Jr. 1985. Morphology, growth, and vasculature of the sympodial rhizome of *Smilacina racemosa* (Liliaceae). *Botanical Gazette* **146**:534–554.

Lankau RA. 2011a. Resistance and recovery of soil microbial communities in the face of *Alliaria petiolata* invasions. *New Phytologist* **189**:536–548.

Lankau RA. 2011b. Interpopulation variation in allelopathic traits informs restoration of invaded landscapes. *Evolutionary Applications* **5**:270–282.

Lankau RA. 2012. Coevolution between invasive and native plants driven by chemical competition and soil biota. *Proceedings of the National Academy of Sciences of the USA* **109**:11240–11245.

Lankau RA, Nuzzo V, Spyreas G, Davis AS. 2009. Evolutionary limits ameliorate the negative impact of an invasive plant. *Proceedings of the National Academy of Sciences of the USA* **106**:15362–15367.

Lankau RA, Bauer JT, Anderson MR, Anderson RC. 2014. Long-term legacies and partial recovery of mycorrhizal communities after invasive plant removal. *Biological Invasions* **16**:1979–1990.

Lapointe L. 1998. Fruit development in *Trillium*: dependence on stem carbohydrate reserves. *Plant Physiology* **117**:183–188.

Larsen LM, Nielsen JK, Sørensen H. 1992. Host plant recognition in monophagous weevils: specialization of *Ceutorhynchus inaffectatus* to glucosinolates from its host plant *Hesperis matronalis*. *Entomologia Experimentalis et Applicata* **64**:49–55.

Leicht-Young SA, Pavlovic NB, Adams JV. 2012. Competitive interactions of garlic mustard (*Alliaria petiolata*) and damesrocket (*Hesperis matronalis*). *Invasive Plant Science and Management* **5**:27–36.

Levine JM, Vilà M, Antonio CMD, Dukes JS, Grigulis K, Lavorel S. 2003. Mechanisms underlying the impacts of exotic plant invasions. *Proceedings of the Royal Society B: Biological Sciences* **270**:775–781.

Lu X, Koide RT. 1994. The effects of mycorrhizal infection on components of plant growth and reproduction. *New Phytologist* **128**:211–218.

McCarthy BC. 1997. Response of a forest understory community to experimental removal of an invasive nonindigenous plant (*Alliaria petiolta*, Brassicaceae). In: Luken JO, Thieret JW, eds. *Assessment and management of plant invasions*. New York: Springer, 117–130.

McDowell NG. 2011. Mechanisms linking drought, hydraulics, carbon metabolism, and vegetation mortality. *Plant Physiology* **155**: 1051–1059.

McDowell NG, Pockman WT, Allen CD, Breshears DD, Cobb N, Kolb T, Plaut J, Sperry J, West A, Williams DG, Yepez EA. 2008. Mechanisms of plant survival and mortality during drought: why do some plants survive while others succumb to drought? *New Phytologist* **178**:719–739.

Meekins JF, McCarthy BC. 1999. Competitive ability of *Alliaria petiolata* (Garlic Mustard, Brassicaceae), an invasive, nonindigenous forest herb. *International Journal of Plant Sciences* **160**:743–752.

Meinhardt KA, Gehring CA. 2012. Disrupting mycorrhizal mutualisms: a potential mechanism by which exotic tamarisk outcompetes native cottonwoods. *Ecological Applications* **22**:532–549.

Morris WF, Pfister CA, Tuljapurkar S, Haridas CV, Boggs CL, Boyce MS, Bruna EM, Church DR, Coulson T, Doak DF, Forsyth S, Gaillard J-M, Horvitz CC, Kalisz S, Kendall BE, Knight TM, Lee CT, Menges ES. 2008. Longevity can buffer plant and animal populations against changing climatic variability. *Ecology* **89**:19–25.

Nakagawa S, Freckleton RP. 2008. Missing inaction: the dangers of ignoring missing data. *Trends in Ecology and Evolution* **23**:592–596.

Newsham KK. 2011. A meta-analysis of plant responses to dark septate root endophytes. *New Phytologist* **190**:783–793.

Olano JM, Menges ES, Martínez E. 2006. Carbohydrate storage in five resprouting Florida scrub plants across a fire chronosequence. *New Phytologist* **170**:99–106.

Pisula NL, Meiners SJ. 2010. Relative allelopathic potential of invasive plant species in a young disturbed woodland. *The Journal of the Torrey Botanical Society* **137**:81–87.

Poulette MM, Arthur MA. 2012. The impact of the invasive shrub *Lonicera maackii* on the decomposition dynamics of a native plant community. *Ecological Applications* **22**:412–424.

Powell LA. 2007. Approximating variance of demographic parameters using the delta method: a reference for avian biologists. *The Condor* **109**:949–954.

Prati D, Bossdorf O. 2004. Allelopathic inhabitation of germination by *Alliaria petiolate* (Brassicaceae). *American Journal of Botany* **91**: 285–288.

R Development Core Team. 2014. *R: a language and environment for statistical computing*. Vienna, Austria.

Rita H, Komonen A. 2008. Odds ratio: an ecologically sound tool to compare proportions. *Annales Zoologici Fennici* **45**:66–72.

Roberts KJ, Anderson RC. 2001. Effect of garlic mustard [*Alliaria petiolata* (Beib. Cavara & Grande)] extracts on plants and arbuscular mycorrhizal (AM) fungi. *The American Midland Naturalist* **146**: 146–152.

Rodgers VL, Stinson KA, Finzi AC. 2008a. Ready or not, garlic mustard is moving in: *Alliaria petiolata* as a member of Eastern North American forests. *BioScience* **58**:426–436.

Rodgers VL, Wolfe BE, Werden LK, Finzi AC. 2008b. The invasive species *Alliaria petiolata* (garlic mustard) increases soil nutrient

availability in northern hardwood-conifer forests. *Oecologia* **157**:459–471.

Rosenthal R, Rosnow R. 1985. *Contrast analysis: focused comparisons in the analysis of variance*. Cambridge: Cambridge University Press.

Rout ME, Callaway RM. 2012. Interactions between exotic invasive plants and soil microbes in the rhizosphere suggest that 'everything is not everywhere'. *Annals of Botany* **110**:213–222.

Ruckli R, Rusterholz HP, Baur B. 2014. Invasion of an annual exotic plant into deciduous forests suppresses arbuscular mycorrhiza symbiosis and reduces performance of sycamore maple saplings. *Forest Ecology and Management* **318**:285–293.

Schuster MJ, Dukes JS. 2014. Non-additive effects of invasive tree litter shift seasonal N release: a potential invasion feedback. *Oikos* **123**:1101–1111.

Schweiger R, Baier MC, Persicke M, Müller C. 2014. High specificity in plant leaf metabolic responses to arbuscular mycorrhiza. *Nature Communications* **5**:Article 3886.

Shannon SM, Bauer JT, Anderson WE, Reynolds HL. 2014. Plant-soil feedbacks between invasive shrubs and native forest understory species lead to shifts in the abundance of mycorrhizal fungi. *Plant and Soil* **382**:317–328.

Shefferson RP. 2009. The evolutionary ecology of vegetative dormancy in mature herbaceous perennial plants. *Journal of Ecology* **97**:1000–1009.

Smith LM, Reynolds HL. 2012. Positive plant-soil feedback may drive dominance of a woodland invader, *Euonymus fortunei*. *Plant Ecology* **213**:853–860.

Smith LM, Reynolds HL. 2014. Light, allelopathy, and post-mortem invasive impact on native forest understory species. *Biological Invasions* **16**:1131–1144.

Smith SE, Read DJ. 2008. *Mycorrhizal symbiosis*, 3rd edn. London: Academic Press.

Stinson KA, Campbell SA, Powell JR, Wolfe BE, Callaway RM, Thelen GC, Hallett SG, Prati D, Klironomos JN. 2006. Invasive plant suppresses the growth of native tree seedlings by disrupting belowground mutualisms. *PLoS Biology* **4**:e140.

Tierens K, Thomma BPH, Brouwer M, Schmidt J, Kistner K, Porzel A, Mauch-Mani B, Cammue BPA, Broekaert WF. 2001. Study of the role of antimicrobial glucosinolate-derived isothiocyanates in resistance of *Arabidopsis* to microbial pathogens. *Plant Physiology* **125**:1688–1699.

Tukey JW. 1991. The philosophy of multiple comparisons. *Statistical Science* **6**:100–116.

Tylianakis JM, Didham RK, Bascompte J, Wardle DA. 2008. Global change and species interactions in terrestrial ecosystems. *Ecology Letters* **11**:1351–1363.

van der Heijden MGA, Bardgett RD, van Straalen NM. 2008. The unseen majority: soil microbes as drivers of plant diversity and productivity in terrestrial ecosystems. *Ecology Letters* **11**:296–310.

Van Diepen LTA, Lilleskov EA, Pregitzer KS. 2011. Simulated nitrogen deposition affects community structure of arbuscular mycorrhizal fungi in northern hardwood forests. *Molecular Ecology* **20**:799–811.

Vaughn SF, Berhow MA. 1999. Allelochemicals isolated from tissues of the invasive weed garlic mustard (*Alliaria petiolata*). *Journal of Chemical Ecology* **25**:2495–2504.

Vitousek PM, Walker LR. 1989. Biological invasion by *Myrica Faya* in Hawai'i: plant demography, nitrogen fixation, ecosystem effects. *Ecological Monographs* **59**:247–265.

Vogelsang KM, Bever JD. 2009. Mycorrhizal densities decline in association with nonnative plants and contribute to plant invasion. *Ecology* **90**:399–407.

Weidenhamer JD. 2006. Distinguishing allelopathy from resource competition: the role of density. In: Reigosa MJ, Pedrol N, Gonzalez L, eds. *Allelopathy: a physiological process with ecological implications*. The Netherlands: Springer, 85–103.

Weir TL, Park SW, Vivanco JM. 2004. Biochemical and physiological mechanisms mediated by allelochemicals. *Current Opinion in Plant Biology* **7**:472–479.

Whigham DF. 2004. Ecology of woodland herbs in temperate deciduous forests. *Annual Review of Ecology, Evolution, and Systematics* **35**:583–621.

Wixted KL, McGraw JB. 2010. Competitive and allelopathic effects of garlic mustard (*Alliaria petiolata*) on American ginseng (*Panax quinquefolius*). *Plant Ecology* **208**:347–357.

Wolfe BE, Klironomos JN. 2005. Breaking new ground: soil communities and exotic plant invasion. *BioScience* **55**:477–487.

Wolfe BE, Rodger VL, Stinson KA, Pringle A. 2008. The invasive plant *Alliaria petiolata* (garlic mustard) inhibits ectomycorrhizal fungi in its introduced range. *Journal of Ecology* **96**:777–783.

Wright DP, Read DJ, Scholes JD. 1998. Mycorrhizal sink strength influences whole plant carbon balance of *Trifolium repens* L. *Plant, Cell and Environment* **21**:881–891.

Wyka T. 1999. Carbohydrate storage and use in an alpine population of the perennial herb, *Oxytropis sericea*. *Oecologia* **120**: 198–208.

Zhu X-C, Song F-B, Liu S-Q, Liu T-D. 2011. Effects of arbuscular mycorrhizal fungus on photosynthesis and water status of maize under high temperature stress. *Plant and Soil* **346**:189–199.

Zuleta A, Sambucetti ME. 2001. Inulin determination for food labeling. *Journal of Agricultural and Food Chemistry* **49**: 4570–4572.

Zuur A, Ieno EN, Walker N, Saveliev AA, Smith GM. 2009. *Mixed effects models and extensions in ecology with R*. New York: Springer Science & Business Media.

9

Local dominance of exotic plants declines with residence time: a role for plant–soil feedback?

Tanja A.A. Speek[1,2,3]*, Joop H.J. Schaminée[4,5], Jeltje M. Stam[6], Lambertus A.P. Lotz[1], Wim A. Ozinga[4] and Wim H. van der Putten[2,3]

[1] Plant Research International, Wageningen University and Research Centre, Droevendaalsesteeg 1, 6708 PB Wageningen, The Netherlands
[2] Laboratory of Nematology. Wageningen University and Research Centre, Droevendaalsesteeg 1, 6708 PB Wageningen, The Netherlands
[3] Department of Terrestrial Ecology, Netherlands Institute of Ecology (NIOO-KNAW), Droevendaalsesteeg 10, 6708 PB Wageningen, The Netherlands
[4] Centre for Ecosystem Studies, Wageningen University and Research Centre, Droevendaalsesteeg 3a, 6708 PB Wageningen, The Netherlands
[5] Department of Ecology, Aquatic Ecology and Environmental Biology Research Group, Radboud University Nijmegen, Heyendaalseweg 135, 6525 AJ Nijmegen, The Netherlands
[6] Laboratory of Entomology, Wageningen University and Research Centre, Droevendaalsesteeg 1, 6708 PB Wageningen, The Netherlands

Associate Editor: Inderjit

Abstract. Recent studies have shown that introduced exotic plant species may be released from their native soil-borne pathogens, but that they become exposed to increased soil pathogen activity in the new range when time since introduction increases. Other studies have shown that introduced exotic plant species become less dominant when time since introduction increases, and that plant abundance may be controlled by soil-borne pathogens; however, no study yet has tested whether these soil effects might explain the decline in dominance of exotic plant species following their initial invasiveness. Here we determine plant–soil feedback of 20 plant species that have been introduced into The Netherlands. We tested the hypotheses that (i) exotic plant species with a longer residence time have a more negative soil feedback and (ii) greater local dominance of the introduced exotic plant species correlates with less negative, or more positive, plant–soil feedback. Although the local dominance of exotic plant species decreased with time since introduction, there was no relationship of local dominance with plant–soil feedback. Plant–soil feedback also did not become more negative with increasing time since introduction. We discuss why our results may deviate from some earlier published studies and why plant–soil feedback may not in all cases, or not in all comparisons, explain patterns of local dominance of introduced exotic plant species.

Keywords: Exotic species; introduced species; local dominance; macro ecology; residence time; soil-borne enemy.

* Corresponding author's e-mail address: tanjaspeek@gmail.com

Introduction

An important challenge for invasion ecologists is to predict the course of invasions of introduced exotic species. This requires insight in the factors that may control the abundance and dominance of species in both their native and new ranges. It has been well established that regional distribution of exotic plant species increases with residence time (Pyšek *et al.* 2004; Hamilton *et al.* 2005; Wilson *et al.* 2007; Milbau and Stout 2008; Bucharova and van Kleunen 2009; Gassó *et al.* 2009). It has also been argued that increased residence time may result in lower local dominance and invasiveness (Carpenter and Cappuccino 2005; Hawkes 2007; Speek *et al.* 2011). Local dominance of introduced exotic plant species may be, at least in part, driven by interactions with soil biota, including effects of soil-borne enemies and symbionts (Inderjit and van der Putten 2010). The question that we address in the present study is how residence time and local dominance of exotic plant species may relate to enemy impact of the soil biota. Ultimately, this information may be used to enhance predictions on the course of invasiveness of introduced exotic plant species.

A possible explanation for lower local dominance of introduced exotic plant species with a long residence time is that enemy species may increasingly adapt and accumulate when time of exposure to the new hosts increases (Hawkes 2007; Diez *et al.* 2010; Dostál *et al.* 2013). Both aboveground (Bentley and Whittaker 1979; Gange and Brown 1989) and belowground (van der Putten *et al.* 1993; Klironomos 2002; Mangan *et al.* 2010; Johnson *et al.* 2012) enemies may control local plant dominance. Release from natural enemies by introduction to a new range has been proposed to enhance the performance of species and, therefore, their invasiveness (Elton 1958; Keane and Crawley 2002). This 'enemy release hypothesis' (Keane and Crawley 2002) has been supported by surveys showing that introduced plant species have fewer enemies in their novel than native range (e.g. Mitchell and Power 2003).

Thus far, the majority of research on enemy release of exotic plant species has been dedicated to aboveground enemies. However, an increasing amount of studies is showing that introduced exotic plant species can be released from native soil-borne enemies as well (Reinhart *et al.* 2003, 2010; Callaway *et al.* 2004; Gundale *et al.* 2014). Introduced exotic plant species suffer less from soil-enemies of the invaded range than congeners that are native in that range (Maron and Vilà 2001; Agrawal *et al.* 2005; Engelkes *et al.* 2008).

The change in performance of exotic species with progressing residence time has been described for several invaders (Simberloff and Gibbons 2004). Loss of exotic dominance might be caused by evolutionary adaptation of enemies in the new range to the introduced plant species (Müller-Schärer *et al.* 2004). Such adaptive potential may be deduced from reported higher frequencies of specialist compared with generalist herbivores (Andow and Imura 1994), higher exposure (Mitchell *et al.* 2010) and higher impact (Hawkes 2007) of enemies on crop and exotic plant species in relation to increasing residence time. Similarly, in New Zealand plant–soil feedback of 12 exotic plant species related negatively to their residence time (Diez *et al.* 2010) and in the Czech Republic giant hogweed (*Heracleum mantegazzianum*) developed negative feedback effects from the soil biota in fields that had been colonized for some decades (Dostál *et al.* 2013). However, in these latter studies, increased enemy exposure has not yet been related to local dominance of the exotic plant species, which is the key aim of the present study.

A recent analysis established that exotic plant species with a long residence time in The Netherlands have lower local dominance than recently introduced species (Speek *et al.* 2011). Until now, no study has related such patterns in local dominance to plant–soil feedback effects. Therefore, in the present study, we determine how residence time, local dominance and soil pathogen effects to exotic species may relate to each other. We tested soil pathogen effects by plant–soil feedback approach (Bever *et al.* 1997), which is a way to experimentally integrate all positive and negative interactions between plants and the soil biota. We first tested the hypothesis that species with a longer residence time have a more negative plant–soil feedback (Diez *et al.* 2010). Then, we tested the hypothesis that species with a more positive plant–soil feedback have a higher local dominance (Klironomos 2002).

Methods

Data on plants, their residence time and local dominance

Data on residence time were derived from information on the period of naturalization according to the standard list of the Dutch flora (Tamis *et al.* 2004). Data on local dominance were derived from the Dutch Vegetation Database (Schaminée *et al.* 2007), containing over 500 000 vegetation records including data on local species cover in plots varying from 1 by 1 m^2 to 10 by 10 m^2. Plot sizes used for recording depended on the characteristics of vegetation, for example largest plot sizes were used for forests. Data on plant species cover were used to calculate local dominance as [the number of vegetation records with that species having >10 per cent ground cover/the total number of vegetation records with that plant species] $\times$ 100 %

(Speek *et al.* 2011). Therefore, local dominance expresses the frequency of how often a plant species has a minimum cover of 10 %, when present in the vegetation record. In order to exclude recorder bias, for example due to avoiding taking records of vegetation heavily invaded by exotic plant species, we used expert judgment to check and where necessary adjust the cover data (Speek *et al.* 2011).

Soil feedback experiment

We used a selection of 20 introduced exotic plant species in The Netherlands for a plant–soil feedback experiment (Supplements). The selection of 20 plant species was based on a number of criteria. First, we excluded woody species, because the length of the plant–soil feedback is too limited for capturing a substantial part of the life cycle of trees. We then selected as many as possible plant species from riverine areas in order to be able to use the same soil origin for all plant species. Finally, the selection was limited as the seeds of some plant species did not germinate. Seeds had been collected by specialized seed companies that collect seeds locally, or by ourselves or colleagues.

Of the 20 plant species, 14 occur in the Millingerwaard (Dirkse *et al.* 2007), a riverine floodplain area of 700 hectares. Millingerwaard is a nature reserve in the riverine floodplain of the river Waal, which is in the southern branch of the Rhine river in The Netherlands (51°87′N, 6°01′E). Three other species occur near or in other riverine areas in The Netherlands and the remaining three occur outside riverine areas. We collected soil from the Millingerwaard area, instead of from a larger variety of sites, as soil from a variety of sites would have introduced additional variation due to soil type, fertility, pH etc. All plant species were forbs that varied in local dominance from 5 to 38 % and in residence time from 75 to 400 years.

Seeds were germinated on glass beads placed in demineralized water. Germination was carried out in transparent plastic containers of $17 \times 12 \times 5$ cm that were placed under conditions of 16 h 22 °C in the light (day) and 8 h 10 °C in the dark (night). *Xanthium strumarium* seeds were germinated at a higher temperature: 16 h 32 °C and 8 h 20 °C. Germinated seedlings were stored at 4 °C and 10/14 h light/dark until transplantation in soil, to ensure equal sizes at start of the experiment. Soil was collected from five random locations in Millingerwaard. Soil to be used as inoculum was collected in October 2010, prior to the first phase of the experiment. Soil from the five sampling locations was sieved (mesh size 5 mm) to remove coarse roots, stones and other large particles, and subsequently homogenized. The bulk soil was collected in January 2010, sterilized by gamma irradiation (25 kGray) and stored in sealed plastic bags at 4 °C until use.

The sensitivity of exotic plant species to soil-borne enemies was determined in a so-called two-phase plant–soil feedback experiment (Bever *et al.* 1997). In the first phase, which started from one pooled sample, the seedlings were grown to condition the field soil. In that phase, soil biota that can grow on resources provided by that particular plant species are enumerated (Grayston *et al.* 1998; Kowalchuk *et al.* 2002). In the second phase, we kept all replicates of own soil separate. In order to do so, the soil of each pot was split into two halves: one half was used as own soil, whereas the other half was mixed with halves of all other replications and species, to be used as away soils. The replicates of the mixed soil were not kept intact, because there was no relationship between replicate 1 conditioned by species A or B. Comparing plant performances in own and mixed soils enabled us to make a home (own) versus away (mixed) comparison, which is a less sensitive and ecologically more realistic method of detecting plant–soil feedback effects than a comparison of non-sterilized versus sterilized soil (Kulmatiski *et al.* 2008). In the final analysis, plant species was the unit of replication.

For the first—conditioning—phase, bulk soil and inoculum were mixed at a ratio of 4 : 1, with a total of 1200 g soil per pot on a dry weight basis. Pots of 1.3 L were used. For the second—feedback—phase, 'own soil' and 'mixed soil' were homogenized with sterilized bulk soil at a ratio of 1 : 1 in order to keep pot volumes equal between the two feedback phases. For each plant species, we had five independent replicates with own and five with mixed soil. Every pot contained three seedlings, except *Amaranthus retroflexus* that was planted as two seedlings per pot due to poor germination of the seeds. Dead seedlings were replaced until the first week after transplanting. Greenhouse conditions were maintained at 60 % RH, day temperature 21 °C, night temperature 16 °C. Daylight was supplemented with lamps (SON-T Agro, 225 μmol^{-1} m^{-2}), to ensure a minimum of 16 h light per day.

Before planting, the water content in each pot was set at 20 % (w/w). Plants were supplied with water three times a week and once a week the water content was re-set to 20 % by weighing. Plants received 10 mL of 0.5 strength Hoagland per pot in weeks 2, 3 and 4, and 20 mL in weeks 5 and 6 after transplanting in order to meet the increasing demand. Plants were harvested 6 weeks after planting. The length of growth was the same for both phases, which is relatively short, but ample for testing feedback responses (van der Putten *et al.* 1988). When harvesting, shoots of the three (or two) plants per pot were clipped at ground level, pooled, dried in paper bags at 75 °C until constant weight and weighed, so that biomass data per pot were obtained.

Statistical analysis

The effect of soil feedback on shoot and root biomass was calculated as ln[(biomass in own soil)/(biomass in mixed soil)] (Brinkman et al. 2010). We assigned pairs of own soil with mixed soil randomly. To analyse whether residence time or local dominance could explain mean shoot and root feedback responses, we used linear models. The unit of replication was the plant species. For residence time we used models with a normal distribution, for local dominance we used models with a binomial distribution and a logit link, with binomial totals set to 50 % (the highest value in our dataset).

We analysed which traits and other factors related best to residence time by a model selection procedure within a linear model with a normal distribution. Thus, we selected the best minimal adequate model with the lowest Akaike Information Criterion value from all possible subsets. Although time and dominance were related, the relation of a trait or other factor to residence time may not necessarily imply that there is a relation with local dominance as well. Therefore, the factors in the best minimal adequate

model were added to a generalized linear model with residence time explaining local dominance. By adding each factor separately, we analysed which one significantly changed the model. Factors that affected the model were likely to be a better explanation for variation in local dominance than residence time. For explaining local dominance we used a binomial distribution with a logit link, binomial totals set at 50 and accounting for overdispersion. All analyses were done in Genstat, version 14.

Results

Opposite to our hypothesis, we found neither a significant relationship between residence time and plant–soil feedback of the exotic plant species, nor for shoots ($F = 0.10$, $t_{18} = -0.32$, $P = 0.751$, Fig. 1) and for roots ($F = 0.41$, $t_{18} = -0.64$, $P = 0.529$). Local plant dominance also did not relate to the feedback effect on shoots ($F = 0.09$, $t_{18} = -0.31$, $P = 0.763$) or roots ($F = 0.73$, $t_{18} = -0.85$, $P = 0.404$). Excluding species from riverine habitats, which may not be responsive to soil biota from that habitat, or

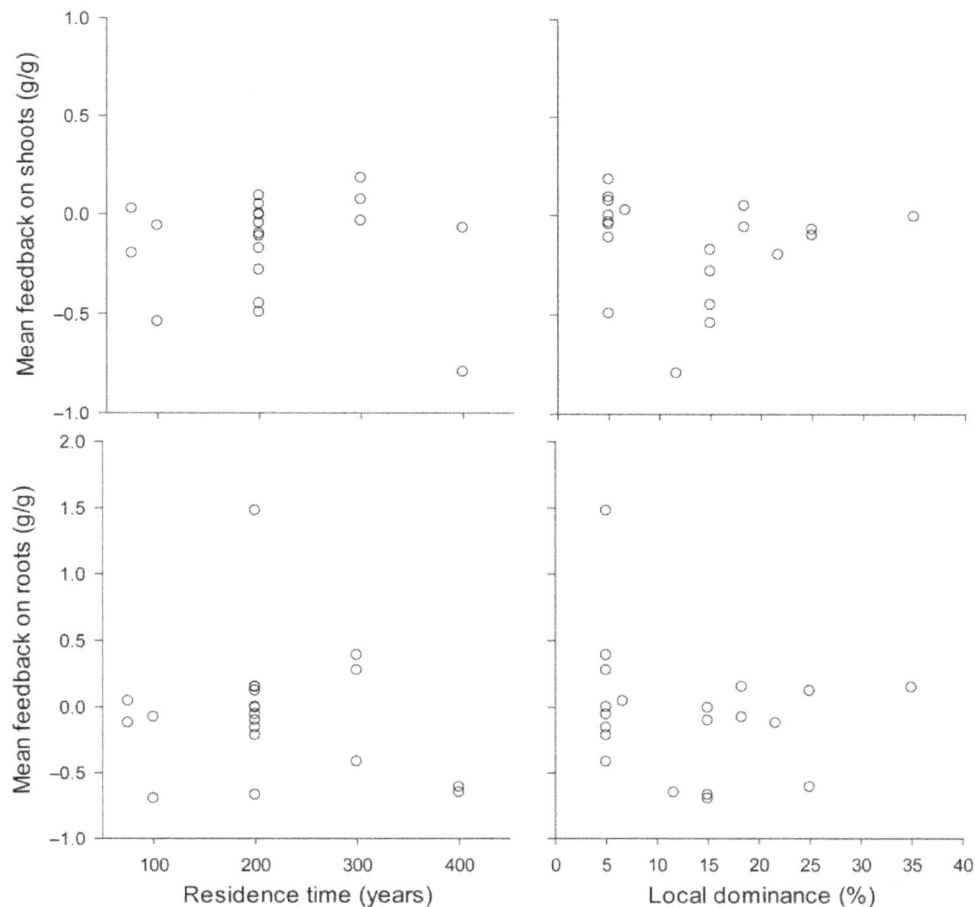

Figure 1. Mean soil feedback effect on the biomass of shoots and roots in relation to the residence time or the local dominance of naturalized exotic plant species in The Netherlands. Each circle represents a different plant species.

Fabaceae species, which may have a different feedback due to symbiosis with rhizobia did not alter the significance of the results (data not shown). Therefore, our hypotheses that species with a longer residence time have a more negative plant–soil feedback, and that species with a less negative or more positive plant–soil feedback have a higher local dominance were not supported.

Discussion

In our study we tested the hypotheses that species with a longer residence time have a more negative plant–soil feedback and that species with a less negative, or more positive plant–soil feedback have a higher local dominance. We used an experimental approach to measure soil-borne enemy impact by plant–soil feedback approach. However, opposite to a study from New Zealand (Diez et al. 2010), and to a study on introduced H. mantegazzianum in the Czech Republic (Dostál et al. 2013) we did not find such a relationship between time since introduction of 20 exotic plant species in the Dutch flora and plant–soil feedback.

There are several possible explanations for these results. Our results could indicate that not all introduced exotic plant species develop negative plant–soil feedback when time since introduction increases. In the field, other ecological processes may be influencing community composition and aboveground interactions can either increase or decrease with the strength of the below-ground interactions. Another possible explanation concerns the choice of soils for the plant–soil feedback experiment. We have chosen soils from areas where most exotic plant species may occur, but we did not use soils from the root zone of particular populations. This approach has led to marked differences in plant–soil feedback between natives and exotics (van Grunsven et al. 2007; Engelkes et al. 2008); however, it has resulted in scattered results when testing soil responses across an entire native range (van Grunsven et al. 2010).

The results may also depend on the relatively short conditioning and testing phases of 6 weeks each. Test phases of 6 weeks can detect feedback effects (van der Putten et al. 1988). Longer test periods may even result in pot limitations, which may obfuscate results. Conditioning for 6 weeks will have been relatively short, but to our experience this is possible when adding soil inocula to sterilized soil, as has been done in the present study.

Our use of pooled soils as 'away' treatment may have provided a conservative estimate of plant–soil feedback effects, because of reducing variances. Nevertheless, since we did not find significant relationships with time since abandonment, or local dominance, our results show that even with a highly sensitive test still no

relationship could be detected between time since introduction, or local dominance and plant–soil feedback. Mixing soils from all plant species to produce 'away' soils could theoretically have led to single pathogens dominating the entire away soil community. However, a previous addition study using a variety of amounts of soil inocula showed that soil feedback effects increased gradually with the amount of inoculum added (van der Putten et al. 1988), which does not point at a disproportional role of pathogens from single plant species in the away soil mixtures.

Plant–soil biota interactions are highly local (Levine et al. 2006; Bezemer et al. 2010; Genung et al. 2012), and adaptation of soil organisms to new plant species does not take place at a national, but at a local scale through direct interactions between plant roots and the soil biota (Schweitzer et al. 2008; Lankau et al. 2009; Lau and Lennon 2011, 2012). As the feedback was estimated at a regional scale, also the local dominance was measured at a regional scale (first occurrence in The Netherlands). Using first occurrence in a larger region as an estimate of residence time could result in an overestimation of the local residence time. On the other hand, the study from New Zealand (Diez et al. 2010) also used data on residence time for the entire country and not specifically for the sites from which the soil has been collected.

We expected plant–soil feedback to be negatively related to local dominance (Klironomos 2002; Mangan et al. 2010). However, in our study we did not observe such an inverse relationship. A possible explanation is that previous studies by Klironomos (2002) and Mangan et al. (2010) on dominance-feedback relationships have been based on native species, and that these relations may differ when considering exotic species. Moreover, we used dominance estimates averaged across the entire Netherlands (Speek et al. 2011), which differs from the local dominance estimates as used in other studies (e.g. Klironomos 2002). National estimates (in the case of The Netherlands concerning an area of appr. 150 × 300 km) will not provide accurate information about the local dominance of exotics in the riverine ecosystem where the soil for testing plant–soil feedback originated from. Therefore, it is possible that soil origin and plant dominance data were not well linked to each other, or that a relationship between plant–soil feedback and dominance works out differently for exotic plant species than for natives. Alternatively, our study may add to other examples where plant dominance does not relate to plant–soil feedback (Reinhart 2012).

An alternative explanation for the rejection of our hypotheses could be that the evolutionary dynamics leading to increased enemy pressure on exotic plant species is not strong enough to result in a change in mean local dominance. Meta-analyses have shown that a general pattern of decreased enemy numbers on exotic species in the novel range was not reflected in a general pattern of higher plant performance (Chun *et al.* 2010). Adaptation can occur both at the soil species level but also at the plant species level. This adaptation at two fronts is likely to result in a mixed general outcome. Moreover, while local dominance has been assumed to increase after introduction to a new range (Keane and Crawley 2002), recent work has shown that most species have the same dominance in both their introduced and native ranges (Firn *et al.* 2011). Clearly, local dominance is a complex trait, with a high variation both between and within species that can be influenced by a large number of ecological processes.

Conclusions

We found no support for the hypothesis that the negative relationship between residence time and local dominance of exotic species in The Netherlands is caused by an increase in negative plant–soil feedback. It may be that data on residence time, dominance, enemy exposure and impact need to be collected all from the same area, or that different choices in plant–soil feedback approach need to be made (e.g. longer conditioning and/or feedback phases, a more sensitive 'away' soil treatment). Alternatively, it might be better to track single species across an introduction gradient (Lankau *et al.* 2009; Lankau 2011). It could also mean that not all introduced exotic plant species develop negative plant–soil feedback when time since introduction increases or that the hypothesized effect of increasing enemy pressure on dominance of introduced exotic plant species might not be strong enough to emerge from examining a large diversity of species across a variation of locations. Therefore even though we are aware of weaknesses of our paper (aspects of the experimental design that were not ideal, for example sampling of soil from one location that did not include all of the study species, pooling 'away' soils, method of pairing of home and away pots to calculate response ratios), our results may add to the debate on change in invasiveness of exotic plant species after introduction.

Sources of Funding

T.A.A.S. and L.A.P.L. were funded by the former Dutch Ministry of Agriculture, Nature and Food Quality, FES-programme 'Versterking Infrastructuur Plantgezondheid'. W.H.v.d.P. was supported by ALW-Vici grant (number 389 865.05.002). W.A.O. was supported by the Dutch Science Foundation (NWO Biodiversity Works).

Contributions by the Authors

All authors contributed to the experimental set up and commented on the manuscript, T.A.A.S., J.M.S. and W.H.v.d.P. performed the experiment, T.A.A.S. performed all analyses, J.H.J.S. and W.A.O. provided data on local dominance of plant species, T.A.A.S. and W.H.v.d.P. wrote the paper.

Acknowledgements

We thank Staatsbosbeheer Regio Oost for allowing permission to work in Millingerwaard and Ciska Raaijmakers for invaluable help during the experiment.

Supporting Information

The following additional information is available in the online version of this article –

Table S1. Plant species naturalized in The Netherlands that were used in soil–plant feedback experiments. Occurrence in Millingerwaard (area where soil was collected) is based on maps in Dirkse *et al.* 2007. + does occur in Millingerwaard; 0 does not occur in Millingerwaard but does occur in other floodplains in The Netherlands; − does not occur in Millingerwaard or other floodplains in The Netherlands.

Literature Cited

Agrawal AA, Kotanen PM, Mitchell CE, Power AG, Godsoe W, Klironomos J. 2005. Enemy release? An experiment with congeneric plant pairs and diverse above- and belowground enemies. *Ecology* 86:2979–2989.

Andow DA, Imura O. 1994. Specialization of phytophagous arthropod communities on introduced plants. *Ecology* 75:296–300.

Bentley S, Whittaker JB. 1979. Effects of grazing by a chrysomelid beetle, *Gastrophysa viridula*, on competition between *Rumex obtusifolius* and *Rumex crispus*. *Journal of Ecology* 67:79–90.

Bever JD, Westover KM, Antonovics J. 1997. Incorporating the soil community into plant population dynamics: the utility of the feedback approach. *Journal of Ecology* 85:561–573.

Bezemer TM, Fountain MT, Barea JM, Christensen S, Dekker SC, Duyts H, van Hal R, Harvey JA, Hedlund K, Maraun M, Mikola J, Mladenov AG, Robin C, de Ruiter PC, Scheu S, Setälä H, Šmilauer P, van der Putten WH. 2010. Divergent composition but similar function of soil food webs of individual plants: plant species and community effects. *Ecology* 91:3027–3036.

Brinkman EP, van der Putten WH, Bakker E-J, Verhoeven KJF. 2010. Plant-soil feedback: experimental approaches, statistical analyses and ecological interpretations. *Journal of Ecology* 98: 1063–1073.

Bucharova A, van Kleunen M. 2009. Introduction history and species characteristics partly explain naturalization success of North

American woody species in Europe. *Journal of Ecology* **97**: 230–238.

Callaway RM, Thelen GC, Rodriguez A, Holben WE. 2004. Soil biota and exotic plant invasion. *Nature* **427**:731–733.

Carpenter D, Cappuccino N. 2005. Herbivory, time since introduction and the invasiveness of exotic plants. *Journal of Ecology* **93**: 315–321.

Chun YJ, van Kleunen M, Dawson W. 2010. The role of enemy release, tolerance and resistance in plant invasions: linking damage to performance. *Ecology Letters* **13**:937–946.

Diez JM, Dickie I, Edwards G, Hulme PE, Sullivan JJ, Duncan RP. 2010. Negative soil feedbacks accumulate over time for non-native plant species. *Ecology Letters* **13**:803–809.

Dirkse GM, Hochstenback SMH, Reijerse AI, Bijlsma R-J, Cerff D. 2007. *Flora van Nijmegen en Kleef 1800–2006: Catalogus van soorten met historische vindplaatsen en recente verspreiding.* Mook.

Dostál P, Müllerová J, Pyšek P, Pergl J, Klinerová T. 2013. The impact of an invasive plant changes over time. *Ecology Letters* **16**: 1277–1284.

Elton C. 1958. *The ecology of invasions by animals and plants.* London: Methuen.

Engelkes T, Morriën E, Verhoeven KJF, Bezemer TM, Biere A, Harvey JA, McIntyre LM, Tamis WLM, van der Putten WH. 2008. Successful range-expanding plants experience less above-ground and below-ground enemy impact. *Nature* **456**:946–948.

Firn J, Moore JL, MacDougall AS, Borer ET, Seabloom EW, HilleRisLambers J, Harpole WS, Cleland EE, Brown CS, Knops JMH, Prober SM, Pyke DA, Farrell KA, Bakker JD, O'Halloran LR, Adler PB, Collins SL, D'Antonio CM, Crawley MJ, Wolkovich EM, La Pierre KJ, Melbourne BA, Hautier Y, Morgan JW, Leakey ADB, Kay A, McCulley R, Davies KF, Stevens CJ, Chu C-J, Holl KD, Klein JA, Fay PA, Hagenah N, Kirkman KP, Buckley YM. 2011. Abundance of introduced species at home predicts abundance away in herbaceous communities. *Ecology Letters* **14**:274–281.

Gange AC, Brown VK. 1989. Insect herbivory affects size variability in plant populations. *Oikos* **56**:351–356.

Gassó N, Sol D, Pino J, Dana ED, Lloret F, Sanz-Elorza M, Sobrino E, Vilà M. 2009. Exploring species attributes and site characteristics to assess plant invasions in Spain. *Diversity and Distributions* **15**: 50–58.

Genung MA, Bailey JK, Schweitzer JA. 2012. Welcome to the neighbourhood: interspecific genotype by genotype interactions in Solidago influence above- and belowground biomass and associated communities. *Ecology Letters* **15**:65–73.

Grayston SJ, Wang SQ, Campbell CD, Edwards AC. 1998. Selective influence of plant species on microbial diversity in the rhizosphere. *Soil Biology and Biochemistry* **30**:369–378.

Gundale MJ, Kardol P, Nilsson MC, Nilsson U, Lucas RW, Wardle DA. 2014. Interactions with soil biota shift from negative to positive when a tree species is moved outside its native range. *New Phytologist* **202**:415–421.

Hamilton MA, Murray BR, Cadotte MW, Hose GC, Baker AC, Harris CJ, Licari D. 2005. Life-history correlates of plant invasiveness at regional and continental scales. *Ecology Letters* **8**: 1066–1074.

Hawkes CV. 2007. Are invaders moving targets? The generality and persistence of advantages in size, reproduction, and enemy release in invasive plant species with time since introduction. *The American Naturalist* **170**:832–843.

Inderjit, van der Putten WH. 2010. Impacts of soil microbial communities on exotic plant invasions. *Trends in Ecology and Evolution* **25**:512–519.

Johnson DJ, Beaulieu WT, Bever JD, Clay K. 2012. Conspecific negative density dependence and forest diversity. *Science* **336**:904–907.

Keane RM, Crawley MJ. 2002. Exotic plant invasions and the enemy release hypothesis. *Trends in Ecology and Evolution* **17**: 164–170.

Klironomos JN. 2002. Feedback with soil biota contributes to plant rarity and invasiveness in communities. *Nature* **417**:67–70.

Kowalchuk GA, Buma DS, de Boer W, Klinkhamer PGL, van Veen JA. 2002. Effects of above-ground plant species composition and diversity on the diversity of soil-borne microorganisms. *Antonie van Leeuwenhoek* **81**:509–520.

Kulmatiski A, Beard KH, Stevens JR, Cobbold SM. 2008. Plant-soil feedbacks: a meta-analytical review. *Ecology Letters* **11**: 980–992.

Lankau RA. 2011. Resistance and recovery of soil microbial communities in the face of *Alliaria petiolata* invasions. *New Phytologist* **189**:536–548.

Lankau RA, Nuzzo V, Spyreas G, Davis AS. 2009. Evolutionary limits ameliorate the negative impact of an invasive plant. *Proceedings of the National Academy of Sciences of the USA* **106**: 15362–15367.

Lau JA, Lennon JT. 2011. Evolutionary ecology of plant-microbe interactions: soil microbial structure alters selection on plant traits. *New Phytologist* **192**:215–224.

Lau JA, Lennon JT. 2012. Rapid responses of soil microorganisms improve plant fitness in novel environments. *Proceedings of the National Academy of Sciences of the USA* **109**: 14058–14062.

Levine JM, Pachepsky E, Kendall BE, Yelenik SG, Lambers JHR. 2006. Plant-soil feedbacks and invasive spread. *Ecology Letters* **9**: 1005–1014.

Mangan SA, Schnitzer SA, Herre EA, Mack KML, Valencia MC, Sanchez EI, Bever JD. 2010. Negative plant-soil feedback predicts tree-species relative abundance in a tropical forest. *Nature* **466**: 752–755.

Maron JL, Vilà M. 2001. When do herbivores affect plant invasion? Evidence for the natural enemies and biotic resistance hypotheses. *Oikos* **95**:361–373.

Milbau A, Stout JC. 2008. Factors associated with alien plants transitioning from casual, to naturalized, to invasive. *Conservation Biology* **22**:308–317.

Mitchell CE, Power AG. 2003. Release of invasive plants from fungal and viral pathogens. *Nature* **421**:625–627.

Mitchell CE, Blumenthal D, Jarošik V, Puckett EE, Pyšek P. 2010. Controls on pathogen species richness in plants' introduced and native ranges: roles of residence time, range size and host traits. *Ecology Letters* **13**:1525–1535.

Müller-Schärer H, Schaffner U, Steinger T. 2004. Evolution in invasive plants: implications for biological control. *Trends in Ecology and Evolution* **19**:417–422.

Pyšek P, Richardson DM, Williamson M. 2004. Predicting and explaining plant invasions through analysis of source area floras: some critical considerations. *Diversity and Distributions* **10**:179–187.

Reinhart KO. 2012. The organization of plant communities: negative plant–soil feedbacks and semiarid grasslands. *Ecology* **93**: 2377–2385.

Local dominance of exotic plants declines with residence time: a role for plant–soil...

103

Reinhart KO, Packer A, van der Putten WH, Clay K. 2003. Plant-soil biota interactions and spatial distribution of black cherry in its native and invasive ranges. *Ecology Letters* **6**: 1046–1050.

Reinhart KO, Tytgat T, van der Putten WH, Clay K. 2010. Virulence of soil-borne pathogens and invasion by *Prunus serotina*. *New Phytologist* **186**:484–495.

Schaminée JHJ, Hennekens SM, Ozinga WA. 2007. Use of the ecological information system SynBioSys for the analysis of large datasets. *Journal of Vegetation Science* **18**:463–470.

Schweitzer JA, Bailey JK, Fischer DG, LeRoy CJ, Lonsdorf EV, Whitham TG, Hart SC. 2008. Plant-soil-microorganism interactions: Heritable relationship between plant genotype and associated soil microorganisms. *Ecology* **89**:773–781.

Simberloff D, Gibbons L. 2004. Now you see them, now you don't!—population crashes of established introduced species. *Biological Invasions* **6**:161–172.

Speek TAA, Lotz LAP, Ozinga WA, Tamis WLM, Schaminée JHJ, van der Putten WH. 2011. Factors relating to regional and local success of exotic plant species in their new range. *Diversity and Distributions* **17**:542–551.

Tamis WLM, van der Meijden R, Runhaar J, Bekker RM, Ozinga WA, Odé B, Hoste I. 2004. Standaardlijst van de Nederlandse flora 2003. *Gorteria* **30**:101–195.

van der Putten WH, van Dijk C, Troelstra SR. 1988. Biotic soil factors affecting the growth and development of *Ammophila arenaria*. *Oecologia* **76**:313–320.

van der Putten WH, van Dijk C, Peters BAM. 1993. Plant-specific soil-borne diseases contribute to succession in foredune vegetation. *Nature* **362**:53–56.

van Grunsven RHA, van der Putten WH, Bezemer TM, Tamis WLM, Berendse F, Veenendaal EM. 2007. Reduced plant-soil feedback of plant species expanding their range as compared to natives. *Journal of Ecology* **95**:1050–1057.

van Grunsven RHA, van der Putten WH, Bezemer TM, Berendse F, Veenendaal EM. 2010. Plant-soil interactions in the expansion and native range of a poleward shifting plant species. *Global Change Biology* **16**:380–385.

Wilson JRU, Richardson DM, Rouget M, Procheş S, Amis MA, Henderson L, Thuiller W. 2007. Residence time and potential range: crucial considerations in modelling plant invasions. *Diversity and Distributions* **13**:11–22.

10

Dominant plant taxa predict plant productivity responses to CO$_2$ enrichment across precipitation and soil gradients

Philip A. Fay[1]*, Beth A. Newingham[2,6], H. Wayne Polley[1], Jack A. Morgan[3], Daniel R. LeCain[3], Robert S. Nowak[4] and Stanley D. Smith[5]

[1] Grassland, Soil, and Water Laboratory, USDA-ARS, 808 E Blackland Rd., Temple, TX 76502, USA
[2] College of Natural Resources, University of Idaho, PO Box 441133, Moscow, ID 83844, USA
[3] Rangeland Resources Research Unit, USDA-ARS, 1701 Centre Avenue, Fort Collins, CO 80526, USA
[4] Department of Natural Resources and Environmental Science/MS 186, University of Nevada Reno, 1664 North Virginia, Reno, NV 89557, USA
[5] School of Life Sciences, University of Nevada, Las Vegas, 4505 S. Maryland Parkway, Las Vegas, NV 89154, USA
[6] Present address: Great Basin Rangelands Research, USDA-ARS, 920 Valley Rd., Reno, NV 89512, USA

Guest Editor: Elise S. Gornish

Abstract. The Earth's atmosphere will continue to be enriched with carbon dioxide (CO$_2$) over the coming century. Carbon dioxide enrichment often reduces leaf transpiration, which in water-limited ecosystems may increase soil water content, change species abundances and increase the productivity of plant communities. The effect of increased soil water on community productivity and community change may be greater in ecosystems with lower precipitation, or on coarser-textured soils, but responses are likely absent in deserts. We tested correlations among yearly increases in soil water content, community change and community plant productivity responses to CO$_2$ enrichment in experiments in a mesic grassland with fine- to coarse-textured soils, a semi-arid grassland and a xeric shrubland. We found no correlation between CO$_2$-caused changes in soil water content and changes in biomass of dominant plant taxa or total community aboveground biomass in either grassland type or on any soil in the mesic grassland ($P > 0.60$). Instead, increases in dominant taxa biomass explained up to 85 % of the increases in total community biomass under CO$_2$ enrichment. The effect of community change on community productivity was stronger in the semi-arid grassland than in the mesic grassland, where community biomass change on one soil was not correlated with the change in either the soil water content or the dominant taxa. No sustained increases in soil water content or community productivity and no change in dominant plant taxa occurred in the xeric shrubland. Thus, community change was a crucial driver of community productivity responses to CO$_2$ enrichment in the grasslands, but effects of soil water change on productivity were not evident in yearly responses to CO$_2$ enrichment. Future research is necessary to isolate and clarify the mechanisms controlling the temporal and spatial variations in the linkages among soil water, community change and plant productivity responses to CO$_2$ enrichment.

Keywords: Central Plains grasslands; climate change; community change; Mojave Desert; primary productivity; rangelands; threshold responses.

* Corresponding author's e-mail address: philip.fay@ars.usda.gov

Introduction

Continued enrichment of the Earth's atmosphere with carbon dioxide (CO_2) is certain through the coming century and is expected to alter climate and terrestrial ecosystem structure and function. In water-limited ecosystems, one major mechanism by which CO_2 enrichment may cause ecosystem change is by increasing soil water (Morgan et al. 2004b; Nowak et al. 2004a). CO_2 enrichment can increase soil water by reducing stomatal conductance and leaf- and canopy-scale transpiration in both C_3 and C_4 species (Nowak et al. 2004b; Ainsworth and Long 2005). Lower stomatal conductance and plant transpiration under CO_2 enrichment in turn decreases soil water depletion, which results in higher soil water and potentially reduced water limitation of ecosystem processes compared with ambient CO_2.

Increased soil water with CO_2 enrichment generally increases the productivity of water-limited plant communities. Greater productivity increases often occur with lower precipitation across precipitation levels typical of grassland ecosystems (Owensby et al. 1996; Morgan et al. 2004b; Nowak et al. 2004b; Körner 2006; Polley et al. 2011), because the increased soil water under CO_2 enrichment is more likely to relieve water limitation (Nowak et al. 2004b). However, in severely water-limited systems such as deserts, CO_2 enrichment may not increase productivity if reduced leaf transpiration is offset by increased leaf area because soil water may not increase. In addition, slow growth rates in drought-adapted desert plants limit the potential to increase community productivity in the short-term (Nowak et al. 2004b; Newingham et al. 2013).

Plant community change is another mechanism by which CO_2 enrichment may increase community productivity (Bradley and Pregitzer 2007; Gornish and Tylianakis 2013). Community change is most likely to alter community productivity when change involves the dominant plant taxa (Smith and Knapp 2003). Smith et al. (2009) suggested that altered resource availability drives the ecosystem change primarily by causing community change. If true, increased soil water with CO_2 enrichment would result in changes in dominant taxa abundance, which would in turn increase productivity. However, increased soil water alone can also result in community productivity increases (Fig. 1).

The contributions of increased soil water and community change to increased community productivity with CO_2 enrichment are also expected to differ with soil texture. At precipitation levels typical of grasslands, soil water will be lower on coarse-textured than fine-textured soils, increasing water limitation of community productivity, and thus the importance of soil water increases in

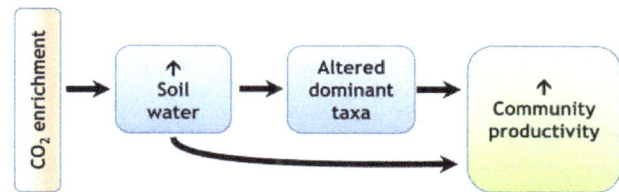

Figure 1. Model by which increased soil water resulting from atmospheric CO_2 enrichment may increase plant community productivity in water-limited ecosystems. The increase in soil water may predict the increase in community productivity, or the effect of soil water increase may be mediated by change in the abundance of dominant plant taxa, which in turn predicts in the increase in community productivity.

regulating productivity. For example, over 4 years of CO_2 enrichment, the increase in plant aboveground biomass was best predicted by soil water on a coarse-textured soil, but increases in biomass were best predicted by the abundance of a dominant grass species on a fine-textured soil (Fay et al. 2012), which accounted for up to 60 % of the effect of CO_2 enrichment (Polley et al. 2012b).

To investigate how water-limited plant communities respond to CO_2 enrichment across varying precipitation amounts and soils, we examined soil water, plant community change and plant productivity responses to CO_2 enrichment in five multi-year CO_2 enrichment experiments from mesic and semi-arid grasslands and in a xeric shrubland in the central and western USA. The experiments span a 6-fold range of mean annual precipitation (MAP), which provides a test of variation in CO_2 responses with precipitation. The mesic grassland experiment contained fine and coarse-textured soils, providing a test of texture effects on CO_2 responses. Interannual variability in precipitation is often high in water-limited systems, and so yearly community productivity increases from CO_2 enrichment were correlated with increases in soil water and community change.

Methods

Sites and experiments

Mesic grassland. Two CO_2 enrichment experiments were conducted at Temple, TX, USA, in the southern portion of the North American tallgrass prairie, a region of humid subtropical climate with warm summers and abundant precipitation (Table 1). The Prairie CO_2 Gradient (PCG, Fig. 2A, Johnson et al. 2000) experiment manipulated CO_2 using two chambers (1 m tall, 1 m wide and 60 m long) placed on an intact formerly grazed pasture. The C_4 perennial grass *Bothriochloa ischaemum* accounted for 51 % of aboveground biomass, accompanied by C_3 perennial forbs, primarily *Solanum dimidiatum* and *Ratibida columnaris* (Polley et al. 2003). In one chamber,

Table 1. Location, climate and CO_2 treatments for the four experiments. MAP, mean annual precipitation; MAT, mean annual temperature.

Experiment	CO_2 treatments (μmol mol^{-1})	Location, lat., long.	Elevation (m)	MAP (mm)	MAT (max, °C)	MAT (min, °C)
PCG	200–550	Texas, USA; 31°05′, −97°20′	186	914	35	3
LYCOG	250–500	Texas, USA; 31°05′, −97°20′	186	914	35	3
OTC	360 and 720	Colorado, USA; 40°40′, −104°45′	1555	320	16	0.5
PHACE	385 and 600	Wyoming, USA; 41°11′, −104°54′	1930	384	18	−3
NDFF	375 and 550	Nevada, USA; 36°46′, −115°57′	970	148	38	−5

Figure 2. CO_2 manipulation experiments included in the study. (A) Carbon dioxide gradient chambers used in PCG and LYCOG experiments in mesic grassland (Photo: P. Fay, USDA-ARS). (B) Open-top chambers used in short-grass steppe experiment (Photo: S. Cox, USDA-ARS). (C) Free-air CO_2 enrichment technology used in mixed-grass prairie PHACE experiment (Photo: S. Cox, USDA-ARS). (D) FACE plot used in xeric shrubland experiment (L. Fenstermaker, Desert Research Institute).

CO_2-enriched air was introduced at the chamber entrance, and photosynthesis by the enclosed vegetation depleted the air of CO_2 as it transited the length of the chamber. Air flow rates were controlled to create a 550–350 μmol mol^{-1} gradient in CO_2. Similarly, ambient air was supplied to the second chamber, which was controlled to create a 365–200 μmol mol^{-1} gradient. The chambers were divided into 5 m sections, and air was cooled and dehumidified in each section to maintain air temperature and vapour pressure deficit near ambient conditions. The CO_2 gradients were maintained during the growing seasons of 1997–2000.

The Lysimeter CO_2 Gradient (LYCOG) experiment was the successor to the PCG experiment and adapted the PCG chambers to enclose 60 intact soil monoliths (1.5 m^3) from a silty clay, a sandy loam and a heavy clay soil, arranged in a stratified random design along the two chambers (Fig. 2A; Polley et al. 2008; Fay et al. 2009). The monoliths supported well-established, constructed communities of native prairie. The C$_4$ grasses *Sorghastrum nutans, Bouteloua curtipendula, Schizachyrium scoparium* and *Tridens albescens* accounted for 80 % of total aboveground biomass. The remainder was C$_3$ forbs (*Solidago canadensis, Salvia azurea*) and one legume (*Desmanthus illinoensis*; Fay et al. 2012; Polley et al. 2012a). Other species were regularly removed by hand or selective glyphosate application. Air temperature and vapour pressure deficit were controlled as for PCG; CO_2 concentration was maintained at 500–380 and 380–250 μmol mol^{-1} during 2006–10. In both experiments, irrigation regimes were representative of average growing season precipitation for the site.

Semi-arid grassland. Two CO_2 enrichment experiments were conducted in a cold semi-arid steppe region of northeastern Colorado and southeastern Wyoming, characterized by cooler summers than the other sites (Table 1). Semi-arid grasslands are more water-limited than mesic grasslands, are dominated by a different, more drought tolerant assemblage of species and are predicted to be more responsive in plant productivity to CO_2 enrichment than mesic grasslands (Morgan et al. 2004b). An open-top chamber (OTC, Fig. 2B) experiment was conducted during 1996–2001 in short-grass steppe near Fort Collins, Colorado. Basal cover of the vegetation was ∼25 %. Two perennial C_3 grasses *Pascopyrum smithii* and *Hesperostipa comata* accounted for 60 % of aboveground biomass, accompanied by a C_4 grass, *Bouteloua gracilis*, and forbs (Morgan et al. 2001, 2004a). Six circular plots of 15.5 m^2 were fitted with OTCs which were randomly assigned to maintain an ambient CO_2 level of 360 μmol mol^{-1} or an elevated CO_2 level of 720 μmol mol^{-1} during the growing season.

The Prairie Heating and CO_2 Experiment (PHACE, Fig. 2C) was conducted in a northern mixed prairie near Cheyenne, Wyoming (Parton et al. 2007). Cool-season C_3 grasses, mostly *P. smithii* and *H. comata*, account for 75 % of aboveground biomass. C_4 grasses, almost exclusively *B. gracilis*, and sedges, forbs and small shrubs account for the rest (Morgan et al. 2011). PHACE treatments were factorial combinations of two levels of CO_2; ambient = 385 μmol mol^{-1} and elevated = 600 μmol mol^{-1} achieved using Free-Air CO_2 Enrichment (FACE) technology, and ambient or warmed temperature regimes (+1.5/3.0 day/night), with five replications each. Treatments were randomly assigned to twenty 3.3 m diameter circular plots. Here we consider data from 2006 to 2009 from ambient temperature treatments.

Xeric shrubland. The Nevada Desert FACE Facility (NDFF, Fig. 2D) was located in the Mojave Desert in southern Nevada, USA, on the Nevada National Security Site, formerly the Nevada Test Site. The climate is mid-latitude desert with the lowest annual precipitation of the sites (Table 1) and was thus predicted to be the least responsive to CO_2 enrichment due to severe water limitation. Most precipitation occurs during the winter. The NDFF used the FACE technology with three ambient plots averaging 375 μmol mol^{-1}, three elevated plots averaging 550 μmol mol^{-1} CO_2 and three non-blower control plots without FACE infrastructure; all plots were 23 m in diameter. The NDFF experiment was conducted from April 1997 to June 2007. Experimental plots were in a *Larrea tridentata*—*Ambrosia dumosa* desert scrub community, which consisted of C_3 shrubs, forbs, annuals and a C_4 bunchgrass (Jordan et al. 1999).

Sampling and data analysis

We used previously published data on soil water content, and dominant taxa and total community aboveground biomass to estimate soil water change, community change and community productivity change in each year of CO_2 enrichment. All sites measured vertical profiles of volumetric soil water content (Nelson et al. 2004; Nowak et al. 2004a; Parton et al. 2007; Fay et al. 2009). In LYCOG soil water potential was calculated from soil water content using soil water characteristic curves for each soil (Fay et al. 2012). At the grassland sites, dominant taxa and community biomass were estimated annually from the dry mass of current-year standing aboveground biomass. Aboveground biomass in the xeric shrubland experiment was determined by a single harvest of standing aboveground biomass after 10 years of CO_2 enrichment (Newingham et al. 2013), and responses of dominant species were determined as in Newingham et al. (2014).

The response ratio β (Amthor and Koch 1996; eq. 1) was computed from the dominant taxa biomass, community biomass, and growing season mean volumetric soil water content at enriched (C_E) and ambient (C_A) CO_2 each year.

$$\beta = \frac{(Y_E/Y_A) - 1}{\ln(C_E/C_A)} \qquad (1)$$

The calculation of β standardizes the responses for different levels of CO_2 used among the experiments.

Soil water content β was calculated for the soil depth that best predicted aboveground biomass: 0–135 cm (PCG), 0–40 cm (LYCOG), 70–100 cm (OTC) and 5–25 cm (PHACE). We calculated β in the mesic grassland CO_2 gradients from values of biomass and soil water content at 500 and 390 μL L^{-1} CO_2. These values were estimated from linear regressions of biomass and soil water content against CO_2 for each year. In LYCOG, regressions were fit across all three soils and for each soil separately. The β values were not calculated for the xeric shrubland because there were no responses to CO_2 enrichment in growing season soil water content and no on-going responses in total or dominant taxa aboveground biomass. Correlations among the β for community biomass, soil water content and dominant taxa were tested with linear regression analysis. The magnitude of community biomass β with no community change was estimated from the y-intercept of community biomass β/dominant taxa β regressions.

Results

Precipitation gradient

Mesic grassland. Mean community biomass β was 0.52 in PCG and 1.19 in LYCOG (Fig. 3A), corresponding to

◇ Xeric shrubland   ○ LYCOG - Mesic
● PCG - Mesic   ■ PHACE - Semiarid
□ OTC - Semiarid

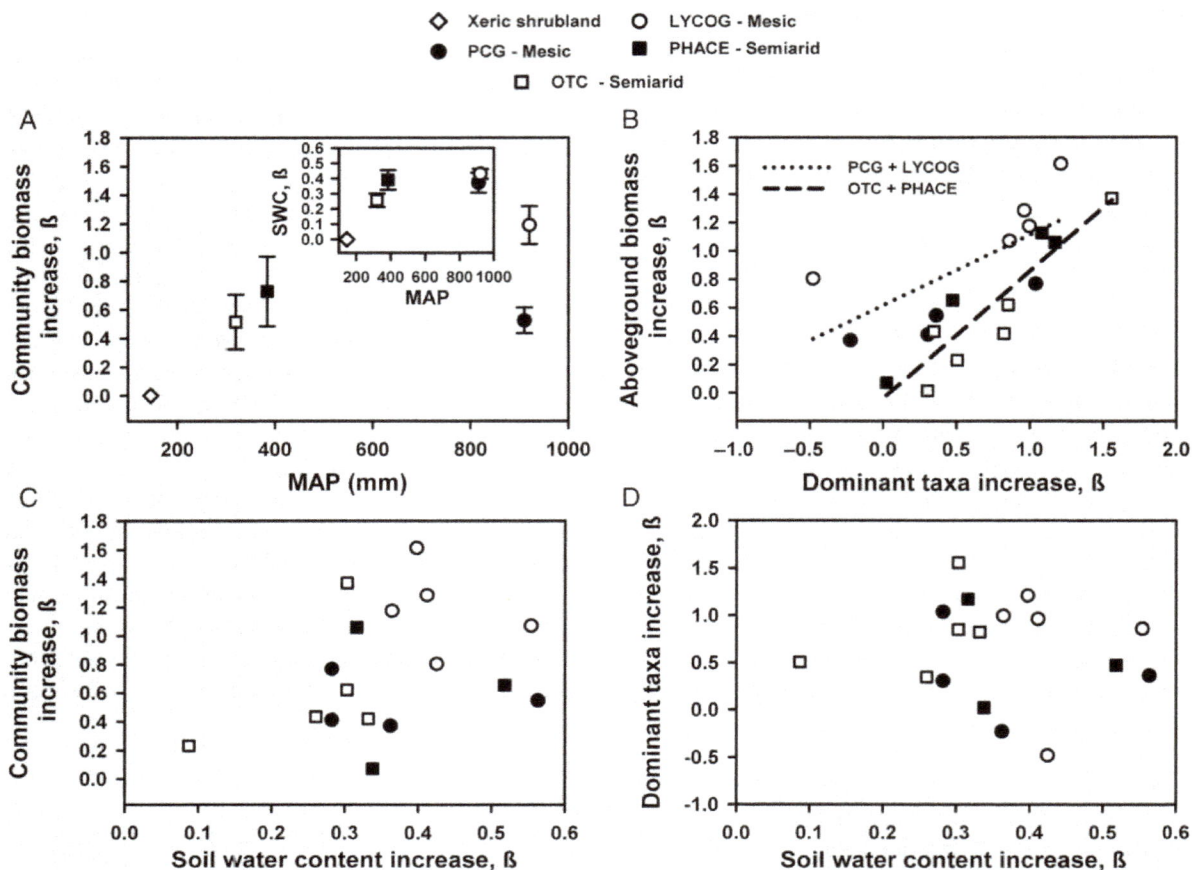

Figure 3. Relationships among community aboveground biomass, MAP, dominant plant taxa biomass and soil water content (SWC) responses to CO_2 enrichment in xeric shrubland, mesic grassland and semi-arid grassland CO_2 enrichment experiments. (A) Community biomass β, mean $\pm$ SE across all years of CO_2 manipulation for each experiment, plotted by site MAP. Inset is the corresponding mean $\pm$ SE soil water content β. (B) Relationship between annual values of β for community biomass and dominant plant taxa, with linear regressions fit separately to mesic grassland and semi-arid grassland experiments. (C and D) Relationships of community biomass β and dominant plant taxa β to soil water content β.

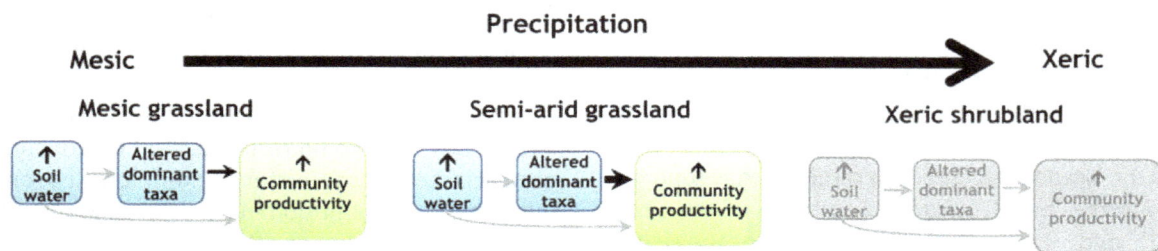

Figure 4. Summary of correlations among changes in soil water, abundance of dominant plant taxa and plant community biomass increases with CO_2 enrichment in mesic grassland, semi-arid grassland and xeric shrubland. Solid arrows denote positive correlations among variables, and thicker lines denote stronger relationships. Grey boxes indicate no significant CO_2 enrichment effects, and grey lines indicate no correlations.

community biomass increases with CO_2 enrichment of 15 and 29 %, respectively. Mean soil water content β was 0.37 for PCG and 0.43 for LYCOG (Fig. 3A, inset), corresponding to 9–10 % increases in soil water content.

Carbon dioxide enrichment altered the abundance of dominant grasses in both mesic grassland CO_2 experiments. In LYCOG, C_4 grass β ranged from −0.9 to 1.2.

In PCG, C_4 grass β ranged from −0.23 to 1.0 (Fig. 3B), corresponding to −20 to 90 % changes in C_4 grass biomass. Across both experiments, community biomass β increased with the C_4 grass β ($R^2 = 0.43$, $P = 0.03$, Fig. 3B). The community biomass β at the y-intercept was 0.6, equating to a 15 % increase in community biomass with CO_2 enrichment in the absence of change in C_4 grass biomass. The values of

β for community and C_4 grass biomass were not correlated with soil water content β ($P > 0.69$, Fig. 3C and D).

Semi-arid grassland. Mean community biomass β was 0.51 in OTC and 0.73 in PHACE (Fig. 3A), corresponding to 31 and 35 % increases, respectively, in community biomass under CO_2 enrichment. Mean soil water content β averaged 0.26 for OTC and 0.39 for PHACE (Fig. 3A, inset), corresponding to 17–18 % increases in soil water content.

Carbon dioxide enrichment increased the abundance of the dominant plant taxa in both semi-arid grassland experiments. C_3 grass β averaged 0.65 in the OTC experiment and 0.30 in the PHACE experiment (Fig. 3B), corresponding to 30–50 % increases with CO_2 enrichment. Community biomass β strongly increased with the C_3 grass β (slope = 0.90, $R^2 = 0.81$, $P = 0.0002$, Fig. 3B), with a slope nearly twice that of the community–C_4 grass β relationship for mesic grassland (slope = 0.50, Figs 3B and 4). The community biomass β at the y-intercept was 0, indicating that there was no community biomass increase with CO_2 enrichment in the absence of change in C_3 grass biomass. As in the mesic grassland, β values for community biomass and C_3 grass biomass were not correlated with soil water content β across the two experiments ($P > 0.59$, Figs 3C and D and 4).

Xeric shrubland. The xeric shrubland responses to CO_2 enrichment were a dramatic departure from those of the mesic and semi-arid grasslands. Carbon dioxide enrichment at the NDFF had no effect on soil water content (Nowak et al. 2004a) or on cumulative total aboveground and belowground biomass (Newingham et al. 2013) and productivity of dominant annual plant taxa (Smith et al. 2014; Fig. 4) following 10 years of CO_2 enrichment. Thus, there was no basis for testing correlations among β values for soil water content, dominant taxa and aboveground biomass in this system.

Soils gradient

The total community biomass increase resulting from CO_2 enrichment varied considerably among soil types in the LYCOG experiment. Mean community biomass β was 1.0 on the fine-textured clay soil and 1.4 on the silty clay and sandy loam soils (Fig. 5A), corresponding to 23–35 % increases in aboveground biomass with CO_2 enrichment. Soil water content mean β was 0.1–0.2 (range 0.03–0.31) in the clay and silty clay soils, increasing to 0.85 (range 0.50–1.04) on the sandy loam soil (Fig. 5A), which also had the highest mean soil water potential (Fig. 5A, inset). These values of β corresponded to 1–26 % increases in soil water content with CO_2 enrichment. Community biomass β was not correlated with soil water content β on any soil

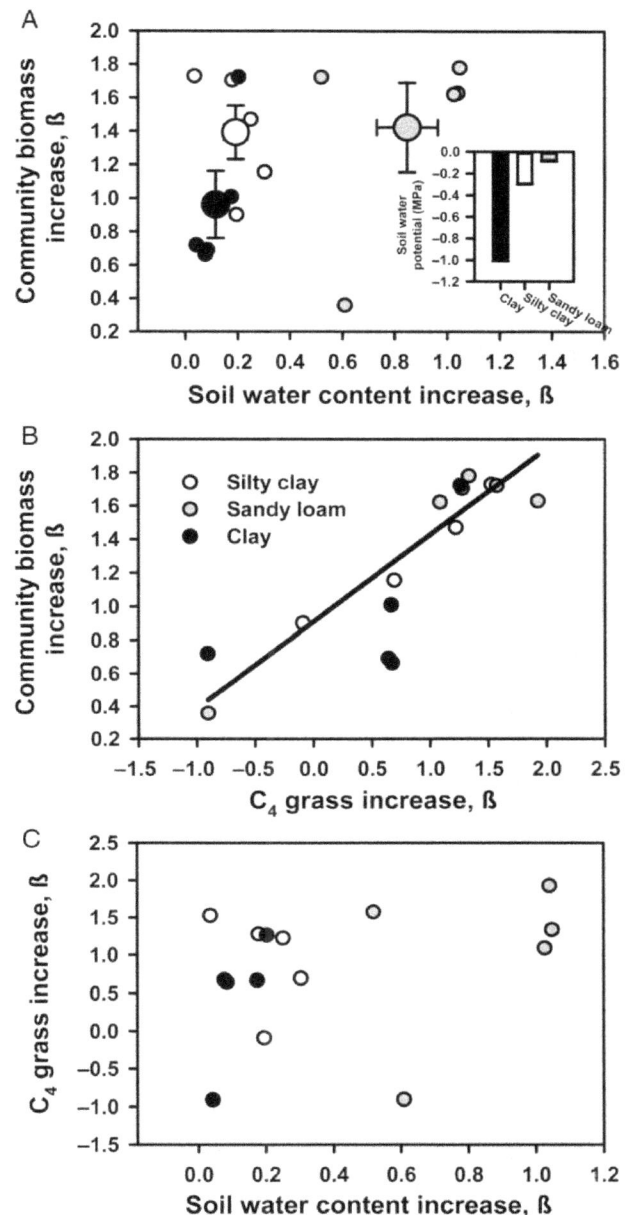

Figure 5. Relationships among community aboveground biomass, dominant plant taxa biomass and soil water content responses to CO_2 enrichment in the three soils in the LYCOG experiment in mesic grassland. (A) Community biomass β, mean ± SE across all years of CO_2 manipulation in relation to soil water content β. Large symbols denote means ± SEs across years for each soil. Inset is the corresponding mean ± SE soil water potential for each soil type. (B) Community biomass β relationship to C_4 grass β, with linear regressions fit to silty clay and sandy loam soils. (C) C_4 grass β relationship to soil water content β.

individually ($0.06 < P < 0.41$, Figs 5A and 6) or across all soils ($P = 0.09$).

The value of β for C_4 grasses ranged from −1.0 to 2.0 across all soils (Fig. 5B), corresponding to −41 to 130 %

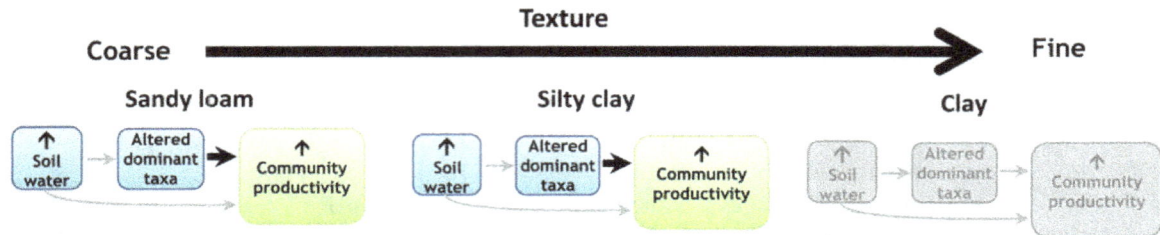

Figure 6. Summary of correlations among changes in soil water, abundance of dominant plant taxa and plant community biomass increases with CO_2 enrichment on three soils in mesic grassland. Solid arrows denote positive correlations among variables, and thicker lines denote stronger relationships. Grey boxes indicate no significant CO_2 enrichment effects, and grey lines indicate no correlations.

changes in C_4 biomass with CO_2 enrichment. The community biomass β increased with C_4 grass β on the silty clay and sandy loam soils (slope = 0.52, $R^2 = 0.90$, $P = <0.0001$). The community biomass β on these two soils was 0.9 at the y-intercept for C_4 grass β, equating to a 2 % increase in community biomass with CO_2 enrichment in the absence of change in C_4 grass biomass. Community biomass β was not correlated with C_4 grass β on the clay soil ($P = 0.31$, Fig. 5B). Dominant grass β was not correlated with soil water content β for any soil ($0.28 < P < 0.58$; Figs 5C and 6).

Discussion

Carbon dioxide enrichment can increase the productivity of communities solely by increasing soil water or by mediating community change. Soil water increases were expected to cause larger community productivity responses with lower precipitation and on coarser-textured soils in the grasslands, but responses were expected to decline sharply in the extreme aridity of the xeric shrubland. Our findings indicate that in the grasslands, yearly soil water content increases never predicted community change or community productivity. Instead, community change caused a larger community productivity response in semi-arid than that in mesic grasslands, and the predicted sharp decline in responsiveness to CO_2 was supported in the xeric shrubland.

The absence of cumulative increases in community productivity with CO_2 enrichment in the xeric shrubland may be explained by the absence of increased soil water and community change. There was no sustained increase in soil water content (Nowak et al. 2004a) because CO_2 enrichment only increased photosynthesis and water-use efficiency and decreased stomatal conductance in occasional wet years (Nowak et al. 2001; Naumburg et al. 2003; Housman et al. 2006; Aranjuelo et al. 2011). There was also no sustained community change because the dominant perennial species were unaffected by CO_2 enrichment (Newingham et al. 2014), and responses were confined to native and exotic annuals which transiently increased with CO_2 enrichment only during high rainfall years (Smith et al. 2000, 2014).

Thus, the mechanisms proposed as the fundamental drivers of productivity increases with CO_2 enrichment in water-limited ecosystems were largely absent in this arid ecosystem.

In the grasslands, the absent correlations between soil water content change, community change and community productivity increases suggests that effects of increased soil water on plant productivity were not well represented by the growing season mean soil moisture response. Increased soil water content may not always translate into greater plant growth because plant growth exhibits a threshold response to soil water (Lambers et al. 2008). Soil water variability, such as an increased duration or severity of soil water deficit, in current or previous years may lower community productivity (Polley et al. 2002; Fay et al. 2003, 2011; Heisler-White et al. 2008; Hovenden et al. 2014; Reichmann and Sala 2014), and the CO_2 effect on soil water during drought periods may be a better predictor of productivity responses than the mean response in soil water over the growing season.

The lack of correlation between yearly soil water change and community responses differs from previous findings. Averaged over multiple years of CO_2 enrichment, increased community productivity occurred in part because of increased soil water in these grassland experiments (Polley et al. 2003; Nelson et al. 2004; Morgan et al. 2011; Fay et al. 2012). The importance of soil water increases averaged over multiple years but not in any given year suggests that predictors of community change or productivity responses at one temporal scale may not apply at other scales in these grasslands (Peters et al. 2004).

Community change was a stronger predictor of community productivity increases in the drier semi-arid grassland compared with the mesic grassland. The weaker relationship to community change in mesic grassland stemmed from 2 years when CO_2 enrichment increased plant productivity but decreased the dominant taxa (Fig. 3B), indicating that productivity of other species in these mesic grassland communities increased with CO_2 enrichment enough to offset decreases in the dominants. In contrast, the dominant taxa never decreased in the semi-arid grassland. Offsetting responses within communities may dampen the productivity response to CO_2 in mesic grassland (Hooper et al. 2005), and both dominant and sub-dominant taxa can contribute to community productivity responses to CO_2 enrichment.

Two other factors may have contributed to the stronger community change effect on community productivity increases in the semi-arid grassland. First, lower MAP and higher mean soil water content increases with CO_2 enrichment may have resulted in release from water limitation for all community members in the semi-arid grassland compared with the mesic grassland. Second, in the mesic grassland, there was no community change effect on community productivity on the clay soil. The strong community change/productivity responses occurred on the two soils with high mean soil water potential (Fay et al. 2012), suggesting that the overall availability of soil water, not the increase in soil water content, determines community and productivity responses to CO_2 enrichment.

In the semi-arid grassland during years with no community change, there was no increase in community biomass, suggesting that yearly increases in community productivity were almost completely explained by community change. Although community change was a weaker predictor of community productivity change in mesic grassland, it still predicted ~80 % of the productivity response. These results suggest that, in years with no community change, community productivity increased by 15–22 % from other potential causes, such as direct effects of CO_2 enrichment on carbon gain or plant water status (Ainsworth and Long 2005).

Conclusions

Community change predicted most of the changes in community productivity in the grasslands and better predicted community productivity responses in the drier grassland and on soils with higher plant availability of soil water in the mesic grassland. The xeric shrubland makes clear that in the absence of increased soil water or community change, increased community productivity is unlikely. The linkage of community change and productivity to yearly changes in soil water with CO_2 enrichment remains unclear, but it may depend more on the temporal variability in soil water than on the size of the increase. Future research needs to isolate and clarify the temporal and spatial mechanisms controlling the linkages among soil water, community change and plant productivity responses to CO_2 enrichment.

Sources of Funding

Mesic grassland experiments: USDA-ARS Climate Change, Soils & Emissions and Pasture, Forage and Rangeland Systems Programs. *Semi-arid grassland experiments*: NSF-TECO (IBN-9524068), NSF (DEB-9708596, 1021559), the NSF LTER (DEB-9350273), USDA-ARS Climate Change, Soils & Emissions Program, USDA-CSREES (2008-35107-18655), the US DOE NICCR (DE-SC0006973). *Xeric shrubland experiments*: US DOE (DE-FG02-03ER63651), NSF (DEB-0212812) and the Nevada Agricultural Experiment Station. Opinions, findings and conclusions or recommendations expressed in this material are those of the authors and do not necessarily reflect the views of the National Science Foundation. USDA is an Equal Opportunity Employer.

Contributions by the Authors

P.A.F. and B.A.N. conceived the manuscript and led the writing, P.A.F. analysed the data and the remaining authors contributed data, edited the manuscript and approved the submission.

Acknowledgements

The authors thank the many technicians, students and collaborators whose fieldwork made these data available and whose preceding research provided the foundation for this paper.

Literature Cited

Ainsworth EA, Long SP. 2005. What have we learned from 15 years of free-air CO_2 enrichment (FACE)? A meta-analytic review of the responses of photosynthesis, canopy properties and plant production to rising CO_2. New Phytologist **165**:351–372.

Amthor JS, Koch GW. 1996. Biota growth factor β: stimulation of terrestrial ecosystem net primary production by elevated atmospheric CO_2. In: Koch GW, Roy J, eds. Carbon dioxide and terrestrial ecosystems. San Diego, CA: Academic Press, 399–414.

Aranjuelo I, Ebbets AL, Evans RD, Tissue DT, Nogués S, van Gestel N, Payton P, Ebbert V, Adams WW III, Nowak RS, Smith SD. 2011. Maintenance of C sinks sustains enhanced C assimilation during long-term exposure to elevated [CO_2] in Mojave Desert shrubs. Oecologia **167**:339–354.

Bradley KL, Pregitzer KS. 2007. Ecosystem assembly and terrestrial carbon balance under elevated CO_2. Trends in Ecology and Evolution **22**:538–547.

Fay PA, Carlisle JD, Knapp AK, Blair JM, Collins SL. 2003. Productivity responses to altered rainfall patterns in a C_4-dominated grassland. Oecologia **137**:245–251.

Fay PA, Kelley AM, Procter AC, Jin VL, Hui D, Jackson RB, Johnson HB, Polley HW. 2009. Primary productivity and water balance of grassland vegetation on three soils in a continuous CO_2 gradient: Initial results from the Lysimeter CO_2 Gradient Experiment. Ecosystems **12**:699–714.

Fay PA, Blair JM, Smith MD, Nippert JB, Carlisle JD, Knapp AK. 2011. Relative effects of precipitation variability and warming on tallgrass prairie ecosystem function. Biogeosciences **8**:3053–3068.

Fay PA, Jin VL, Way DA, Potter KN, Gill RA, Jackson RB, Polley HW. 2012. Soil-mediated effects of subambient to increased carbon dioxide on grassland productivity. *Nature Climate Change* **2**: 742–746.

Gornish ES, Tylianakis JM. 2013. Community shifts under climate change: Mechanisms at multiple scales. *American Journal of Botany* **100**:1422–1434.

Heisler-White JL, Knapp AK, Kelly EF. 2008. Increasing precipitation event size increases aboveground net primary productivity in a semi-arid grassland. *Oecologia* **158**:129–140.

Hooper DU, Chapin FS, Ewel JJ, Hector A, Inchausti P, Lavorel S, Lawton JH, Lodge DM, Loreau M, Naeem S, Schmid B, Setälä H, Symstad AJ, Vandermeer J, Wardle DA. 2005. Effects of biodiversity on ecosystem functioning: a consensus of current knowledge. *Ecological Monographs* **75**:3–35.

Housman DC, Naumburg E, Huxman TE, Charlet TN, Nowak RS, Smith SD. 2006. Increases in desert shrub productivity under elevated carbon dioxide vary with water availability. *Ecosystems* **9**: 374–385.

Hovenden MJ, Newton PCD, Wills KE. 2014. Seasonal not annual rainfall determines grassland biomass response to carbon dioxide. *Nature* **511**:583–586.

Johnson HB, Polley HW, Whitis RP. 2000. Elongated chambers for field studies across atmospheric CO_2 gradients. *Functional Ecology* **14**:388–396.

Jordan DN, Zitzer SF, Hendrey GR, Lewin KF, Nagy J, Nowak RS, Smith SD, Coleman JS, Seemann JR. 1999. Biotic, abiotic and performance aspects of the Nevada Desert Free-Air CO_2 Enrichment (FACE) facility. *Global Change Biology* **5**:659–668.

Körner C. 2006. Plant CO_2 responses: an issue of definition, time and resource supply. *New Phytologist* **172**:393–411.

Lambers H, Chapin FS, Chapin FS, Pons TL. 2008. *Plant physiological ecology*, 2nd edn. New York: Springer.

Morgan JA, LeCain DR, Mosier AR, Milchunas DG. 2001. Elevated CO_2 enhances water relations and productivity and affects gas exchange in C_3 and C_4 grasses of the Colorado shortgrass steppe. *Global Change Biology* **7**:451–466.

Morgan JA, Mosier AR, Milchunas DG, Lecain DR, Nelson JA, Parton WJ. 2004a. CO_2 enhances productivity, alters species composition, and reduces digestibility of shortgrass steppe vegetation. *Ecological Applications* **14**:208–219.

Morgan JA, Pataki DE, Körner C, Clark H, Del Grosso SJ, Grünzweig JM, Knapp AK, Mosier AR, Newton PCD, Niklaus PA, Nippert JB, Nowak RS, Parton WJ, Polley HW, Shaw MR. 2004b. Water relations in grassland and desert ecosystems exposed to elevated atmospheric CO_2. *Oecologia* **140**:11–25.

Morgan JA, Lecain DR, Pendall E, Blumenthal DM, Kimball BA, Carrillo Y, Williams DG, Heisler-White J, Dijkstra FA, West M. 2011. C_4 grasses prosper as carbon dioxide eliminates desiccation in warmed semi-arid grassland. *Nature* **476**:202–205.

Naumburg E, Housman DC, Huxman TE, Charlet TN, Loik ME, Smith SD. 2003. Photosynthetic responses of Mojave Desert shrubs to free air CO_2 enrichment are greatest during wet years. *Global Change Biology* **9**:276–285.

Nelson JA, Morgan JA, Lecain DR, Mosier AR, Milchunas DG, Parton WA. 2004. Elevated CO_2 increases soil moisture and enhances plant water relations in a long-term field study in semi-arid shortgrass steppe of Colorado. *Plant and Soil* **259**: 169–179.

Newingham BA, Vanier CH, Charlet TN, Ogle K, Smith SD, Nowak RS. 2013. No cumulative effect of 10 years of elevated [CO_2] on per-

ennial plant biomass components in the Mojave Desert. *Global Change Biology* **19**:2168–2181.

Newingham BA, Vanier CH, Kelly LJ, Charlet TN, Smith SD. 2014. Does a decade of elevated [CO_2] affect a desert perennial plant community? *New Phytologist* **201**:498–504.

Nowak RS, DeFalco LA, Wilcox CS, Jordan DN, Coleman JS, Seemann JR, Smith SD. 2001. Leaf conductance decreased under free-air CO_2 enrichment (FACE) for three perennials in the Nevada desert. *New Phytologist* **150**:449–458.

Nowak RS, Zitzer SF, Babcock D, Smith-Longozo V, Charlet TN, Coleman JS, Seemann JR, Smith SD. 2004a. Elevated atmospheric CO_2 does not conserve soil water in the Mojave Desert. *Ecology* **85**: 93–99.

Nowak RS, Ellsworth DS, Smith SD. 2004b. Functional responses of plants to elevated atmospheric CO_2—do photosynthetic and productivity data from FACE experiments support early predictions? *New Phytologist* **162**:253–280.

Owensby CE, Ham JM, Knapp A, Rice CW, Coyne PI, Auen LM. 1996. Ecosystem-level responses of tallgrass prairie to elevated CO_2. In: Koch GW, Mooney HA, eds. *Carbon dioxide and terrestrial ecosystems*. San Diego, CA: Academic Press, 147–162.

Parton WJ, Morgan JA, Wang G, Del Grosso S. 2007. Projected ecosystem impact of the Prairie heating and CO_2 enrichment experiment. *New Phytologist* **174**:823–834.

Peters DPC, Pielke RA, Bestelmeyer BT, Allen CD, Munson-McGee S, Havstad KM. 2004. Cross-scale interactions, nonlinearities, and forecasting catastrophic events. *Proceedings of the National Academy of Sciences of the USA* **101**:15130–15135.

Polley HW, Johnson HB, Derner JD. 2002. Soil- and plant-water dynamics in a C_3/C_4 grassland exposed to a subambient to superambient CO_2 gradient. *Global Change Biology* **8**:1118–1129.

Polley HW, Johnson HB, Derner JD. 2003. Increasing CO_2 from subambient to superambient concentrations alters species composition and increases above-ground biomass in a C_3/C_4 grassland. *New Phytologist* **160**:319–327.

Polley HW, Johnson HB, Fay PA, Sanabria J. 2008. Initial response of evapotranspiration from tallgrass prairie vegetation to CO_2 at subambient to elevated concentrations. *Functional Ecology* **22**: 163–171.

Polley HW, Morgan JA, Fay PA. 2011. Application of a conceptual framework to interpret variability in rangeland responses to atmospheric CO_2 enrichment. *The Journal of Agricultural Science* **149**:1–14.

Polley HW, Jin VL, Fay PA. 2012a. CO_2-caused change in plant species composition rivals the shift in vegetation between mid-grass and tallgrass prairies. *Global Change Biology* **18**: 700–710.

Polley HW, Jin VL, Fay PA. 2012b. Feedback from plant species change amplifies CO_2 enhancement of grassland productivity. *Global Change Biology* **18**:2813–2823.

Reichmann LG, Sala OE. 2014. Differential sensitivities of grassland structural components to changes in precipitation mediate productivity response in a desert ecosystem. *Functional Ecology* **28**: 1292–1298.

Smith MD, Knapp AK. 2003. Dominant species maintain ecosystem function with non-random species loss. *Ecology Letters* **6**:509–517.

Smith MD, Knapp AK, Collins SL. 2009. A framework for assessing ecosystem dynamics in response to chronic resource alterations induced by global change. *Ecology* **90**:3279–3289.

Smith SD, Huxman TE, Zitzer SF, Charlet TN, Housman DC, Coleman JS, Fenstermaker LK, Seemann JR, Nowak RS. 2000. Elevated CO_2 increases productivity and invasive species success in an arid ecosystem. *Nature* **408**:79–82.

Smith SD, Charlet TN, Zitzer SF, Abella SR, Vanier CH, Huxman TE. 2014. Long-term response of a Mojave Desert winter annual plant community to a whole-ecosystem atmospheric CO_2 manipulation (FACE). *Global Change Biology* **20**: 879–892.

11

Root contact responses and the positive relationship between intraspecific diversity and ecosystem productivity

Lixue Yang[1,2], Ragan M. Callaway[1]* and Daniel Z. Atwater[3]

[1] Division of Biological Sciences and the Institute on Ecosystems, The University of Montana, Missoula, MT 59812, USA
[2] School of Forestry, Northeast Forestry University, Harbin 150040, China
[3] Department of Plant Pathology, Physiology, and Weed Science, Virginia Tech University, Blacksburg, VA 24061, USA

Associate Editor: James F. Cahill

Abstract. High species and functional group richness often has positive effects on ecosystem function including increasing productivity. Recently, intraspecific diversity has been found to have similar effects, but because traits vary far less within a species than among species we have a much poorer understanding of the mechanisms by which intraspecific diversity affects ecosystem function. We explored the potential for identity recognition among the roots of different *Pseudoroegneria spicata* accessions to contribute to previously demonstrated overyielding in plots with high intraspecific richness of this species relative to monocultures. First, we found that when plants from different populations were planted together in pots the total biomass yield was 30 % more than in pots with two plants from the same population. Second, we found that the elongation rates of roots of *Pseudoroegneria* plants decreased more after contact with roots from another plant from the same population than after contact with roots from a plant from a different population. These results suggest the possibility of some form of detection and avoidance mechanism among more closely related *Pseudoroegneria* plants. If decreased growth after contact results in reduced root overlap, and reduced root overlap corresponds with reduced growth and productivity, then variation in detection and avoidance among related and unrelated accessions may contribute to how ecotypic diversity in *Pseudoroegneria* increases productivity.

Keywords: Ecosystem productivity; identity recognition; intraspecific genetic diversity; *Pseudoroegneria spicata*; root interactions.

Introduction

Biodiversity and ecosystem functioning are often positively linked (Balvanera *et al.* 2006; Cardinale *et al.* 2007, 2011). Among the most prominent explanations for why diversity improves ecosystem function is that different species complement or facilitate each other in chemical, spatial or temporal resource use (Eisenhauer

* Corresponding author's e-mail address: ray.callaway@mso.umt.edu

2012), and that more species-rich communities experience less suppression by pathogenic fungi than monocultures (Maron et al. 2011; Schnitzer et al. 2011; Kulmatiski et al. 2012). Furthermore, recent studies have shown that the root growth of some species increases in species mixtures compared with monocultures causing belowground overyielding, potentially due to reduced effects of plant pathogens in diverse assemblages (Mommer et al. 2010; de Kroon et al. 2012).

Most studies have focussed on the roles of species and functional group diversity on ecosystem functioning; however, a few recent studies have found that intraspecific diversity can have similar effects on ecosystem function (Crutsinger et al. 2006; Fridley and Grime 2010; Cook-Patton et al. 2011; Crawford and Rudgers 2012; Schöb et al. 2015). Similarly, Atwater and Callaway (2015) found that intraspecific diversity of *Pseudoroegneria spicata* (Pursh) Á. Löve increased productivity to a similar degree as generally reported in the literature for species diversity. This overyielding was primarily due to complementary interactions among accessions; however, there was no evidence that overyielding was related to resource depletion, a commonly cited mechanism for overyielding in species-diverse communities. Thus the mechanisms that lead to overyielding by diverse assemblages of *Pseudoroegneria* remain poorly understood.

A possible, but to our knowledge unexplored, mechanism for variation in productivity in ecotypic monocultures is different degrees of overlap among the roots of individual plants (see Schenk et al. 1999; Novoplansky 2009). Spatially segregated root systems have been documented among conspecifics at the scale of whole root systems and individual roots (Schenk et al. 1999), and segregation appears to be affected in some cases by the changes in root growth of one plant in response to contact with the roots of another plant, or identity recognition (Mahall and Callaway 1992; Cahill et al. 2010; Cahill and McNickle 2011). For example, root segregation may provide competitive advantages for resources or space for some individuals over others, functioning effectively as the establishment of territories (Schenk et al. 1999). It is unknown how spatial root segregation might affect the growth of individual plants, and how this ramifies to community productivity. Furthermore, root responses to other roots can be highly complex, including both decreased and increased growth rates after contact.

There is substantial evidence for different forms of identity recognition among the roots of individual plants, among individuals within populations, among populations and among species (Mahall and Callaway 1991, 1992; Krannitz and Caldwell 1995; Schenk et al. 1999; Gersani et al. 2001; Callaway 2002; Falik et al. 2003; Semchenko et al. 2007; Metlen et al. 2009; Novoplansky

2009; Bhatt et al. 2011). Mahall and Callaway (1996) explored self-non-self recognition among different populations of *Ambrosia dumosa* and found that the roots of plants demonstrated sharp declines in growth after contact with the roots of another plant from the same population. This decline did not occur when roots contacted the roots of an individual from a distant population. Others have reported various forms of identity recognition and subsequent change in root behaviour (de Kroon and Visser 2003; Donohue 2003; Cheplick and Kane 2004; Gruntman and Novoplansky 2004; Dudley and File 2007) and many of these studies have explored the potential for such recognition to affect interactions among and within species (Falik et al. 2006; see review by Novoplansky 2009). Recently, Semchenko et al. (2014) found that root exudates can communicate information about genetic relatedness, population origin and species identity. However, to our knowledge, no studies have explored how the responses of roots to other roots might contribute to how ecotypic diversity increases ecosystem function.

Because of this lack of research, and the wide range of ways that roots respond to each other and to resources, it is hard to predict whether contact and avoidance (i.e. territorial-like responses) or increased growth response to contact (e.g. Gersani et al. 2001; Novoplansky 2009; Fang et al. 2013) might increase overall productivity (see Milla et al. 2009). It is reasonable to predict that decreased root overlap might result in the improved growth of individuals in a community but not greater productivity overall. On the other hand, high root overlap has the potential to increase productivity via more complete root exploration of the soil and resource uptake (de Kroon and Visser 2003; de Kroon et al. 2012). Thus exploring correlations between root–root responses and productivity have the potential to resolve important general questions about the ecosystem ramifications of root behaviour.

Here we explored potential mechanisms for the positive relationship between ecotypic diversity of *P. spicata* and productivity reported by Atwater and Callaway (2015) in which plots with high intraspecific richness overyielded relative to monocultures. Specifically, we tested the hypotheses that (i) the total biomass of two interacting plants from different populations would be more than that for two plants from the same population and (ii) the growth rates of *Pseudoroegneria* roots would decrease more when contacting roots from other individuals from the same populations than when contacting roots from individuals from other populations.

Methods

We used the same populations of *P. spicata* studied by Atwater and Callaway (2015). We obtained seeds of

Pseudoroegneria from 12 sites throughout western North America, and with one exception, seeds were field collected in Montana or acquired from true-bred lines managed by the USDA Plant Germplasm Introduction and Testing Research Station in Pullman, Washington, USA. The one exception was a high-yielding wild-selected cultivar from south-eastern Washington, 'Goldar', which we purchased because of problems with seed viability of some of the naturally collected accessions.

To compare intra-genotype and inter-genotype interactions in control conditions, in May 2012 we sowed seeds from each of the 11 genotypes into 200 mL rocket pots to establish treatments in which plants were grown alone or with another individual. Six individuals from each population (accession) were grown alone (total $n = 66$), eight individuals were grown with another individual from the same population (total $n = 88$) and one individual from each population was grown with one other plant from each of the other populations (total $n = 55$). The pots were re-organized randomly on the bench once per week and watered every 2 days. Seedlings were harvested in August 2013, dried at 60 °C and total biomass was weighed. We analysed the effects of treatment on root mass and total biomass using linear mixed models with treatment (grown alone, with a plant from the same population or with a plant from another population) as a fixed effect and the identity of each interacting plant (i.e. its population) as random effects. We specified alone-vs-same population and alone-vs-other population as contrasts, and also calculated differences between biomass with same- and other-population as *post hoc* contrasts.

In a second experiment we monitored the growth rates of roots of individual *Pseudoroegneria* plants when they grew into the rhizospheres of other plants and made contact with roots of the same accession or different accession in the same chambers used and described in Mahall and Callaway (1991). These chambers were 20.5 × 12.5 × 2 cm, inside dimensions and were filled with 30 grit silica sand. Chambers were oriented at a 45° angle so that geotropic roots would grow down against Plexiglas viewing windows that could be covered and uncovered with shutters that excluded light. These chambers were placed in a greenhouse and the experiment ran from early February to late April. We initiated the experiment with 10 chambers containing intra-populations pairs and 14 chambers containing inter-population pairs but after mortality this replication was reduced to $n = 8$ and $n = 12$. Our inter-population pairs did not include all possible combinations, and these were chosen randomly with the condition that no pair combination was repeated. Two times during the experiment the chambers were saturated with a solution of

1.2 g L^{-1} of water-soluble fertilizer (15-2-20, The Scotts Company, Marysville, OH, USA). Sand in the chambers was kept continuously moist. After plants were established in 'target' and 'test' chambers, these chambers were connected so that roots of a test plant would grow into the rhizosphere and roots of a target plant. Elongation rates of all test-plant roots were visible and measurable through the Plexiglas along which they grew. Every 2 days the locations of root tips of the test plants were measured and elongation rates were measured as the distance travelled over those 2 days. Most roots of the test plants ultimately made contact with the roots of target plants and we calibrated our comparisons of growth rates of intra-population and inter-population pairs by the day they made contact. This calibration allowed us to compare growth rates prior to contact to growth rates after contact. We analysed root growth using linear mixed models with treatment (growth with a plant from the same population or from another population) and sampling date as fixed effects and focal plant population, target plant population as random effects. This design enabled us to treat daily measurements of repeated measures and account for effects of target and test plant (Faraway 2005). Because a different combination of target and test plants was used in each root chamber, the effects of chamber were redundant with effects of target and test plant and were not included in the final model. We used rate of root growth over 2 days from each time point as the dependent variable. All analyses were done using package 'lme4' in R version 3.1.2 (R Core Team 2012; Bates *et al.* 2014). Parameter significance was estimated using Satterthwaite approximation with package 'lmerTest' (Kunetsova *et al.* 2013).

Results

When *Pseudoroegneria* individuals from the same population were planted together the total biomass yield in pots (both plants combined) was not greater than that of plants grown alone (Fig. 1). In other words, intra-population competition suppressed the growth of the two individuals to the point where total yield was the same regardless of the number of plants in per pot ($t_{same-vs-control} = 0.014$; df = 147.9; $P = 0.989$). In contrast, when plants from different populations were planted together the total biomass yield of pots was 30 % more than in pots with two plants from the same population (*post hoc*; $t_{same-vs-other} = 4.238$; df = 154.6; $P < 0.0001$) and 27 % more than in pots with individual *Pseudoroegneria* plants ($t_{other-vs-control} = 4.133$; df = 155.0; $P < 0.0001$). We found a similar but slightly stronger pattern for root biomass yield with a 37 % increase in pots with two plants from different populations

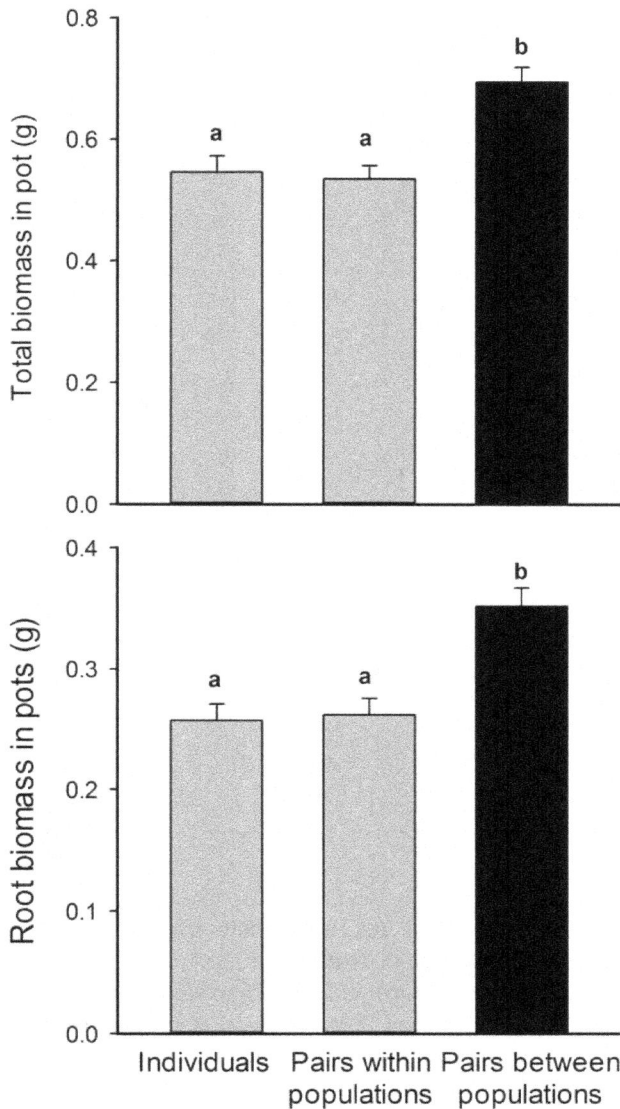

Figure 1. Total and root biomass in pots with either a P. spicata plant grown alone, grown with another plant of the same population or with a plant from a different population. Means that share a letter are not significantly different and error bars represent 1 SE.

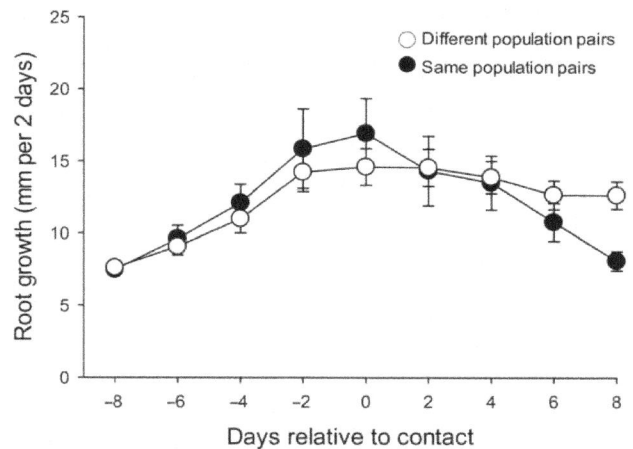

Figure 2. Rates of elongation of roots of P. spicata plants grown towards and making contact with roots of other P. spicata plants from either the same population or a different population. Elongation rates were standardized in time by aligning at Day 0 the days on which their contact with a target root was recorded (see Mahall and Callaway 1996). Error bars represent 1 SE.

relative to pots with two plants from the same population (*post hoc*; $t_{same-vs-other} = 4.228$; df $= 154.5$; $P < 0.0001$) and a 34 % increase relative to pots with one plant ($t_{other-vs-control} = 4.566$; df $= 155.0$; $P < 0.0001$). As for biomass yield, total root yield was not different when one plant was grown alone versus when two plants of the same population grew together ($t_{same-vs-control} = 0.526$; df $= 148.3$; $P = 0.599$). When we allowed a focal accession × treatment interaction in the model, this did not improve fit for total mass or root mass (log-likelihood comparison: $P = 0.757$, $P = 0.999$), and the random effects of focal accession were estimated to be very low (at least five orders of magnitude smaller than the

residual variance). In sum, intra- and inter-population competition suppressed the growth of individual plants, but the total yield of the plants in inter-population combinations was less suppressed than it was in intraspecific combinations.

In the root chamber experiment, root growth rates in both treatments increased steadily until root contact, at which point they either stabilized or significantly slowed, depending on whether the competitors were from the same or different populations (Fig. 2; Table 1). At the days of contact and for 2 days prior to contact, plants with same-population neighbours grew 20 % faster than plants with neighbours from different populations ($P < 0.0043$). After contact the roots of test plants growing with a target plant from the same population rapidly decreased in growth rate. Differences in growth rates between same-population and different-population target plants were statistically significant 6 and 8 days after contact ($P < 0.0418$). By 8 days after contact, test plants growing with neighbours from the same population were growing 36 % slower than the roots of test plants making contact with a plant from a different population (Fig. 2; Table 1).

Discussion

We found that apparent relatedness (from the same population) of *Pseudoroegneria* plants affected root growth both before and after initial root contact. Initial root growth was faster in target plants growing with a conspecific competitor from the same population, compared with those competing with a member of a different population. Immediately after initial root contact this

Table 1. Results of linear mixed model of rate of root growth (mm per 2 days) against days since initial contact and treatment (grown with same population or with other population), with target and focal plant identity as random effects. A positive value of *B* for treatment effects means that plants with same-population neighbours grew faster than plants with other-population neighbours. Eight days prior to contact was used as the reference category because treatment effects had not yet begun to appear.

	B	df	T	P
Day − 6	1.230 ± 0.610	665.8	2.018	0.0440
Day − 4	2.896 ± 0.610	665.8	4.749	<0.0001
Day − 2	6.041 ± 0.610	665.8	9.909	<0.0001
Day of contact	6.630 ± 0.610	665.8	10.875	<0.0001
Day + 2	6.659 ± 0.610	665.8	10.921	<0.0001
Day + 4	5.954 ± 0.610	665.8	9.766	<0.0001
Day + 6	5.191 ± 0.610	665.8	8.515	<0.0001
Day + 8	4.837 ± 0.610	665.8	7.933	<0.0001
Treatment	−0.274 ± 0.716	328.0	−0.382	0.7025
Treatment × Day − 6	0.676 ± 0.952	665.8	0.710	0.4780
Treatment × Day − 4	1.682 ± 0.952	665.8	1.768	0.0776
Treatment × Day − 2	2.743 ± 0.952	665.8	2.882	0.0041
Treatment × Day 0	2.729 ± 0.952	665.8	2.867	0.0043
Treatment × Day + 2	0.023 ± 0.952	665.8	0.024	0.9811
Treatment × Day + 4	−0.129 ± 0.952	665.8	−0.136	0.8920
Treatment × Day + 6	−1.941 ± 0.952	665.8	−2.039	0.0418
Treatment × Day + 8	−4.462 ± 0.952	665.8	−4.688	<0.0001

pattern reversed, with rates of root growth sharply falling off in plants interacting with a plant from the same population. This suggests the possibility of some form of contact detection and avoidance among *Pseudoroegneria* plants that depends to some degree on relatedness. In a similar study, Krannitz and Caldwell (1995) measured elongation rates of *Pseudoroegneria* roots as they encountered conspecific roots, roots of the closely related *Agropyron desertorum* (*Pseudoroegneria* used to be classified as *Agropyron*) or roots of the unrelated *Artemisia tridentata*. Unlike our results, they did not find decreased growth after contact for conspecific interactions, but they did find that *Pseudoroegneria* roots sharply decreased elongation after contact with *Agropyron* roots and that this effect differed among *Pseudoroegneria* genotypes. *Agropyron* roots were not affected by *Pseudoroegneria* roots. These results were corroborated by Huber-Sannwald et al. (1996), who found that *Pseudoroegneria* appeared to recognize and respond differently to competition with conspecifics versus competition with *Agropyron*.

Root interactions have been widely discussed as potential drivers of the relationship between plant *species* diversity and ecosystem functioning (Wilson and Tilman 2002; Cahill 2003; Rajaniemi 2003; Rajaniemi et al. 2003; Mommer et al.

2010; de Kroon et al. 2012). But there are two conflicting ideas for how root interactions might function in this context. By far the most cited mechanism is that of niche partitioning; the idea that different species more fully occupy belowground niche space, more completely utilizing resources and thus increasing productivity. However, Mommer et al. (2010) and de Kroon et al. (2012) provided experimental evidence for an alternative mechanism. They found that the roots of grassland species overyielded in the presence of the roots of other species, substantially increasing root density without any evidence for spatial partitioning. Furthermore, this overyielding was connected to more beneficial soil biota in diversity species mixtures, linking their results to a body of other recent work demonstrating how soil biota have powerful overall effects on the species diversity–productivity relationship (Maron et al. 2011; Schnitzer et al. 2011). Our results, but for intraspecific diversity, are broadly supportive of the findings reported by de Kroon et al. (2012), in that we found that greater root overlap and root production among different accessions of *Pseudoroegneria* correlated with greater productivity in the field (Atwater and Callaway 2015) and in pots by mixes of accessions (Fig. 1). Put another way, the decrease in root growth after contact and the lower root growth demonstrated in intra-accession interactions corresponded with lower productivity in low accession-diversity plots in the field (Atwater and Callaway 2015). However, intraspecific relatedness in general can correspond with a wide range of ecosystem function. For example, Biernaskie (2011) found that differences in yield among mixes of identical genotypes of *Ipomea hederacea* was greater than that of mixed genotypes, suggesting that productivity might be higher for related individuals. However, to our knowledge, our study is the first to connect root–root interactions to intraspecific diversity–ecosystem function relationships. If increased root overlap increases the overall exploration of soil space and resource acquisition, then our results suggest that mixtures of population-accession might increase resource acquisition and growth due to increased root overlap.

In a similar, but interspecific example, Padilla *et al.* (2013) found that monocultures of *Festuca rubra* had higher root densities and a faster rate of soil nitrate depletion than monocultures of *Plantago lanceolata*, which indicated that the former was a superior competitor for nutrients. However, in experiments they found that *Festuca* was an inferior competitor to *Plantago*. In these experiments *Plantago* overyielded in root growth. They argued that that competitive superiority occurred through root growth stimulation by a competitor prior to nutrient depletion instead of superior ability to deplete nutrients. Inhibition after contact may also function as detection and avoidance mechanisms that might reduce competition among established and closely related plants, a form of territoriality (Schenk *et al.* 1999). This may explain the initially rapid root proliferation of *Pseudoroegneria* growing with neighbours from the same population. If more closely related neighbours are stronger competitors (Cahill *et al.* 2008), and if this pattern holds within species as well as among species (as in Cheplick and Kane 2004), increased root proliferation in the presence of members of the same population could be important for establishing competitive superiority. In our study, rapid initial root proliferation could be a form of territorial expansion. However, a dramatic reduction in proliferation following root contact suggests that there is a cost to further proliferation into territory already occupied by a related competitor. It is not immediately clear why this should be the case, or why we observed a different pattern in plants growing with a less related neighbour.

It is important to note that we used accessions of *Pseudoroegneria* (or potentially ecotypes) from across a very large portion of the regional distribution of the species. The seeds of each accession were pooled, and not single seed descent families. Thus we do not yet know whether enough genetic variation exists within populations to create important effects on overyielding in natural populations. However, by using such a wide breadth of *Pseudoroegneria* accessions we have explored the *potential* of intraspecific diversity–ecosystem function relationships and provided a working hypothesis for a mechanism.

Our results add to the growing body of evidence for how roots may respond to contact with other roots individual plants, among individuals within populations, among populations and among species (Mahall and Callaway 1991, 1992; Krannitz and Caldwell 1995; Schenk *et al.* 1999; Gersani *et al.* 2001; Callaway 2002; Falik *et al.* 2003; Semchenko *et al.* 2007; Metlen *et al.* 2009; Novoplansky 2009). However, the mechanisms by which closely related species might respond to each other are not known. Badri *et al.* (2012) found that more defence and stress-related proteins were released from roots when a specific control genotype of *Arabidopsis* was grown alone than when it was co-cultured with another homozygous individual or with an unrelated plant. They pointed out that their results suggested that plants can detect and respond to ecotypic variation in neighbours. In contrast, Gruntman and Novoplansky (2004) reported that self-/non-self-discrimination among *Buchloe dactyloides* plants was mediated by physiological coordination among roots developing on the same plant (also see Mahall and Callaway 1996). Furthermore, we do not know whether differences in root growth between different accessions are related to between-kin aspects of altruism or simply increased competition between non-related individuals.

In summary, our results link a growing body of literature on the capacity of roots from different species and different accessions to detect and respond to each other to how diverse mixtures of species or accessions enhance ecosystem function. We do not know whether similar mechanisms might also operate in species-diverse systems, but it warrants investigation.

Sources of Funding

R.M.C. thanks the Montana Institute on Ecosystems and NSF EPSCoR Track-1 EPS-1101342 (INSTEP 3) for support.

Contributions by the Authors

D.Z.A. developed the conceptual background of intraspecific diversity and ecosystem functions, R.M.C. developed the conceptual background of root–root interactions. R.M.C. and L.Y. designed the experiment, L.Y. carried out the experiment and all authors wrote the paper.

Acknowledgements

We thank the Plant Germplasm Introduction and Testing

Research Station, and Curator Vicki Bradley, in Pullman, WA, for supplying *Pseudoroegneria* seeds.

Literature Cited

Atwater DZ, Callaway RM. 2015. Testing the mechanisms of diversity-dependent overyielding in a grass species. *Ecology*, doi:10.1890/15-0889.1.

Badri DV, De-la-Peña C, Lei Z, Manter DK, Chaparro JM, Guimarães RL, Sumner LW, Vivanco JM. 2012. Root secreted metabolites and proteins are involved in the early events of plant-plant recognition prior to competition. *PLoS ONE* 7:e46640.

Balvanera P, Pfisterer AB, Buchmann N, He JS, Nakashizuka T, Raffaelli D, Schmid B. 2006. Quantifying the evidence for biodiversity effects on ecosystem functioning and services. *Ecology Letters* 9:1146–1156.

Bates D, Maechler M, Bolker B, Walker S. 2014. lme4: linear mixed-effects models using Eigen and S4. R package version 1.1-7. http://CRAN.R-project.org/package=lme4 (16 April 2015).

Bhatt MV, Khandelwal A, Dudley SA. 2011. Kin recognition, not competitive interactions, predicts root allocation in young *Cakile edentula* seedling pairs. *New Phytologist* 189:1135–1142.

Biernaskie JM. 2011. Evidence for competition and cooperation among climbing plants. *Proceedings of the Royal Society B: Biological Sciences* 278:1989–1996.

Cahill JF. 2003. Lack of relationship between below-ground competition and allocation to roots in 10 grassland species. *Journal of Ecology* 91:532–540.

Cahill JF Jr, McNickle GG. 2011. The behavioral ecology of nutrient foraging by plants. *Annual Review of Ecology, Evolution, and Systematics* 42:289–311.

Cahill JF, Kembel SW, Lamb EG, Keddy PA. 2008. Does phylogenetic relatedness influence the strength of competition among vascular plants? *Perspectives in Plant Ecology, Evolution and Systematics* 10:41–50.

Cahill JF Jr, McNickle GG, Haag JJ, Lamb EG, Nyanumba SM, St. Clair CC. 2010. Plants integrate information about nutrients and neighbors. *Science* 328:1657.

Callaway RM. 2002. The detection of neighbors by plants. *Trends in Ecology and Evolution* 17:104–105.

Cardinale BJ, Wright JP, Cadotte MW, Carroll IT, Hector A, Srivastava DS, Loreau M, Weis JJ. 2007. Impacts of plant diversity on biomass production increase through time because of species complementarity. *Proceedings of the National Academy of Sciences of the USA* 104:18123–18128.

Cardinale BJ, Matulich KL, Hooper DU, Byrnes JE, Duffy E, Gamfeldt L, Balvanera P, O'Connor MI, Gonzalez A. 2011. The functional role of producer diversity in ecosystems. *American Journal of Botany* 98:572–592.

Cheplick GP, Kane KH. 2004. Genetic relatedness and competition in *Triplasis purpurea* (Poaceae): resource partitioning or kin selection. *International Journal of Plant Sciences* 165:623–630.

Cook-Patton SC, McArt SH, Parachnowitsch AL, Thaler JS, Agrawal AA. 2011. A direct comparison of the consequences of plant genotypic and species diversity on communities and ecosystem function. *Ecology* 92:915–923.

Crawford KM, Rudgers JA. 2012. Plant species diversity and genetic diversity within a dominant species interactively affect plant community biomass. *Journal of Ecology* 100:1512–1521.

Crutsinger GM, Collins MD, Fordyce JA, Gompert Z, Nice CC, Sanders NJ. 2006. Plant genotypic diversity predicts community structure and governs an ecosystem process. *Science* 313:966–968.

de Kroon H, Visser EJW. 2003. *Root ecology*. Berlin: Springer.

de Kroon H, Hendriks M, van Ruijven J, Ravenek J, Padilla FM, Jongejans E, Visser EJW, Mommer L. 2012. Root responses to nutrients and soil biota: drivers of species coexistence and ecosystem productivity. *Journal of Ecology* 100:6–15.

Donohue K. 2003. The influence of neighbor relatedness on multilevel selection in the Great Lakes Sea Rocket. *The American Naturalist* 162:77–92.

Dudley SA, File AL. 2007. Kin recognition in an annual plant. *Biology Letters* 3:435–438.

Eisenhauer N. 2012. Aboveground–belowground interactions as a source of complementarity effects in biodiversity experiments. *Plant and Soil* 351:1–22.

Falik O, Reides P, Gersani M, Novoplansky A. 2003. Self/non-self discrimination in roots. *Journal of Ecology* 91:525–531.

Falik O, de Kroon H, Novoplansky A. 2006. Physiologically mediated self/non-self root discrimination in *Trifolium repens* has mixed effects on plant performance. *Plant Signaling and Behavior* 1:116–121.

Fang S, Clark RT, Zheng Y, Iyer-Pascuzzi AS, Weitz JS, Kochian LV, Edelsbrunner H, Liao H, Benfey PN. 2013. Genotypic recognition and spatial responses by rice roots. *Proceedings of the National Academy of Sciences of the USA* 110:2670–2675.

Faraway JJ. 2005. *Extending the linear model with R: generalized linear, mixed effects and nonparametric regression models*. Boca Raton, FL: Chapman & Hall/CRC, 312 pp.

Fridley JD, Grime JP. 2010. Community and ecosystem effects of intraspecific genetic diversity in grassland microcosms of varying species diversity. *Ecology* 91:2272–2283.

Gersani M, Brown JS, O'Brien EE, Maina GM, Abramsky Z. 2001. Tragedy of the commons as a result of root competition. *Journal of Ecology* 89:660–669.

Gruntman M, Novoplansky A. 2004. Physiologically mediated self/non-self discrimination in roots. *Proceedings of the National Academy of Sciences of the USA* 101:3863–3867.

Huber-Sannwald E, Pyke DA, Caldwell MM. 1996. Morphological plasticity following species-specific recognition and competition in two perennial grasses. *American Journal of Botany* 83:919–931.

Krannitz PG, Caldwell MM. 1995. Root growth responses of three Great Basin perennials to intra- and interspecific contact with other roots. *Flora* 190:161–167.

Kulmatiski A, Beard KH, Heavilin J. 2012. Plant-soil feedbacks provide an additional explanation for diversity-productivity relationships. *Proceedings of the Royal Society B: Biological Sciences* 279:3020–3026.

Kunetsova A, Brockhoff PB, Christensen RHB. 2013. lmerTest: tests for random and fixed effects for linear mixed models (lmer objects of lme4 package). http://CRAN.R-project.org/package=lmerTest (16 April 2015).

Mahall BE, Callaway RM. 1991. Root communication among desert shrubs. *Proceedings of the National Academy of Sciences of the USA* 88:874–876.

Mahall BE, Callaway RM. 1992. Root communication mechanisms and intracommunity distributions of two Mojave Desert shrubs. *Ecology* 73:2145–2151.

Mahall BE, Callaway RM. 1996. Effects of regional origin and genotype on intraspecific root communication in the desert shrub *Ambrosia dumosa* (Asteraceae). *American Journal of Botany* **83**: 93–98.

Maron JL, Marler M, Klironomos JN, Cleveland CC. 2011. Soil fungal pathogens and the relationship between plant diversity and productivity. *Ecology Letters* **14**:36–41.

Metlen KL, Aschehoug ET, Callaway RM. 2009. Plant behavioural ecology: dynamic plasticity in secondary metabolites. *Plant, Cell and Environment* **32**:641–653.

Milla R, Forero DM, Escudero A, Iriondo JM. 2009. Growing with siblings: a common ground for cooperation or for fiercer competition among plants? *Proceedings of the Royal Society B: Biological Sciences* **276**:2531–2540.

Mommer L, van Ruijven J, de Caluwe H, Smit-Tiekstra AE, Wagemaker CAM, Ouborg NJ, Bögemann GM, van der Weerden GM, Berendse F, de Kroon H. 2010. Unveiling belowground species abundance in a biodiversity experiment: a test of vertical niche differentiation among grassland species. *Journal of Ecology* **98**:1117–1127.

Novoplansky A. 2009. Picking battles wisely: plant behaviour under competition. *Plant, Cell and Environment* **32**:726–741.

Padilla FM, Mommer L, de Caluwe H, Smit-Tiekstra AE, Wagemaker CAM, Ouborg NJ, de Kroon H. 2013. Early root overproduction not triggered by nutrients decisive for competitive success belowground. *PLoS ONE* **8**:e55805.

Rajaniemi TK. 2003. Evidence for size asymmetry of belowground competition. *Basic and Applied Ecology* **4**:239–247.

Rajaniemi TK, Allison VJ, Goldberg DE. 2003. Root competition can cause a decline in diversity with increased productivity. *Journal of Ecology* **91**:407–416.

R Core Team. 2012. R: a language and environment for statistical computing. Vienna, Austria: R Foundation for Statistical Computing. ISBN 3-900051-09-0. http://www.R-project.org/ (16 April 2015).

Schenk HJ, Callaway RM, Mahall BE. 1999. Spatial root segregation: are plants territorial? *Advances in Ecological Research* **28**: 145–180.

Schnitzer SA, Klironomos JN, HilleRisLambers JH, Kinkel LL, Reich PB, Xiao K, Rillig MC, Sikes BA, Callaway RM, Mangan SA, van Nes EH, Scheffer M. 2011. Soil microbes drive the classic plant diversity–productivity pattern. *Ecology* **92**:296–303.

Schöb C, Kerle S, Karley AJ, Morcillo L, Pakeman RJ, Newton AC, Brooker RW. 2015. Intraspecific genetic diversity and composition modify species-level diversity–productivity relationships. *New Phytologist* **205**:720–730.

Semchenko M, Hutchings MJ, John EA. 2007. Challenging the tragedy of the commons in root competition: confounding effects of neighbour presence and substrate volume. *Journal of Ecology* **95**:252–260.

Semchenko M, Saar S, Lepik A. 2014. Plant root exudates mediate neighbour recognition and trigger complex behavioural changes. *New Phytologist* **204**:631–637.

Wilson SD, Tilman D. 2002. Quadratic variation in old-field species richness along gradients of disturbance and nitrogen. *Ecology* **83**:492–504.

12

Disentangling root system responses to neighbours: identification of novel root behavioural strategies

Pamela R. Belter and James F. Cahill Jr*
Department of Biological Sciences, University of Alberta, Edmonton, AB, Canada T6G 2E9

Associate Editor: Inderjit

Abstract. Plants live in a social environment, with interactions among neighbours a ubiquitous aspect of life. Though many of these interactions occur in the soil, our understanding of how plants alter root growth and the patterns of soil occupancy in response to neighbours is limited. This is in contrast to a rich literature on the animal behavioural responses to changes in the social environment. For plants, root behavioural changes that alter soil occupancy patterns can influence neighbourhood size and the frequency or intensity of competition for soil resources; issues of fundamental importance to understanding coexistence and community assembly. Here we report a large comparative study in which individuals of 20 species were grown with and without each of two neighbour species. Through repeated root visualization and analyses, we quantified many putative root behaviours, including the extent to which each species altered aspects of root system growth (e.g. rooting breadth, root length, etc.) in response to neighbours. Across all species, there was no consistent behavioural response to neighbours (i.e. no general tendencies towards root over-proliferation nor avoidance). However, there was a substantial interspecific variation showing a continuum of behavioural variation among the 20 species. Multivariate analyses revealed two novel and predominant root behavioural strategies: (i) size-sensitivity, in which focal plants reduced their overall root system size in response to the presence of neighbours, and (ii) location-sensitivity, where focal plants adjusted the horizontal and vertical placement of their roots in response to neighbours. Of these, size-sensitivity represents the commonly assumed response to competitive encounters—reduced growth. However, location sensitivity is not accounted for in classic models and concepts of plant competition, though it is supported from recent work in plant behavioural ecology. We suggest that these different strategies could have important implications for the ability of a plant to persist in the face of strong competitors, and that location sensitivity may be a critical behavioural strategy promoting competitive tolerance and coexistence.

Keywords: Coexistence; competition; habitat use and separation; plant behaviour; plant foraging; plant strategies; root ecology.

* Corresponding author's e-mail address: cahillj@ualberta.ca

Introduction

The close proximity of neighbours, combined with strongly overlapping resource requirements, results in competition for limiting resources being a commonly experienced ecological interaction among plants. Competition can greatly reduce individual fitness and alter evolutionary trajectories (Keddy 2001). At the community level, competitive interactions can lead to competitive exclusion, may alter community structure among co-occurring species (Lamb et al. 2009) and can influence plant invasions (Levine 2001; Gurevitch et al. 2011, Bennett et al. 2014). Thus, competition has the potential to alter fundamental aspects influencing the evolution, persistence and coexistence of species in natural and managed landscapes. Despite the importance of competition at many organizational scales, and despite it being an inherently social interaction, only recently have ecologists explicitly focused on understanding plant competition through a behavioural lens (e.g. Gersani et al. 2001; Cahill and McNickle 2011; McNickle and Brown 2014). Here, we build upon behavioural concepts and approaches to better understand how plants alter root growth in the context of social interactions.

In many systems, particularly herbaceous communities such as grasslands, the majority of plant biomass is belowground (Schenk and Jackson 2002). Additionally, when measured, root competition is often a more severe limitation to plant growth than is competition aboveground (Casper and Jackson 1997). Nonetheless, our understanding of plant responses to neighbouring shoots is substantially more advanced (e.g. Smith and Whitelam 1997) than our understanding of plant responses to neighbouring roots (Cahill and McNickle 2011). Better information of how plants alter growth patterns and modify patterns of soil occupancy in response to neighbouring roots should advance our understanding of the causes and consequences of competition and coexistence. By using concepts drawn from the field of behaviour, what a plant does in response to some change in the biotic or abiotic environment (Silvertown and Gordon 1989), one can draw upon a rich conceptual foundation to understand deterministic and plastic growth patterns in plants.

There is substantial evidence that many species of plants have the capacity to alter patterns of root placement in response to neighbours (Schenk 2006; reviewed in Cahill and McNickle 2011). The general patterns found include spatial segregation of neighbouring root systems (Baldwin and Tinker 1972; Brisson and Reynolds 1994; Caldwell et al. 1996; reviewed in Schenk et al. 1999; Holzapfel and Alpert 2003), over-proliferation of roots in the area of potential interaction (Gersani et al. 2001; Maina et al. 2002; Padilla et al. 2013), along with

examples of no response (Litav and Harper 1967; Semchenko et al. 2007). Behavioural responses to neighbours appear species specific, and can change as a function of neighbour identity (Mahall and Callaway 1991; Falik et al. 2003; Bartelheimer et al. 2006; Fang et al. 2013). Despite the strong evidence that plants exhibit complexity and contingency in how they occupy and explore the soil environment (Mommer et al. 2012), the research performed to date is predominately a series of individual studies with idiosyncratic methods and measures, species selections and variable results. Lacking has been a broadly comparative approach to understanding how plants respond to the roots of neighbours (McNickle and Brown 2014), analogous to efforts to understand how plant roots respond to the spatial distribution of soil nutrients (Campbell et al. 1991).

How a plant modifies its occupation of the soil environment in response to a neighbour has important implications for competition for limiting soil resources. Root segregation could result in habitat differentiation, leading to a lack of a 'shared' resource pool, and thus enhancing coexistence (Silvertown 2004). In contrast, plants which tend to aggregate roots at the zone of interaction may exaggerate the spatial overlap of soil depletion zones, leading to enhanced competitive interactions (Gersani et al. 2001). Though there is no existing theory describing which kinds of species are more or less likely to be segregators, aggregators or non-responders in the context of root interactions, there is a theory available in the context of how plants alter root placement and foraging behaviour in response to patchily distributed soil resources. Campbell et al. (1991) predicted that 'large scale foragers' (plants with large root systems) will exhibit little ability to precisely place roots in nutrient patches, while smaller scale foragers will have greater ability to finely adjust root distribution. A phylogenetically controlled meta-analysis did not find support for this prediction (Kembel and Cahill 2005). Instead, Kembel and colleagues (2005, 2008) found that foraging precision in relation to nutrients was positively associated with a number of traits typically associated with weediness and ruderal life-history strategies. How size, competitiveness and other plant traits are associated with plant responsiveness to neighbours is unknown.

In this study we experimentally test three specific questions. (i) Are there general patterns in an individual's root behaviour to neighbouring plants among 20 co-occurring grassland species? (ii) Is a plant's root behaviour contingent upon neighbour identity? (iii) What other plants traits are associated with root behavioural strategies? To answer these questions, we visualized roots using a window box apparatus, allowing for root identification and quantification.

Methods

Species selection

Focal plant species selection. We recognize that there is no single optimum combination of species to be included within a comparative study. As we were predominantly interested in questions related to co-existence, we chose species which potentially co-occur within the native rough fescue (*Festuca hallii* (Vasey) Piper) grasslands near Edmonton, Alberta, Canada. The rough fescue grasslands have been described elsewhere (Lamb and Cahill 2008), with the majority of the biomass consisting of grasses and the majority of diversity being found among the eudicots. In particular, Asteraceae and Poaceae are highly represented in terms of diversity and abundance (Bennett *et al.* 2014), and thus we emphasized species belonging to these two families here.

In total, we included 20 species belonging to six families: Asteraceae (10 species); *Achillea millefolium* L., *Artemesia frigid* Willd., *Artemesia ludoviciana* Nutt., *Erigeron glabellus* Nutt., *Gaillardia aristata* Pursh, *Heterotheca villosa* (Pursh) Shinners, *Solidago missouriensis* Nutt., *Symphyotrichum ericoides* (L.) G.L. Nesom, *Symphyotrichum falcatum* (Lindl.) G.L. Nesom, and *Symphyotrichum laeve* (L.) Á. Löve & D. Löve; Poaceae (five species); *Bouteloua gracilis* (Kunth) Lag. ex Griffiths, *Bromus inermis* Leyss., *Elymus glaucus* Buckley, *Koeleria macrantha* (Ledeb.) Schult., and *Poa pratensis* L; Rosaceae (two species); *Drymocallis arguta* Pursh, and *Geum triflorum* Pursh; Brassicaceae (one species); *Descurainia sophia* (L.) Webb ex Prantl; Fabaceae (one species); *Astragalus agrestis* Douglas ex G. Don; Polygonaceae (one species); *Rumex crispus* L. These species have all been used in other studies conducted by the Cahill lab (Wang *et al.* 2010), grow under growth room conditions and are found in the native grasslands in the area (Bennett and Cahill 2013). These species are representative of the larger species pool at this field site, and as they were not chosen for specific aspects of their growth or abundance, species identity is a 'random effect'.

Seed was field-collected from multiple, naturally occurring plants at the University of Alberta Roy Berg Kinsella Research Ranch located near Kinsella, Alberta, Canada (53°05N, 111°33W).

Neighbour plant species selection. Given the large number of species used in this study, along with the substantial time required to visualize and enumerate root growth (below), it was not feasible to conduct a fully pairwise set of competition trials including all species combinations. Instead, we chose to use a phytometer-based approach (*sensu* Wang et al. 2010).

We chose two species not found in this field site, *Phleum pratense* L., Poaceae, and *Lactuca sativa* L. cv. Esmeralda M.I.., Asteraceae, to serve as neighbour species to our 20 focal species. Our intent in selecting these species was to obtain a generic measure of focal plant response to neighbours, rather than one for which there was potentially a long and co-evolved history. We also chose to include one eudicot and one monocot to limit, for stronger phylogenetic representation. We recognize that results may differ if other species were chosen.

Experimental design

One individual of each focal species was grown under three neighbour treatments: *P. pratense* neighbour, *L. sativa* neighbour and no neighbour (alone). Due to limits in the rate of processing window boxes for visualization, and the size of our growth room, we used temporal, rather than spatial, replication. Each trial consisted of a single replicate of each focal species (20) × neighbour (3) combination; 60 window boxes in total. Replicates were grown between April 2012 and January 2013, with a trial lasting 30–40 days. Due to varying germination success, as well as occasionally limited root visibility in the photos, each species–neighbour combinations was replicated 2–7 times, with most combinations replicated at least three times.

Window box design, soil conditions and planting

To enhance our ability to visualize roots, we used a window box design that forces plants to grow in a nearly two-dimensional plane. We recognize that though this general approach has been used previously (e.g. Mahall and Callaway 1991), it results in highly artificial growth conditions. Nonetheless, we believe that the standardization of growth conditions afforded is critical to initial efforts in undertaking a comparative study of root responsiveness.

Plants were grown in window boxes made of two 215 by 280 mm Plexiglas sheets (one black, one clear) and side spacers (13 mm wide by 5 mm deep) which separated the two Plexiglas sheets creating the soil space (Fig. 1). This configuration was held together with binder clips along the sides. Approximately 30 mm of polyester batting fibre and a horizontal bamboo skewer were arranged at the bottom of each window box to prevent soil leakage yet allowing for drainage. This configuration provided ~5 × 190 × 250 mm of soil space for plant growth.

Window boxes were filled with a homogeneous soil composition of 3 : 1 sand : topsoil mix, amended with ~2 % manure by volume. Though we did not perform nutrient analyses on these soils, prior work with similar soils (Wang *et al.* 2010) suggests plant growth would be nutrient limited, particularly nitrogen. Mineral nutrient

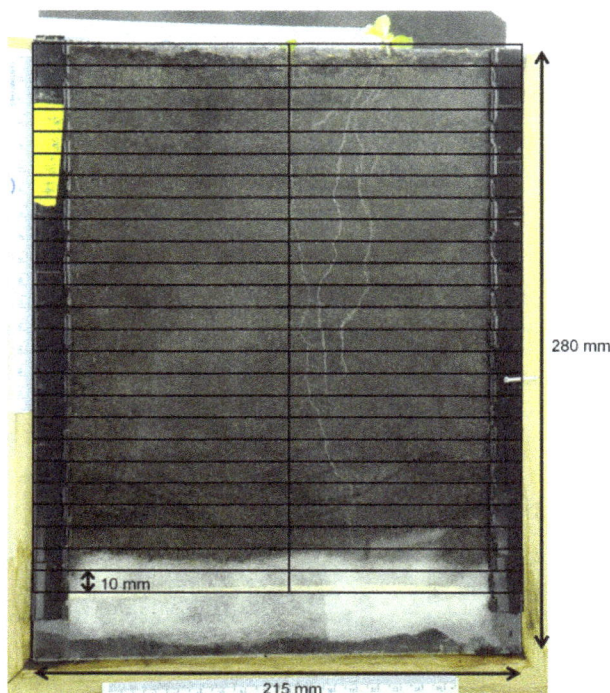

Figure 1. Schematic of experimental window boxes. Soil space available to the plants is ~5 × 190 × 250 mm. For competition treatments the centre plant is the focal species with the neighbour planted to the right, halfway between the focal plant and box edge. No neighbour plant would be present in the control alone treatment. Overlaid grid shows the depth intervals added for image processing with the centre line delineating the right and left side of focal plant for measures of horizontal asymmetry towards a neighbour (to the right).

limitation is also common to the nitrogen-limited soils of the local grasslands from which these seeds were collected (Lamb and Cahill 2008). The top 1 cm of each window box was filled with peat moss (Sun Gro Horticulture Canada Ltd.) to help retain soil moisture.

A single focal species was planted as seed into the centre of each window box, 9.5 cm from each edge. When neighbour plants were used, neighbour seeds were germinated on moist filter paper and bare root transplanted halfway between the focal plant and window box edge. Adding the neighbour plant after germination of the focal plant allowed us to ensure two equally aged seedlings, despite different time-to-germination among species.

Growth, visualization and harvest

The experiment was conducted under controlled environmental conditions (16 : 8 h light : dark cycle at 24 °C) within a growth room at the University of Alberta Biotron. Window boxes were placed in racks, set to a 40° angle, with the clear side facing down and away from the light source. The angled growing position encouraged more

root-Plexiglass contact, enhancing visualization of root growth. To reduce root exposure to light, the clear Plexiglass was covered with a black plastic sheet when roots were not being visualized.

Root visualization and harvest. Roots were visualized every 3 days following the germination of the focal plant, for a total of 10 picture sessions. Visualization consisted of photographs taken using a Nikon D80 with shutter priority, shutter speed of 1/30 s and 50 mm focal length. Camera settings, distance and lighting were constant across visualization sessions and replicate trials.

After 27 days of growth, the window boxes were opened, and the neighbour and target plants were removed. Roots and shoots of each plant were separated, rinsed free of soil, dried (48 h at 70 °C) and individually weighed.

Image analysis and response variables. All photos were inspected to ensure roots of both individuals (if present) were visible. In the few cases where this was not the case, those replicates were removed from further analysis. Using ArcGIS (v10.1; ESRI) the images were digitized by tracing all roots with lines, and coding each root as belonging to either the focal or neighbour plant. To assist with subsequent analyses, we digitally subdivided each image into twenty-five 10 mm depth intervals. These intervals were further subdivided into 'left' and 'right' cells, oriented with respect to the main vertical axis of each focal plant (Fig. 1).

We constructed 10 measures for each species indicating its overall behavioural responsiveness to the presence of a neighbour. These included four size-related metrics (aboveground biomass, belowground biomass, total biomass, total root length), two measures of habitat occupancy (total root system area, and maximum root system breadth), three architectural measures (depth of maximum root system breadth, horizontal asymmetry in root length, horizontal asymmetry in root system area) and one measure of relative allocation to root growth (root : shoot biomass ratio). Details of each measure are provided in Table 1. Though other metrics could be calculated, we believe that this suite of measures broadly describes root system architectural responses to neighbours.

Statistical analysis

Multi-species tests. Linear mixed models were used to analyse general patterns in the effects of neighbours on the focal plants, for each of the 10 response variables. Models incorporated planting treatment as a fixed factor (plants grown alone, with *Lactuca sativa* neighbour, or with *Phleum pratense* neighbour) and focal species as a random factor. To meet the assumptions of normality, proportion variables were arcsine transformed; all other

Table 1. Description of the 10 response measures describing aspects of plant root systems.

Behaviour	Description
Aboveground biomass	Dry mass (g) of all aboveground tissues (g)
Belowground biomass	Dry mass (g) of all belowground tissues (g)
Total biomass	Combined dry mass (g) of aboveground and belowground plant tissues (g)
Root : shoot ratio	Ratio of belowground biomass to aboveground biomass for a given individual
Total root length	Total length of roots (mm) traced using ArcGIS software and attributed to a given individual plant
Total root system area	A convex hull is created around all of the roots of each individual plant. The area of this convex hull (mm^2) is considered to be the total root system area 'occupied' by the plant
Maximum root system breadth	The vertical soil space was divided into 10 mm intervals. For each depth interval the distance (mm) between the farthest root points left and right of centre is calculated. The largest of these widths represents maximum width of the root system
Horizontal asymmetry (root length)	Proportion of total root length for a given individual plant that is found to the right of plant centre. When a neighbour is present, this measure corresponds to the proportion of total root length placed towards that neighbour
Horizontal asymmetry (root system area)	Proportion of total root occupation area for a given individual plant that is found to the right of plant centre. When a neighbour is present, this measure corresponds to the proportion of total occupation area towards that neighbour
Depth of maximum root system breadth	The 10 mm depth interval in which the maximum width is found. The depth measure is the lower end of the interval. For example, a depth of 10 mm would indicate the interval between 0 and 10 mm

variables were ln transformed. Analyses were performed using IBM SPSS Statistics (version 20). Models were also run excluding the random factor, allowing the determination of whether accounting for the variation associated with focal species identity altered model fit based on Hurvich and Tsai's criterion (AICc), accounting for small sample sizes.

A priori contrasts [IBM SPSS Statistics (version 20) TEST subcommand in MIXED] were used to determine whether neighbour presence, independent of the identity of the neighbour, altered focal plant response. Only 16 of the 20 focal species had multiple replicates for all three neighbour treatments, and these were included in the analysis.

Species-specific responses. To determine how individual species responded to neighbours, we calculated log-response ratios (*sensu* Cahill 1999; Hedges *et al.* 1999) for each of the response variables for each replicate of focal × neighbour combination:

$$LRR = \ln\left(\frac{V_N}{V_A}\right)$$

where V_N is the response value for the focal plant when a neighbour (either *Lactuca sativa* or *Phleum pratense*) was present and V_A is the response value when the focal plant was grown alone. To calculate LRR, each focal species replicate with a neighbour was paired to an alone plant

of the same species based on trial number and resting angle of the boxes. Individual replicates of alone plants were not paired more than once within a neighbour treatment. Positive LRR values indicate an increased response with the neighbour (e.g. increased target plant root biomass); negative values indicate a reduced response with the neighbour (e.g. reduced target plant biomass). One-sample *t*-tests were used to test whether each mean species response ratio was significantly different from zero (no difference between responses with and without neighbour species). Analyses were performed using IBM SPSS Statistics (version 20).

Multivariate response and trait correlations. We used principal components analysis (PCA) to explore whether there were multivariate correlations in root responses among species, analogous to larger trait-based studies exploring overall plant strategies (e.g. Grime *et al.* 1997). The PCA was performed using a correlation matrix and equamax rotation in IBM SPSS (version 20). Each species consisted of a single row of data, with its mean LRR for each of the six response variables (LRR) serving as the columns: total biomass, total root length, maximum root system breadth, root : shoot ratio, horizontal asymmetry in root length and depth of maximum root system breadth. We used 6, rather than 10, variables to reduce potential redundancies within the data set.

To test whether a species' root system size was correlated with its root system responsiveness to neighbours, and if this responsiveness was associated with the degree of competition experienced, we performed four regressions. Root system responsiveness was explored in both the horizontal and vertical dimensions by using the absolute value of each species' mean LRR horizontal asymmetry in root length and LRR depth of maximum root system width, respectively. The absolute value of each species' mean LRR was used in order to analyse the magnitude of root system responsiveness independent of direction. For correlations between root system responsiveness and root system size, the mean of ln belowground biomass of each species when grown alone was used as the dependent variable of size. To test whether a species' root responsiveness to neighbours was associated with the degree of competitive suppression it experienced, the mean LRR total biomass across both neighbour treatments was used as the independent variable. Regression analysis was performed using IBM SPSS Statistics (version 20).

Results

Root response to neighbours

General patterns. Across all focal species, there was no overall and consistent effect of the presence of a neighbour on any of the 10 response variables **[see Supporting Information—Table S1]**. However, underneath the lack of a central tendency towards a neighbour effect lies a substantial interspecific variation among the focal species. Including focal plant identity as a random factor in the general linear mixed models substantially increased model fit for 8 of the 10 response variables **[see Supporting Information—Table S1]**, explaining between 20 and 80 % of the variation in a given response variable **[see Supporting Information—Table S1]**. Thus, though on average plants exhibited no root behavioural responses to neighbours, substantial variation in responses occurred among the 20 focal species (Figs 2 and 3). We note that there was no indication, in observation of both roots and shoots, that plants were 'pot-bound', nor that space itself was a limiting resource.

Species-specific responses. An interspecific variation in root responsiveness to neighbours is seen by examination of the response ratios of each response variable (Figs 2 and 3) **[see Supporting Information—Tables S2–S11]**. For all response variables, neighbours caused increases, decreases or no change, depending upon focal species identity (Figs 2 and 3). Visual examination of Figs 2 and 3 indicate no clear trends in responses across plant groups (eudicot versus monocot) or within families, nor consistent effects of neighbour identity on root system responses.

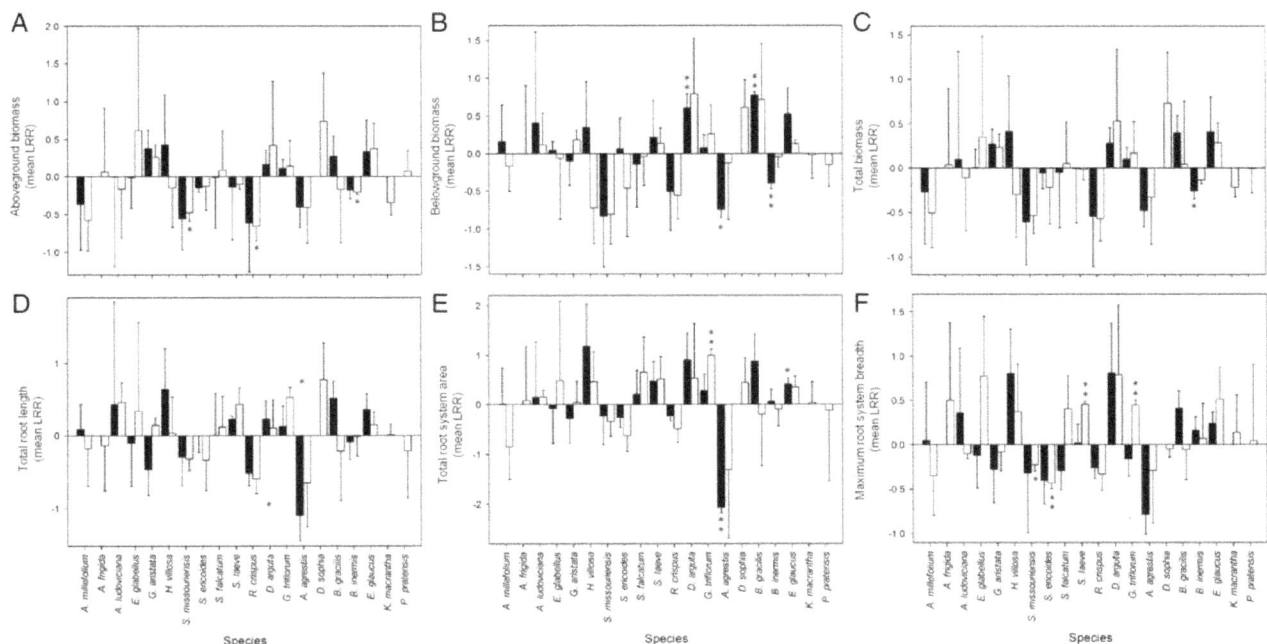

Figure 2. Mean size and habitat occupancy responses (+1 S.E.) of 20 species to neighbour treatment. Graphs show LRR response measures: (A) aboveground biomass, (B) belowground biomass, (C) total biomass, (D) total root length, (E) total root system area, and (F) maximum root system breadth. Closed bars represent the species mean LRR with *Lactuca sativa* neighbour treatment and open bars represent the species mean with *Phleum pratense* neighbour treatment. Asterisks indicate the results of one-sample *t*-tests for a difference from zero (no difference between responses with and without neighbour). *$P < 0.10$ and **$P < 0.05$.

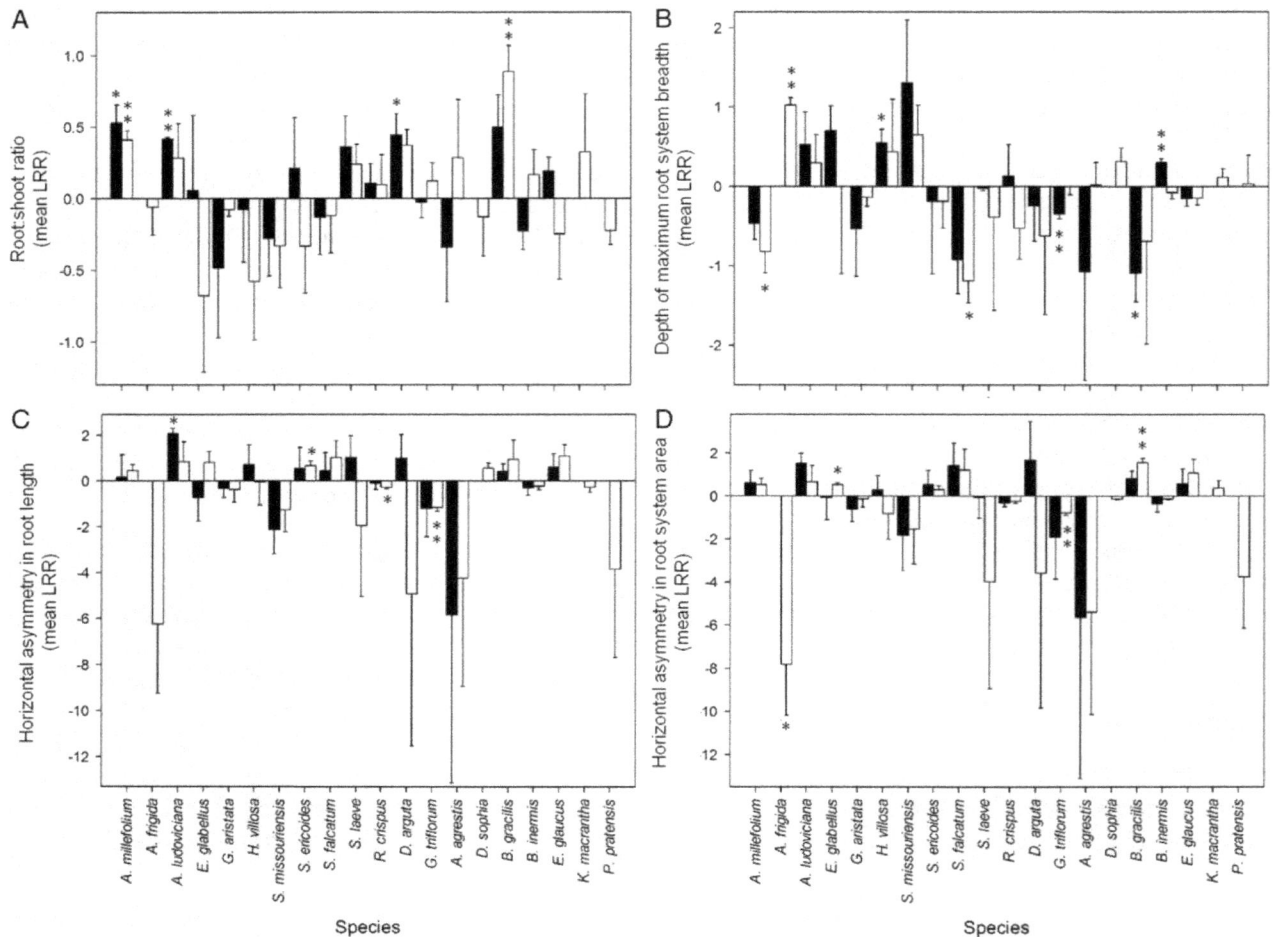

Figure 3. Mean architectural and relative growth allocation responses (+1 S.E.) of 20 species to neighbour treatment. Graphs show LRR response measures: (A) root : shoot ratio, (B) depth of maximum root system breadth, (C) horizontal asymmetry in root length towards neighbour, and (D) horizontal asymmetry in root system area towards neighbour. Closed bars represent the species mean LRR with *Lactuca sativa* neighbour treatment and open bars represent the species mean with *Phleum pratense* neighbour treatment. Asterisks indicate results of one-sample t-tests for a difference from zero (no difference between responses with and without neighbour). *P < 0.10 and **P < 0.05.

However, 20 species is insufficient to conduct formal phylogenetic analyses, precluding estimates of evolutionary conservatism in root behaviour (*sensu* Kembel and Cahill 2005).

Multivariate response. The six response variables used to describe plant responses to neighbours (total biomass, root : shoot ratio, total root length, horizontal asymmetry in root length, maximum root system breadth and depth of maximum breadth) were reduced to two main axes using PCA, explaining 68 % of the variations in the data (Fig. 4). The first axis (39 %) indicates positive correlations among how a plant's total root length, total biomass and maximum root system breadth respond to the presence of a neighbour. Axis two explains an additional 29 % of the variation in the data, and indicates the responsiveness of a plant's root : shoot biomass ratio and horizontal

asymmetry in response to a neighbour are positively correlated with each other, but negatively correlated with a plant's vertical plasticity in response to a neighbour. As before, there was no indication of a consistent difference among monocot and eudicot plant species in how they respond to neighbours.

Trait correlations. There was no significant relationship between a species' root system size (ln belowground biomass of species when grown alone) and its root system responsiveness to neighbours in the horizontal ($R^2 = 0.074$, $F_{1,18} = 1.436$, $P = 0.246$) nor vertical dimensions ($R^2 = 0.150$, $F_{1,18} = 3.166$, $P = 0.092$). Root system responsiveness to neighbours in the horizontal and vertical dimensions was quantified as mean species LRR horizontal asymmetry in root length and LLR depth of maximum root system width, respectively.

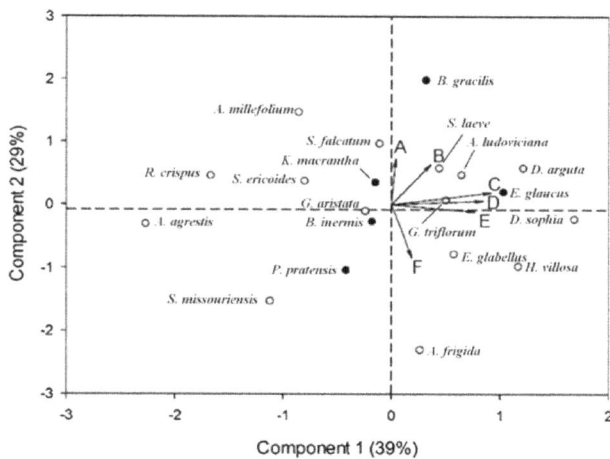

Figure 4. Principal components analysis of six mean response variables (LRR) of 20 species to neighbour treatment. Response variables are: (A) root : shoot ratio, (B) horizontal asymmetry in root length towards neighbour, (C) root length, (D) total biomass, (E) maximum root system breadth, and (F) depth of maximum root system breadth. Component 1 explains 39 % of the variance and Component 2 explains 29 % of the variance.

Similarly, neither horizontal ($R^2 = 0.017$, $F_{1,18} = 0.303$, $P = 0.589$) nor vertical ($R^2 = 0.017$, $F_{1,18} = 0.309$, $P = 0.585$) root responsiveness were associated with the degree of competitive suppression the focal plant experienced (LRR total biomass).

Discussion

General patterns

In previous work, root behavioural responses to neighbours have varied from no response (e.g. Litav and Harper 1967; Semchenko et al. 2007), to segregation (e.g. Baldwin and Tinker 1972; Brisson and Reynolds 1994; Caldwell et al. 1996; Schenk et al. 1999) or over-proliferation (e.g. Gersani et al. 2001; Maina et al. 2002; Padilla et al. 2013). Results presented here (Fig. 1) are consistent with the lack of consistency in these prior findings. We suggest that such behavioural variation is now well demonstrated, and we argue against a strict interpretation of the 'Tragedy of the Commons' prediction of over-proliferation of roots in the zone of competitive encounters (Gersani et al. 2001). Instead, the variation in behaviour observed here, and in prior studies, is consistent with a broader view that multiple adaptive strategies may occur when plants play competitive games (McNickle and Dybzinski 2013; McNickle and Brown 2014). We also note that neither observing behavioural variation in root responses to neighbours, nor modelling fitness differentials associated with different behavioural types is equivalent to demonstrating these behaviours are adaptive. Again, the study of plant foraging behaviour is substantially behind the

understanding of the adaptive value of competitive behaviours aboveground, such as the shade-avoidance response (Dudley and Schmitt 1996). We suggest that more focus on testing the fitness consequences of alternative foraging behaviours is a potentially fruitful area for future research.

Though there was a substantial variation in how plants responded to neighbours, we found no evidence that responses were functionally different among monocots and eudicots. This was surprising, as Kembel and Cahill (2005) found broad differences in the root foraging plasticity of monocot and eudicot species in response to nutrient heterogeneity. Furthermore, both Cahill et al. (2008) and Kiær et al. (2013) showed different competitive effects among monocots and eudicots, and thus we had expected to see clustering of these two groups in terms of behaviour in response to neighbours. We are unable to determine whether our lack of response was due to our relatively limited phylogenetic representation (only 20 species), or whether our results indicate a lack of phylogenetic bias in the tendency to alter root behaviour in response to neighbours.

Similarly, we also found no consistent effect of neighbour identity on root responsiveness to a neighbour. Although previous studies have not always included neighbour identity as a variable for investigation, when they have the comparison is usually between inter- and intra-specific competition (Mahall and Callaway 1991; Bartelheimer et al. 2006) or genotypes of the same species (Callaway and Mahall 2007; Dudley and File 2007; Murphy and Dudley 2009; Fang et al. 2013). Evidence suggests that some plants are able to identify their neighbours at the root level (Chen et al. 2012), and that some species can alter their root responses according to that identity (Mahall and Callaway 1991; Bartelheimer et al. 2006; Callaway and Mahall 2007; Dudley and File 2007; Murphy and Dudley 2009; Fang et al. 2013). It is unclear why we found no similar effect here; though caution that it is difficult to draw strong conclusions, as only two neighbour species were used.

Species-level responses and root behavioural strategies

As mentioned previously, we chose these 20 focal species to be representative of the species that co-occur in a local grassland; they were not chosen to test species-specific hypotheses regarding behavioural responses and strategies. Consequently, each species received relatively little replication, with the strength of the data coming from the comparisons among species. Though these data can be used to test a number of ecologically relevant questions [e.g. are specific root behavioural types associated with high/low abundance in natural system; do specific behavioural types influence a species' response to other ecological challenges (e.g. herbivory)], such questions are well beyond the scope of this manuscript. Instead, we limit our discussion to the two novel behavioural strategies we have identified which are used by plants in response to growing with a neighbour (Fig. 4): size-sensitivity and location-sensitivity.

Size-sensitivity. Nearly 40 % of the variations in species' root responses to neighbours were driven by changes in three size-related traits (total root length, change in maximum root system breadth and change in total biomass; Fig. 4). Not surprisingly, these were all positively correlated and indicate an overall reduction in plant size in response to growth with neighbours (i.e. net effects of competition). It is important to recognize, however, that associated with this reduction in plant size is also a reduction in the area of soil occupied by an individual's root system. Depending upon the allometry of these changes within an individual at the community level, there could be important implications for plant neighbourhood size, biomass distributions in the soil, the degree to which pools of limiting resources are shared among neighbours, as well as resource and host availability for mutualists and other members of the soil community. We suggest that this perspective on the ecological importance of shifts in soil occupancy patterns due to social interactions is overlooked within plant ecology, though widely recognized in the context of animal territoriality, density and resource availability (Hixon 1980).

Location sensitivity. Not all focal species became smaller in response to growth with neighbours, such that there was no main effect of the presence or absence of neighbours for any response variable, including biomass measures **[see Supporting Information—Table S1]**. However, a lack of biomass effect does not equate to a lack of response to neighbours (Fig. 4). We found nearly 30 % of the variations in root responses to neighbours were associated with changes in biomass allocation (R : S ratio) and fine-scale changes in root placement (horizontal asymmetry and depth of maximum root system breadth), rather overall size. These changes indicate a second root system strategy incorporating behavioural plasticity, rather than simply gross biomass responses. We suggest that this is a potentially critical finding, as it highlights that the impacts of neighbours extend further than the traditionally studied resource limitation-biomass reduction paradigm. These data highlight a potential need to begin more robust exploration of the 'non-resource' consequences of neighbours on plant growth and coexistence, analogous to the rapidly increasing research into the non-consumptive effects of predators on prey populations (e.g. Peckarsky et al. 2008).

The ability of plants to modify the fine-scale vertical and horizontal placement of roots in response to neighbours is well established (e.g. Mahall and Callaway 1991; Cahill et al. 2010; Mommer et al. 2010), and has a number of consequences for coexistence, invasion and ecosystem processes. Segregation of the roots of neighbouring plants has long been argued to be a mechanism allowing for species coexistence (Parrish and Bazzaz 1976; Berendse 1983, Craine et al. 2005), due to a reduction in the intensity of competition. The findings here suggest that such a differentiation in micro-scale habitat need not to occur only due to fixed traits of plants (e.g. deep- versus shallow-rooted species), but that behavioural modifications in response to local conditions are not uncommon among plant species. We suggest that reliance on fixed plant traits as a means of understanding the functional ecology of plants can lead to a significant misunderstanding of the mechanisms by which plants can interact with other plants and their environment. We suggest that location-sensitivity behaviours are a potential mechanism that could lead to enhanced coexistence and altered ecosystem functions, even in the face of a strong competitor. It may also be one potential mechanism by which plants are able to tolerate (in a fitness context), growing with aggressive neighbours.

We found no support for the idea that our measures of root responsiveness were related to either plant size (*sensu* the scale and precision ideas of Campbell et al. 1991), nor were they associated with the competition experienced by the focal plants. However, we believe that more work focussed on these root responsive strategies is needed, particularly in the context of fitness consequences, competitive tolerance and avoidance, community assembly and ecosystem function. We also agree with McNickle and Brown (2014) who suggest the accumulation of more and of different types of root trait data allows for novel insights into how plants forage and interact in the soil environment.

We note several limitations in our identification of root responsiveness strategies, including a relatively small number of species (though more than have been used before), nearly two-dimensional growing conditions, short duration of the experiment, use of seedlings rather than mature plants and limited replication within species. How these strategies relate to fitness, the ability to perform in the presence of other ecological processes and non-foraging plant traits is also not known.

Conclusions

Here we used a comparative approach to identify novel behavioural strategies in how plants alter root growth in response to neighbours. Our findings highlight the need to consider species identity when predicting response to neighbours, rather than expect a single dominant strategy of over-proliferation, avoidance or neutrality. Instead, all of these behavioural responses were observed among different species. Though such idiosyncratic responses increase the difficulty of understanding, they do indicate it is critical to understand the biology of the specific species involved in any social interaction. We confirmed prior findings that some species have the potential to alter their fine-scale horizontal and vertical root placement behaviour in response to neighbours, even without showing a negative growth consequence of the 'competitor'. This potentially has important implications for species coexistence, and may be a behavioural trait-filter influencing community assembly and ecosystem function.

Sources of Funding

This work was supported by a Discovery Grant, and Discovery Accelerator, awarded by the Natural Sciences and Engineering Research Council of Canada to J.F.C.

Contributions by the Authors

Both authors conceived of the research and made substantial contributions to the manuscript. P.R.B. conducted the experimentation and analyses.

Acknowledgements

We thank Charlene Nielsen for assistance with the ArcGIS analysis and members of the Cahill lab group for feedback on this research.

Supporting Information

The following additional information is available in the online version of this article –

Table S1. Results of general linear mixed model analysis of the fixed factor neighbour treatment (alone, *Lactuca sativa*, or *Phleum pratense*) on 10 response variables with focal species included as a random factor. A change in the AICc value is obtained when focal species is included as a random factor in the analysis. *A priori* contrast of response to neighbours tests alone (1) versus either *Lactuca sativa* (−0.5) or *Phleum pratense* (−0.5) neighbours.

Table S2. One-sample *t*-tests for the difference between the mean log-response ratio for aboveground biomass (when grown with neighbour) and zero (indicating no response to neighbour).

Table S3. One-sample *t*-tests for the difference between the mean log-response ratio for belowground biomass (when grown with neighbour) and zero (indicating no response to neighbour).

Table S4. One-sample *t*-tests for the difference between the mean log-response ratio for total biomass (when grown with neighbour) and zero (indicating no response to neighbour).

Table S5. One-sample *t*-tests for the difference between the mean log-response ratio for total root length (when grown with neighbour) and zero (indicating no response to neighbour).

Table S6. One-sample *t*-tests for the difference between the mean log-response ratio for root system area (when grown with neighbour) and zero (indicating no response to neighbour).

Table S7. One-sample *t*-tests for the difference between the mean log-response ratio for maximum root system breadth (when grown with neighbour) and zero (indicating no response to neighbour).

Table S8. One-sample *t*-tests for the difference between the mean log-response ratio for root : shoot ratio (when grown with neighbour) and zero (indicating no response to neighbour).

Table S9. One-sample *t*-tests for the difference between the mean log-response ratio for horizontal asymmetry in root length towards neighbour (when grown with neighbour) and zero (indicating no response to neighbour).

Table S10. One-sample *t*-tests for the difference between the mean log-response ratio for horizontal asymmetry in root system area towards neighbour (when grown with neighbour) and zero (indicating no response to neighbour).

Table S11. One-sample *t*-tests for the difference between the mean log-response ratio for depth of maximum root system breadth (when grown with neighbour) and zero (indicating no response to neighbour).

Literature Cited

Baldwin JP, Tinker PB. 1972. A method for estimating the lengths and spatial patterns of two interpenetrating root systems. *Plant and Soil* **37**:209–213.

Bartelheimer M, Steinlein T, Beyschlag W. 2006. Aggregative root placement: a feature during interspecific competition in inland sand-dune habitats. *Plant and Soil* **280**:101–114.

Bennett JA, Cahill JF Jr. 2013. Conservatism of responses to environmental change is rare under natural conditions in a native grassland. *Perspectives in Plant Ecology, Evolution and Systematics* **15**:328–337.

Bennett JA, Stotz GC, Cahill JF Jr. 2014. Patterns of phylogenetic diversity are linked to invasion impacts, not invasion resistance, in a native grassland. *Journal of Vegetation Science* **25**: 1315–1326.

Berendse F. 1983. Interspecific competition and niche differentiation between *Plantago lanceolata* and *Anthoxanthum odoratum* in a natural hayfield. *Journal of Ecology* **71**:379–390.

Brisson J, Reynolds JF. 1994. The effect of neighbors on root distribution in a creosotebush (*Larrea tridentata*) population. *Ecology* **75**:1693–1702.

Cahill JF Jr. 1999. Fertilization effects on interactions between above- and belowground competition in an old field. *Ecology* **80**:466–480.

Cahill JF Jr, McNickle GG. 2011. The behavioral ecology of nutrient foraging by plants. *Annual Review of Ecology, Evolution, and Systematics* **42**:289–311.

Cahill JF Jr, Kembel SW, Lamb EG, Keddy PA. 2008. Does phylogenetic relatedness influence the strength of competition among vascular plants? *Perspectives in Plant Ecology, Evolution and Systematics* **10**:41–50.

Cahill JF Jr, McNickle GG, Haag JJ, Lamb EG, Nyanumba SM, St. Clair CC. 2010. Plants integrate information about nutrients and neighbors. *Science* **328**:1657.

Caldwell MM, Manwaring JH, Durham SL. 1996. Species interactions at the level of fine roots in the field: influence of soil nutrient heterogeneity and plant size. *Oecologia* **106**:440–447.

Callaway RM, Mahall BE. 2007. Plant ecology: family roots. *Nature* **448**:145–147.

Campbell BD, Grime JP, Mackey JML. 1991. A trade-off between scale and precision in resource foraging. *Oecologia* **87**:532–538.

Casper BB, Jackson RB. 1997. Plant competition underground. *Annual Review of Ecology and Systematics* **28**:545–570.

Chen BJW, During HJ, Anten NPR. 2012. Detect thy neighbor: identity recognition at the root level in plants. *Plant Science* **195**:157–167.

Craine JM, Fargione J, Sugita S. 2005. Supply pre-emption, not concentration reduction, is the mechanism of competition for nutrients. *New Phytologist* **166**:933–940.

Dudley SA, File AL. 2007. Kin recognition in an annual plant. *Biology Letters* **3**:435–438.

Dudley SA, Schmitt J. 1996. Testing the adaptive plasticity hypothesis: density-dependent selection on manipulated stem length in *Impatiens capensis*. *The American Naturalist* **147**:445–465.

Falik O, Reides P, Gersani M, Novoplansky A. 2003. Self/non-self discrimination in roots. *Journal of Ecology* **91**:525–531.

Fang S, Clark RT, Zheng Y, Iyer-Pascuzzi AS, Weitz JS, Kochian LV, Edelsbrunner H, Liao H, Benfey PN. 2013. Genotypic recognition and spatial responses by rice roots. *Proceedings of the National Academy of Sciences of the USA* **110**:2670–2675.

Gersani M, Brown JS, O'Brien EE, Maina GM, Abramsky Z. 2001. Tragedy of the commons as a result of root competition. *Journal of Ecology* **89**:660–669.

Grime JP, Thompson K, Hunt R, Hodgson JG, Cornelissen JHC, Rorison IH, Hendry GAF, Ashenden TW, Askew AP, Band SR, Booth RE, Bossard CC, Campbell BD, Cooper JEL, Davison AW, Gupta PL, Hall W, Hand DW, Hannah MA, Hillier SH, Hodkinson DJ, Jalili A, Liu Z, Mackey JML, Matthews N, Mowforth MA, Neal AM, Reader RJ, Reiling K, Ross-Fraser W, Spencer RE, Sutton F, Tasker DE, Thorpe PC, Whitehouse J. 1997. Integrated screening validates primary axes of specialisation in plants. *Oikos* **79**:259–281.

Gurevitch J, Fox GA, Wardle GM, Inderjit, Taub D. 2011. Emergent insights from the synthesis of conceptual frameworks for biological invasions. *Ecology Letters* **14**:407–418.

Hedges LV, Gurevitch J, Curtis PS. 1999. The meta-analysis of response ratios in experimental ecology. *Ecology* **80**:1150–1156.

Hixon MA. 1980. Food production and competitor density as the determinants of feeding territory size. *The American Naturalist* **115**:510–530.

Holzapfel C, Alpert P. 2003. Root cooperation in a clonal plant: connected strawberries segregate roots. *Oecologia* **134**:72–77.

Keddy PA. 2001. *Competition*, 2nd edn. Dordrecht: Kluwer Academic Publishers.

Kembel SW, Cahill JF Jr. 2005. Plant phenotypic plasticity belowground: a phylogenetic perspective on root foraging trade-offs. *The American Naturalist* **166**:216–230.

Kembel SW, de Kroon H, Cahill JF Jr, Mommer L. 2008. Improving the scale and precision of hypotheses to explain root foraging ability. *Annals of Botany* **101**:1295–1301.

Kiær LP, Weisbach AN, Weiner J. 2013. Root and shoot competition: a meta-analysis. *Journal of Ecology.* **101**:1298–1312.

Lamb EG, Cahill JF Jr. 2008. When competition does not matter: grassland diversity and community composition. *The American Naturalist* **171**:777–787.

Lamb EG, Kembel SW, Cahill JF Jr. 2009. Shoot, but not root, competition reduces community diversity in experimental mesocosms. *Journal of Ecology* **97**:155–163.

Levine JM. 2001. Local interactions, dispersal, and native and exotic plant diversity along a California stream. *Oikos* **95**:397–408.

Litav M, Harper JL. 1967. A method for studying spatial relationships between the root systems of two neighbouring plants. *Plant and Soil* **26**:389–392.

Mahall BE, Callaway RM. 1991. Root communication among desert shrubs. *Proceedings of the National Academy of Sciences of the USA* **88**:874–876.

Maina GG, Brown JS, Gersani M. 2002. Intra-plant versus inter-plant root competition in beans: avoidance, resource matching or tragedy of the commons. *Plant Ecology* **160**:235–247.

McNickle GG, Brown JS. 2014. An ideal free distribution explains the root production of plants that do not engage in a tragedy of the commons game. *Journal of Ecology* **102**:963–971.

McNickle GG, Dybzinski R. 2013. Game theory and plant ecology. *Ecology Letters* **16**:545–555.

Mommer L, Van Ruijven J, De Caluwe H, Smit-Tiekstra AE, Wagemaker CAM, Joop Ouborg N, Bögemann GM, Van Der Weerden GM, Berendse F, de Kroon H. 2010. Unveiling belowground species abundance in a biodiversity experiment: a test of vertical niche differentiation among grassland species. *Journal of Ecology* **98**:1117–1127.

Mommer L, van Ruijven J, Jansen C, van de Steeg HM, de Kroon H. 2012. Interactive effects of nutrient heterogeneity and competition: implications for root foraging theory? *Functional Ecology* **26**:66–73.

Murphy GP, Dudley SA. 2009. Kin recognition: competition and cooperation in *Impatiens* (Balsaminaceae). *American Journal of Botany* **96**:1990–1996.

Padilla FM, Mommer L, de Caluwe H, Smit-Tiekstra AE, Wagemaker CAM, Ouborg NJ, de Kroon H. 2013. Early root overproduction not triggered by nutrients decisive for competitive success belowground. *PLoS ONE* **8**:e55805.

Parrish JAD, Bazzaz FA. 1976. Underground niche separation in successional plants. *Ecology* **57**:1281–1288.

Peckarsky BL, Abrams PA, Bolnick DI, Dill LM, Grabowski JH, Luttbeg B, Orrock JL, Peacor SD, Preisser EL, Schmitz OJ, Trussel GC. 2008. Revisiting the classics: considering nonconsumptive effects in textbook examples of predator–prey interactions. *Ecology* **89**: 2416–2425.

Schenk HJ. 2006. Root competition: beyond resource depletion. *Journal of Ecology* **94**:725–739.

Schenk HJ, Jackson RB. 2002. Rooting depths, lateral root spreads and below-ground/above-ground allometries of plants in water-limited ecosystems. *Journal of Ecology* **90**:480–494.

Schenk HJ, Callaway RM, Mahall BE. 1999. Spatial root segregation: are plants territorial? *Advances in Ecological Research* **28**: 145–180.

Semchenko M, John EA, Hutchings MJ. 2007. Effects of physical connection and genetic identity of neighbouring ramets on root-placement patterns in two clonal species. *New Phytologist* **176**:644–654.

Silvertown J. 2004. Plant coexistence and the niche. *Trends in Ecology and Evolution* **19**:605–611.

Silvertown J, Gordon DM. 1989. A framework for plant behavior. *Annual Review of Ecology and Systematics* **20**:349–366.

Smith H, Whitelam GC. 1997. The shade avoidance syndrome: multiple responses mediated by multiple phytochromes. *Plant, Cell and Environment* **20**:840–844.

Wang P, Stieglitz T, Zhou DW, Cahill JF Jr. 2010. Are competitive effect and response two sides of the same coin, or fundamentally different? *Functional Ecology* **24**:196–207.

13

Using an optimality model to understand medium and long-term responses of vegetation water use to elevated atmospheric CO$_2$ concentrations

Stanislaus J. Schymanski[1,2]*, Michael L. Roderick[3,4] and Murugesu Sivapalan[5,6]

[1] Department of Environmental Systems Science, Swiss Federal Institute of Technology Zurich, Universitätstrasse 16, 8092 Zurich, Switzerland
[2] Formerly at: Max Planck Institute for Biogeochemistry, Jena, Germany
[3] Research School of Earth Sciences and Research School of Biology, Australian National University, Canberra 2601, Australia
[4] Australian Research Council Centre of Excellence for Climate System Science, Canberra 2601, Australia
[5] Department of Geography and Geographic Information Science, University of Illinois at Urbana-Champaign, Urbana, Illinois, USA
[6] Department of Civil and Environmental Engineering, University of Illinois at Urbana-Champaign, Urbana, Illinois, USA

Associate Editor: Elise S. Gornish

Abstract. Vegetation has different adjustable properties for adaptation to its environment. Examples include stomatal conductance at short time scale (minutes), leaf area index and fine root distributions at longer time scales (days–months) and species composition and dominant growth forms at very long time scales (years–decades–centuries). As a result, the overall response of evapotranspiration to changes in environmental forcing may also change at different time scales. The vegetation optimality model simulates optimal adaptation to environmental conditions, based on the assumption that different vegetation properties are optimized to maximize the long-term net carbon profit, allowing for separation of different scales of adaptation, without the need for parametrization with observed responses. This paper discusses model simulations of vegetation responses to today's elevated atmospheric CO$_2$ concentrations (eCO$_2$) at different temporal scales and puts them in context with experimental evidence from free-air CO$_2$ enrichment (FACE) experiments. Without any model tuning or calibration, the model reproduced general trends deduced from FACE experiments, but, contrary to the widespread expectation that eCO$_2$ would generally decrease water use due to its leaf-scale effect on stomatal conductance, our results suggest that eCO$_2$ may lead to unchanged or even increased vegetation water use in water-limited climates, accompanied by an increase in perennial vegetation cover.

Keywords: Adaptation; ecohydrology; evapotranspiration; global change; optimality; vegetation.

* Corresponding author's e-mail address: stan.schymanski@env.ethz.ch

Introduction

Elevated atmospheric CO_2 concentrations (eCO_2) are generally expected to lead to reductions in stomatal conductance and hence leaf-scale water use (Wong et al. 1979; Drake et al. 1997). This physiological response has been incorporated into many land surface models, allowing to account for the 'physiological effect' of eCO_2 on surface temperatures in addition to the 'radiative effect' (Sellers et al. 1996; Cao et al. 2010). Several modelling studies have concluded that the physiological effect of eCO_2 on stomata may have resulted in regional and global shifts in the water balance and a general increase in river runoff (e.g. Gedney et al. 2006; Betts et al. 2007; Gopalakrishnan et al. 2011). However, other modelling studies reported that the leaf-scale effect may be offset by concurrent changes in leaf area index, dampening the reduction in vegetation water use due to eCO_2 (Piao et al. 2007; Wu et al. 2012; Niu et al. 2013) and implicated land use change or changes in solar irradiance as possible reasons for increases in continental river runoff (Oliveira et al. 2011). So far, there is only limited empirical evidence for the full range of vegetation responses to eCO_2, but both theoretical considerations and remote sensing data have led some authors to link the observed global increase in perennial vegetation cover ('woody thickening') to increasing atmospheric CO_2 concentrations (C_a) (Bond and Midgley 2000, 2012; Berry and Roderick 2002; Eamus and Palmer 2008; Donohue et al. 2013), suggesting that stomatal closure is indeed only the first step in a long cascade of potential effects of eCO_2. These may include alterations in species compositions, perennial vegetation cover and rooting depths, which come about as the amount of transpiration required to fix a given amount of CO_2 declines with increasing atmospheric CO_2 concentrations. Such alterations are likely to only become obvious after several generations of plants, which, for perennial plants, can take decades to centuries or beyond.

Large-scale free-air CO_2 enrichment (FACE) experiments allow separation of the C_a effect on different plant species from other environmental changes, which is very difficult for remote sensing observations. However, the first FACE experiments were only launched in the 1990s, focussing mainly on temperate ecosystems (Ainsworth and Long 2005), and most of them have come to an end already (Norby and Zak 2011), as they were not intended for the study of long-term vegetation dynamics in response to eCO_2. The present study investigates whether eCO_2 might affect vegetation and the water balance differently in the medium and long term using a previously tested model that incorporates dynamic feedbacks between natural vegetation and the water balance (Schymanski et al. 2009b). Rather than prescribing vegetation response to environmental change, the model is based on the assumption that vegetation self-optimizes to maximize its 'Net Carbon Profit' (i.e. maximizing the difference between carbon acquired by photosynthesis and carbon spent on maintenance of the organs involved in its uptake) and finds the 'optimal' vegetation for given environmental conditions. Here we use this model to investigate the different time scales of vegetation response to eCO_2.

We selected four study sites ranging from dry (water-limited) to wet (energy-limited) conditions in Australia. At each site, we use the model to solve for the optimal vegetation under an assumed climate-CO_2 combination. We use the model runs to ask the following questions:

(1) What would be the difference in predicted annual transpiration rates if only quickly varying vegetation properties (sub-annual scale) were allowed to respond to eCO_2 (medium-term response)?

(2) What would be the difference in predicted annual transpiration rates if all vegetation properties were allowed to respond to increased CO_2 (long-term response)?

(3) Does an increase in atmospheric CO_2 have similar effects on transpiration in all four catchments and climates for both the medium and long-term responses?

Methods

Vegetation optimality model

The model used in this study (vegetation optimality model, VOM) is a coupled water balance and vegetation dynamics model, which does not rely on any input of site-specific vegetation properties or past observations of vegetation response to environmental forcing. This model has been described elsewhere in detail (Schymanski et al. 2008b, 2009b) and the model code is available online (https://github.com/schymans/VOM). In summary, the VOM consists of a physically based multi-layer soil water balance model (0.5 m thick soil layers down to an impermeable bedrock in this study) interfacing with a root water uptake model, which again interfaces with a tissue water balance and leaf gas exchange model. Water fluxes between soil layers and into the fine roots are formulated as functions of water potential gradients and resistances, while leaf gas exchange is simulated as a function of stomatal conductance and leaf-air mole fraction differences. The leaf-internal sink strength for CO_2 is modelled based on a biochemical model of photosynthesis (von Caemmerer 2000), but simplified by omitting carboxylation-limited conditions (see **Supporting Information** or Schymanski 2007; Schymanski et al. 2009b). For the present study, the soil water balance model was also simplified in that the catchment was represented by a rectangular block of soil rather than a linear hillslope as in Schymanski (2007) and Schymanski et al. (2009b). This was found necessary to improve consistency and robustness while parameterizing different catchment geometries (see **Supporting Information** for details).

Optimality, adjustable vegetation properties and associated trade-offs

The VOM approach is based on the assumption that natural vegetation has co-evolved with its environment over a long period of time leading to a composition that is optimally adapted to the conditions. Optimal adaptation is simulated by allowing dynamic adjustments of different vegetation properties at different time scales:

(1) Foliage projective cover and max. rooting depth of perennial plants (decades)
(2) Water-use strategies (decades)
(3) Foliage projective cover of seasonal plants (daily)
(4) Photosynthetic capacity and vertical fine root distributions (daily)
(5) Canopy conductance (hourly)

The different vegetation properties are optimized to maximize the community long-term net carbon profit (NCP), i.e. leaf CO_2 uptake minus respiration costs due to maintenance and turnover of foliage, wood and roots (Schymanski 2007; Schymanski et al. 2009b).

The canopy is represented by two 'big leaves'. One big leaf of invariant size ($M_{A,p}$, m^2 big-leaf area m^{-2} ground area) represents perennial vegetation and another big leaf of varying size ($M_{A,s}$, m^2 big-leaf area m^{-2} ground area) represents seasonal vegetation. As the big leaves are not assumed to transmit any light, no overlap between these two leaves is allowed, so that $M_{A,s} + M_{A,p} \leq 1$. The seasonal vegetation is allowed to vary in its spatial extent ($M_{A,s}$), but has a limited maximum rooting depth ($y_{r,s} = 1$ m), while the perennial component has optimized but invariant $M_{A,p}$ and rooting depth ($y_{r,p}$). Maximum rooting depths are assumed to be invariant in time, but the distribution of roots within each root zone is allowed to vary on a day-by-day basis. The photosynthetic capacity in each big leaf (represented by electron transport capacity, J_{max25}) is also allowed to vary from day to day, while stomatal conductivity (g_s) in each big leaf is allowed to vary on an hourly scale.

The costs and benefits in terms of NCP associated with the optimized parameters in the VOM can be separated into direct and indirect costs and benefits. The direct benefits relate to an increase in photosynthesis, e.g. by increasing big-leaf size, photosynthetic capacity or stomatal conductance. The direct costs relate to increased respiration, e.g. by increased maintenance respiration related to an increased photosynthetic capacity. The indirect benefits relate to carbon gains and losses at a later time, e.g. the consequence of increased stomatal conductance can be a prolonged period of drought-induced stomatal closure and reduced photosynthesis later. Another example is an increase in rooting depth, which has a direct maintenance cost but only an indirect benefit of allowing greater stomatal conductivity and photosynthesis during drought periods. To maximize photosynthetic carbon uptake (A_g) with a limited amount of water, transpiration should be controlled by stomata in such a way that the slope between CO_2 uptake and transpiration ($\partial E_t / \partial A_g$) is kept constant during a day (Cowan and Farquhar 1977; Cowan 1982, 1986; Schymanski et al. 2008a). This slope is denoted by λ_s and λ_p for seasonal and perennial vegetation, respectively. Over longer time periods, the parameters λ_s and λ_p should be sensitive to the availability of soil water and this sensitivity could be seen as a plant physiological response shaped by evolution to suit a given environment (Cowan and Farquhar 1977; Cowan 1982). In the VOM, the sensitivity of λ_s and λ_p to soil water is parametrized as

$$\lambda_s = c_{\lambda f,s} \left(\sum_{i=1}^{i_{r,s}} h_i \right)^{c_{\lambda e,s}} \tag{1}$$

and

$$\lambda_p = c_{\lambda f,p} \left(\sum_{i=1}^{i_{r,p}} h_i \right)^{c_{\lambda e,p}} \tag{2}$$

where h denotes the matric suction head in the soil while $i_{r,s}$ and $i_{r,p}$ denote the deepest soil layer accessed by roots of seasonal and perennial plants, respectively, while the summation is performed over all soil layers (i) within the rooting zone. The parameters $c_{\lambda f,s}$, $c_{\lambda e,s}$, $c_{\lambda f,p}$ and $c_{\lambda e,p}$ are assumed to represent the long-term adaptation of a plant community to its environment and are likely influenced by the species composition of the community.

Separation of medium and long-term responses

Using meteorological data over 30 years, long-term adaptation of vegetation to the environment is modelled by the optimization of six parameters ($M_{A,p}$, $y_{r,p}$, $c_{\lambda f,p}$, $c_{\lambda e,p}$, $c_{\lambda f,s}$ and $c_{\lambda e,s}$) to maximise NCP. The optimization is performed using the shuffled complex evolution (Duan et al. 1993, 1994; Muttil and Liong 2004), which searches the parameter space for the global optimum by re-running the 30-year simulation repeatedly with different parameter values. During each run, electron transport capacity of seasonal ($J_{max25,s}$) and perennial plants ($J_{max25,p}$), vegetated surface area covered by seasonal plants ($M_{A,s}$) and the root surface areas of perennial and seasonal plants ($S_{Ar,p}$ and $S_{Ar,s}$, respectively) are optimized dynamically on a day-by-day basis. For a more detailed description of the optimization algorithms, see Schymanski (2007) and Schymanski et al. (2009b).

The same 30 years of meteorological forcing for each site were used in combination with different atmospheric CO_2 concentrations (C_a = 317, 350 and 380 ppm, representing the observed C_a values in 1960, 1990 and 2005, respectively). The response to eCO_2 was then taken as the difference between the results for 350 or 380 and 317 ppm and simulated responses at different sites were compared to answer Question 3 in the Introduction.

Medium-term responses (Question 1 in the Introduction) were simulated by taking the C_a = 317 ppm simulations and re-running with C_a = 350 ppm and C_a = 380 ppm, while only allowing optimization of those vegetation properties that were assumed to vary at seasonal and shorter time scales (root surface areas, stomatal conductances and electron transport capacities). In other words, medium-term response refers to simulations where those variables marked as 'Constant' in Table 1 were optimized for C_a = 317 ppm, while all other variables were optimized for C_a = 350 ppm or C_a = 380 ppm.

To simulate long-term adaptation (Question 2 in Introduction), optimization of all vegetation parameters in Table 1 was performed independently under each C_a level for each site.

Study sites and site-specific data

The four study sites chosen were all part of the OzFlux network [Ozflux is the Australian and New Zealand Flux Research and Monitoring Network (http://www.dar.csiro.au/lai/ozflux/index.html), which is part of a global network coordinating regional and global analysis of observations from micro-meteorological tower sites (Fluxnet, http://www.fluxnet.ornl.gov/fluxnet/index.cfm)]. The sites span a climatic gradient from semi-arid to humid. The OzFlux sites are long-term monitoring sites for canopy scale CO_2 and water vapour exchange. These sites were Virginia Park (VIR) and Cape Tribulation (CT) in Queensland, Tumbarumba (TUM) in New South Wales and Howard Springs (HS) in the Northern Territory. The geographic locations, vegetation types and key climatic properties of the different sites are summarized in Table 2, while satellite-derived dynamics of foliage projective cover (FPC, fraction of ground area occupied by vertical projection of foliage) is illustrated in Fig. 1. Catchment and soil properties at the different sites are given in Tables 3 and 4.

Meteorological data for the sites were obtained from the Queensland Department of Natural Resources, Mines and Water [SILO Data Drill (http://www.nrm.qld.gov.au/silo)]. The data set contained, among others, daily totals of global solar radiation, precipitation, and class A pan evaporation, daily maxima and minima of air temperature and daily values for atmospheric vapour

Table 1. Optimized vegetation properties in the VOM and their assumed time scales of variation. Subscripts p and s denote perennial and seasonal vegetation, respectively. Canopy conductance is optimized indirectly, as it depends on environmental conditions, J_{max25} and λ, the latter of which is determined by the $c_{\lambda...}$ parameters using Eqs. (1) and (2).

Symbol	Description	Dynamics
$c_{\lambda e,p}$	Exponent of water-use function (perennial veg.)	Constant
$c_{\lambda e,p}$	Exponent of water-use function (seasonal veg.)	Constant
$c_{\lambda f,p}$	Factor of water-use function for (perennial veg.)	Constant
$c_{\lambda f,s}$	Factor of water-use function for (seasonal veg.)	Constant
$G_{s,p}$	Canopy conductance to CO_2 (perennial veg.)	Hourly
$G_{s,s}$	Canopy conductance to CO_2 (seasonal veg.)	Hourly
$J_{max25,p}$	Electron transport capacity at 25 °C (perennial veg.)	Daily
$J_{max25,s}$	Electron transport capacity at 25 °C (seasonal veg.)	Daily
$M_{A,p}$	Fractional cover perennial big leaf	Constant
$M_{A,s}$	Fractional cover seasonal big leaf	Daily
$S_{Ar,p}$	Fine root surface area per soil volume (perennial veg.)	Daily
$S_{Ar,s}$	Fine root surface area per soil volume (seasonal veg.)	Daily
$y_{r,p}$	Maximum rooting depth (perennial veg.)	Constant
λ_p	Slope of $E_t(A_g)$-curve (perennial veg.)	Daily
λ_s	Slope of $E_t(A_g)$-curve (seasonal veg.)	Daily

Table 2. Locations and general conditions of the investigated sites. E_p, net radiation ($I_{n,a}$) divided by latent heat of vaporization (λ_E).

Site	Name	Latitude, longitude	Vegetation	Annual rainfall	Annual E_p ($= I_{n,a}/\lambda_E$)
VIR	Virginia Park	19°53'S, 146°33'E	Open woodland Savanna	580 mm	1810 mm
HS	Howard Springs	12°30S, 131°09'E	Open forest Savanna	1719 mm	1876 mm
TUM	Tumbarumba	35°39'S, 148°09'E	Wet sclerophyll forest	1288 mm	1155 mm
CT	Cape Tribulation	16°06'S, 145°27'E	Tropical rain forest	4097 mm	2085 mm

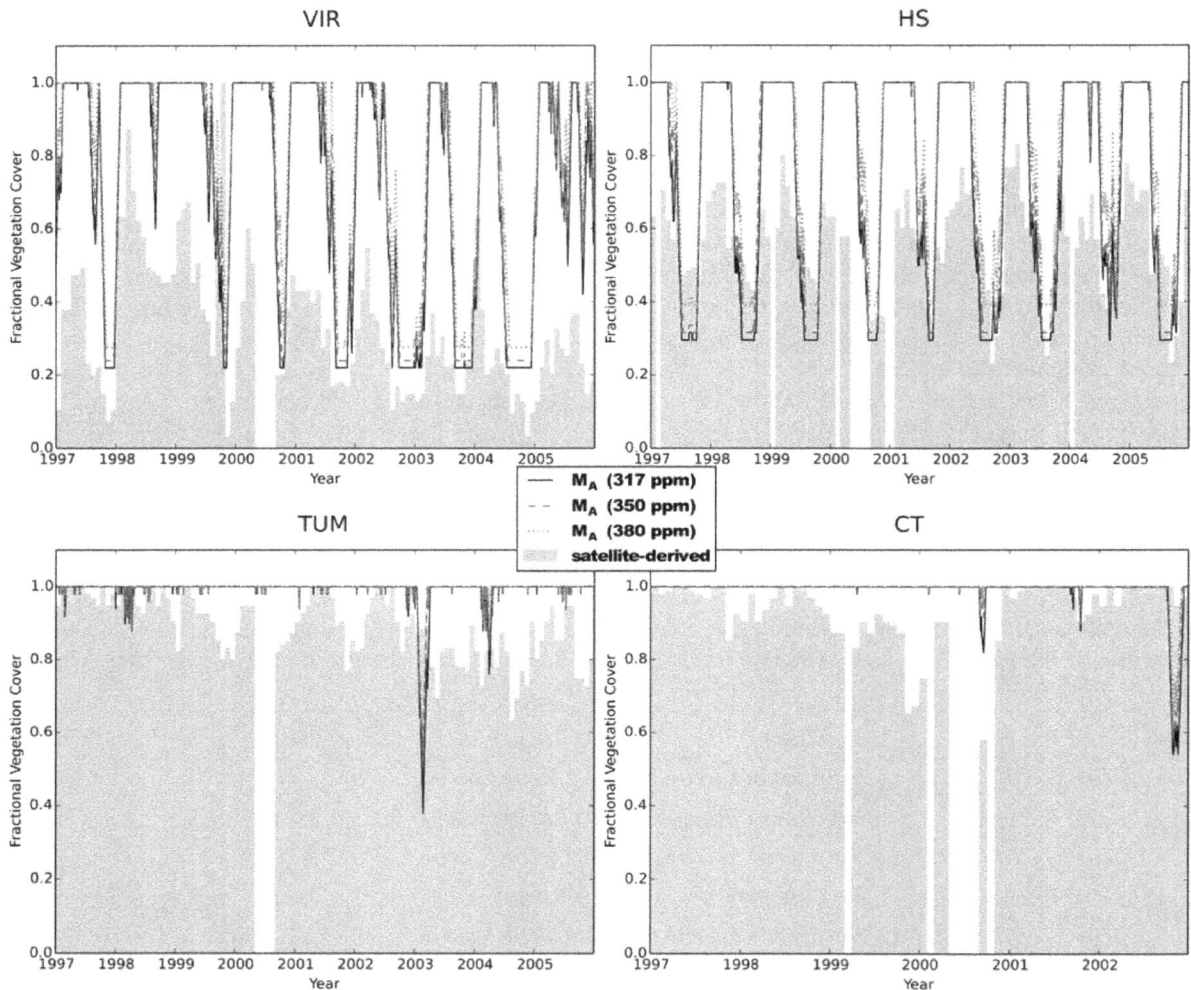

Figure 1. Simulated and satellite-derived FPC at the different sites. Simulation results taken from long-term adaptation runs at 317 (solid lines), 350 (dashed lines) and 380 ppm atmospheric CO_2 concentrations (dotted lines), satellite-derived (AVHRR) estimates of fractional foliage cover (grey shaded) derived from Donohue et al. (2008). Note that gaps in the satellite-derived FPC in year 2000 are due to missing data, not catastrophic events.

pressure, all of which were obtained by interpolation of data from the nearest measurement stations and/or estimated based on proxy data. The methodology used for the compilation of the data set is described in Jeffrey et al. (2001). Daily rainfall was distributed evenly over 24 h, while global irradiance (I_g) and air temperature (T_a) were transformed into hourly values by adding diurnal variation as described in **Supporting Information**. The photosynthetically active photon flux density (I_a, mol quanta m^{-2} s^{-1}) was obtained from global irradiance (I_g, W m^{-2}) using a conversion coefficient of 4.57 × 10^{-6} mol J^{-1} (Thimijan and Heins 1983).

Table 3. Site-specific input data. Z, average soil surface position above bedrock; z_r, average channel elevation above bedrock; γ_0, slope angle near drainage channel.

Site	Soil type	Catchment structure (Z, z_r, γ_0)
VIR	Sandy loam	15 m, 5 m, 2°
HS	Sandy loam	15 m, 10 m, 2°
TUM	Loam	30 m, 5 m, 11.5°
CT	Sandy clay loam	15 m, 5 m, 2°

Table 4. Van Genuchten parameters for the different soil types (Carsel and Parrish 1988). θ_r, residual volumetric water content; θ_s, saturated water content; α_{vG}, inverse of air entry suction; n_{vG}, measure of pore size distribution; K_{sat}, saturated hydraulic conductivity.

Texture	θ_r	θ_s	α_{vG} (m^{-1})	n_{vG}	K_{sat} (m s^{-1})
Sandy loam	0.065	0.41	7.5	1.89	1.228×10^{-5}
Loam	0.078	0.43	3.6	1.56	2.889×10^{-6}
Sandy clay loam	0.1	0.39	5.9	1.48	3.639×10^{-6}

Results

Simulated and observed FPC dynamics

Mean FPC (the sum of the area fractions covered by the perennial and seasonal big leaves, $M_{a,s} + M_{a,p}$) responded positively to eCO$_2$ in all simulations (Table 5), but obviously to a very small extent where FPC was already close to 1 at low C_a (TUM and CT). At these light-limited sites, the model simulated an unexpectedly (and unrealistic) low perennial fractional cover ($M_{A,p}$) of 0.28–0.29 for TUM and 0.24–0.25 for CT. On the other hand, simulated seasonal fractional cover ($M_{A,s}$) was high and largely invariant at these sites, resulting in almost full cover when both seasonal and perennial fractional covers were combined (Fig. 1). Figure 1 also illustrates that the simulations do capture seasonality and inter-annual variability of satellite-derived FPC estimates at the drier sites (e.g. years 1998–99 at VIR or 2003–04 at HS), and the lack of seasonality at the wetter sites (TUM and CT). However, the model predicts full cover on many occasions when satellite-derived FPC is below 0.8 or even 0.5 (at VIR). Note that the simulations under eCO$_2$ suggest a clear increase in perennial FPC (base lines in Fig. 1) at the two drier sites.

Stomatal conductance, roots and evapotranspiration

In all simulations, stomatal conductance decreased in response to elevated CO$_2$ concentrations (C_a) (Table 5). Except for the driest site, the simulated medium-term response of evapotranspiration (E_T, sum of transpiration by perennial and seasonal vegetation plus soil evaporation) was a decrease in the order of 10–80 mm year^{-1} for an increase in C_a of 63 ppm (Fig. 2). The simulated long-term response, in contrast, ranged from a slight decrease (10–25 mm year^{-1}) to increases in E_T by up to 70 mm year^{-1} (Fig. 2) when increasing C_a from 317 to 380 ppm. The largest increase at the HS site was accompanied by an increase in perennial vegetation rooting depth (from 4.5 to 5 m) and $M_{A,p}$ (from 0.3 to 0.39) in the model. For all other sites, simulated rooting depths of perennial vegetation were at 2 m and invariant (data not shown). For the driest site (VIR), simulated perennial FPC increased from 0.22 to 0.28, with a corresponding increase in transpiration by perennial plants ($E_{t,p}$ in the second part of Table 5), but this was accompanied by a decrease in transpiration by seasonal vegetation and soil evaporation ($E_{t,s}$ and E_s, respectively, in the second part of Table 5), resulting in a very small sensitivity of total E_T to C_a at this site. In general, the predictions considering long-term adaptation led to higher E_T rates than those considering medium-term adaptation only (Fig. 2). Simulated root area indices (RAI) had a tendency to increase with C_a in the water-limited catchments and to decrease in the energy-limited catchments. The increases in RAI with C_a at the water-limited sites were much more pronounced in the medium-term adaptation scenario (up to 100 % increase) than for long-term adaptation (up to 25 % increase, Table 5).

Medium-term simulations. Transpiration responses were partly offset by opposite responses in soil evaporation, which was strongly correlated with surface soil moisture (Θ_1, top 50 cm, Fig. 3A). Figure 3A also illustrates that in the medium-term simulations, surface soil moisture decreased at the driest site (VIR), changed very little at the intermediate site (HS) and increased at the energy-limited sites (TUM and CT) in response to increasing C_a. Simulated trends in root area indices (RAI$_p$ and RAI$_s$ for perennial and seasonal vegetation, respectively) were relatively similar to those in transpiration rates, except at HS, where a strong increase in RAI$_s$ coincided with little change in $E_{t,s}$ (Fig. 3A).

Long-term simulations. Simulated soil moisture increased slightly at all sites (all changes <5 %, Table 5). Here, soil evaporation decreased with increasing FPC, in favour of increasing transpiration by perennial plants ($E_{t,p}$), while transpiration by seasonal plants ($E_{t,s}$) did not show very clear trends (Fig. 3B). Root area indices (RAI$_p$ and RAI$_s$) again show similar trends to transpiration rates, with an exception at HS, where the largest increase in $E_{t,p}$ was accompanied by a decrease in RAI$_p$.

Table 5. Simulated responses to increasing atmospheric CO_2 concentrations (C_a). First column in each block gives the actual values (for $C_a = 317$ ppm), while subsequent columns contain deviations (in %) from these values. Negative differences marked in red font. 'Medium-term response', constant vegetation properties (see Table 1) were optimized for $C_a = 317$ ppm; 'Long-term adaptation', all vegetation properties were optimized for the respective C_a. P, precipitation; Q, drainage and runoff; E_T, evapotranspiration (transpiration + soil evaporation)[1]; E_t, transpiration[1]; E_s, soil evaporation[1]; G_s, big-leaf CO_2 stomatal conductance[2]; WUE, water-use efficiency (total A_g/total E_t); iWUE, intrinsic WUE [average (A_g/G_s)]; M_A, fractional cover of big leaf; A_g, CO_2 uptake rate[1]; J_{max25}, leaf electron transport capacity[2]; λ_p and λ_s, median of $\partial E_t/\partial A_g$; RAI, root area index (fine root surface area per ground area); Θ_1, soil saturation degree in top soil layer; Av(Θ), average saturation degree within the rooting zone of perennial vegetation. All magnitudes given as averages (λ_p and λ_s: median values) over last 5 years of simulation. Note that at steady-state, total P = total Q + total E_T. However, in the simulations for VIR, soil water storage (saturated + unsaturated) varies by up to 1000 mm on a decadal scale, and in fact decreased in the last 5 years of the simulation by roughly 500 mm, explaining the mean annual imbalance of 100 mm at this site (see Fig. 2 in the SI). [1]Per m^2 ground area; [2]per m^2 projected leaf area.

Variable C_a	Units ppm	VIR			HS			TUM			CT		
		317	10.4	19.9	317	10.4	19.9	317	10.4	19.9	317	10.4	19.9
Medium-term response													
Total P	mm year^{-1}	401	0.0	0.0	1630	0.0	0.0	1070	0.0	0.0	3280	0.0	0.0
Total Q	mm year^{-1}	101	−1.7	−3.0	335	3.1	7.1	261	7.4	13.2	1320	3.6	6.9
Total E_T	mm year^{-1}	400	1.0	1.5	1320	−0.9	−1.8	883	−1.8	−3.5	1790	−2.5	−4.8
$E_{t,p}$	mm year^{-1}	102	1.2	1.5	611	−1.5	−2.9	227	−3.4	−6.1	456	−3.0	−5.8
$E_{t,s}$	mm year^{-1}	182	6.6	9.9	570	0.3	0.2	458	−2.2	−4.6	1040	−3.3	−6.2
E_s	mm year^{-1}	115	−8.1	−11.9	143	−2.9	−5.5	198	1.0	2.0	296	1.0	1.8
$G_{s,p}$	mmol s^{-1}	117	−2.0	−4.1	364	−2.5	−4.7	404	−5.3	−9.5	926	−4.1	−7.8
$G_{s,s}$	mmol s^{-1}	103	−2.4	−4.1	226	−3.7	−7.0	401	−5.4	−9.7	701	−4.3	−8.0
$M_{A,p}$		0.22	0.0	0.0	0.30	0.0	0.0	0.28	0.0	0.0	0.24	0.0	0.0
$M_{A,s}$		0.42	4.6	6.6	0.44	3.5	6.9	0.71	0.5	1.0	0.76	0.2	0.2
$A_{g,p}$	mmol day^{-1}	87.7	10.5	19.2	208	7.6	13.9	230	4.1	7.4	206	4.8	8.5
$A_{g,s}$	mmol day^{-1}	160	16.0	28.1	254	10.4	19.7	535	5.2	9.4	604	5.3	9.7
WUE_p	mmol mol^{-1}	5.63	9.2	17.5	2.24	9.2	17.3	6.65	7.7	14.3	2.97	8.1	15.2
WUE_s	mmol mol^{-1}	5.78	8.9	16.5	2.92	10.1	19.4	7.67	7.6	14.7	3.82	8.9	17.0
$iWUE_p$	μmol mol^{-1}	151	11.3	21.5	76.1	9.6	17.7	84.1	8.6	16.1	47.3	7.2	13.9
$iWUE_s$	μmol mol^{-1}	156	11.6	22.2	108	12.7	23.6	94.8	8.9	16.6	63.0	8.5	15.6
$J_{max25,p}$	μmol s^{-1}	257	4.9	8.6	351	2.6	4.7	809	1.0	2.1	501	1.2	2.1
$J_{max25,s}$	μmol s^{-1}	218	3.9	8.6	252	2.8	4.2	785	1.2	2.1	491	1.3	2.2
λ_p	mol mol^{-1}	287	−4.4	−7.4	2060	0.2	0.7	953	3.0	6.1	3670	5.2	9.2
λ_s	mol mol^{-1}	207	−3.3	−5.1	809	−1.3	−1.7	1190	3.3	6.5	2360	2.6	4.8
RAI_p	m^2 m^{-2}	0.37	43.8	82.6	0.44	−4.0	−8.9	0.15	−7.8	−14.6	0.094	−4.8	−8.7
RAI_s	m^2 m^{-2}	0.53	45.5	89.8	0.59	38.3	83.6	0.38	−2.4	−1.2	0.17	−7.6	−12.5
Θ_1		0.11	−2.7	−4.9	0.20	−0.3	−0.3	0.41	1.3	2.7	0.54	1.1	2.0
Av(Θ)		0.20	−2.1	−3.3	0.24	0.7	1.7	0.50	0.9	1.8	0.61	0.7	1.4

Continued

Photosynthesis and water-use efficiency

Table 5 and Fig. 3A and B show the following additional trends:

(1) Photosynthetic capacities (J_{max25}) and CO_2 assimilation rates (A_g) increased with C_a in all simulations, with a stronger increase in A_g for seasonal vegetation in the medium term, but a stronger increase for perennial vegetation in the long term. These trends were consistent across all four sites.

(2) Water-use efficiency [both WUE and intrinsic WUE (iWUE)] were generally lower for perennial compared

Table 5. *Continued*

Variable C_a	Units ppm	VIR			HS			TUM			CT		
		317	10.4	19.9	317	10.4	19.9	317	10.4	19.9	317	10.4	19.9
Long-term adaptation													
Total P	mm year^{-1}	401	0.0	0.0	1630	0.0	0.0	1070	0.0	0.0	3280	0.0	0.0
Total Q	mm year^{-1}	101	−2.6	0.9	335	−1.1	−19.9	261	5.5	8.0	1320	1.7	2.3
Total E_T	mm year^{-1}	400	1.2	0.3	1320	0.2	5.9	883	−1.3	−2.1	1790	−1.1	−1.5
$E_{t,p}$	mm year^{-1}	102	10.1	13.1	611	5.9	26.2	227	−0.6	−0.8	456	−4.3	−2.4
$E_{t,s}$	mm year^{-1}	182	1.2	0.3	570	−5.4	−13.5	458	−2.3	−4.2	1040	−0.1	−1.8
E_s	mm year^{-1}	115	−6.6	−11.1	143	−2.1	−3.4	198	0.3	1.2	296	0.4	1.0
$G_{s,p}$	mmol s^{-1}	117	−0.0	−9.6	364	−1.8	−8.1	404	−5.5	−7.5	926	−4.6	−4.4
$G_{s,s}$	mmol s^{-1}	103	−3.6	−6.0	226	−6.1	−7.9	401	−4.4	−6.9	701	−0.7	−1.9
$M_{A,p}$		0.22	9.3	25.6	0.30	7.3	32.8	0.28	3.4	4.7	0.24	−0.4	1.4
$M_{A,s}$		0.42	2.7	5.2	0.44	0.7	−4.3	0.71	−0.5	−0.8	0.76	0.3	−0.2
$A_{g,p}$	mmol day^{-1}	87.7	18.3	39.6	208	14.7	51.0	230	7.4	12.2	206	4.2	9.9
$A_{g,s}$	mmol day^{-1}	160	12.0	22.1	254	6.7	5.6	535	4.3	7.6	604	5.7	9.4
WUE_p	mmol mol^{-1}	5.63	7.5	23.5	2.24	8.3	19.6	6.65	8.1	13.1	2.97	8.8	12.6
WUE_s	mmol mol^{-1}	5.78	10.7	21.6	2.92	12.9	22.1	7.67	6.7	12.3	3.82	5.8	11.4
$iWUE_p$	μmol mol^{-1}	151	10.4	23.9	76.1	10.9	19.3	84.1	8.9	15.7	47.3	8.7	13.9
$iWUE_s$	μmol mol^{-1}	156	11.9	22.7	108	13.2	23.1	94.8	8.2	15.1	63.0	7.2	13.6
$J_{max25,p}$	μmol s^{-1}	257	3.5	3.5	351	1.8	5.0	809	1.0	1.9	501	1.0	2.0
$J_{max25,s}$	μmol s^{-1}	218	3.8	8.1	252	2.2	3.5	785	1.4	2.1	491	1.5	2.3
λ_p	mol mol^{-1}	287	−6.5	−18.4	2060	0.9	−4.6	953	2.7	8.2	3670	2.6	11.1
λ_s	mol mol^{-1}	207	−7.3	−13.3	809	−6.7	−8.4	1190	6.0	13.5	2360	7.5	13.5
RAI_p	m^2 m^{-2}	0.37	21.1	51.2	0.44	9.4	−15.7	0.15	−3.3	−8.1	0.094	−5.4	−5.9
RAI_s	m^2 m^{-2}	0.53	8.6	7.2	0.59	11.8	−15.9	0.38	10.1	2.2	0.17	−3.2	−9.9
Θ_1		0.11	0.7	3.8	0.20	1.1	4.9	0.41	1.1	1.9	0.54	0.5	1.2
$Av(\Theta)$		0.20	−2.2	−0.9	0.24	−0.6	2.5	0.50	0.6	1.0	0.61	0.4	0.7

with seasonal vegetation at each site, and increased linearly with increasing atmospheric CO_2 at all sites with up to 24 % increase at 380 ppm compared with 317 ppm, both under medium and long-term adaptation.

(3) The median values of the slope between CO_2 uptake and transpiration ($\lambda = \partial E_t/\partial A_g$) give an indication whether water is used more or less conservatively. The lower the values of λ, the more water use is limited to times favourable for higher WUE (e.g. high relative humidity of the air and light availability) and hence the more conservative the water use. Simulated λ_p and λ_s (for perennial and seasonal vegetation, respectively) generally decreased with increasing C_a at the water-limited sites (VIR and HS),

while they increase with C_a at the energy-limited sites (TUM and CT). This effect is slightly more pronounced under long-term adaptation but equally clear under medium-term adaptation.

Direct comparison of relative changes in response to eCO_2 under medium and long-term adaptation scenarios reveals clear differences only for drainage (Q), total evapotranspiration (E_T), transpiration by perennial and seasonal vegetation ($E_{t,p}$ and $E_{t,s}$, respectively), FPC, CO_2 assimilation rate (A_g), the water-use strategy indicator λ and RAI, as summarized in Table 6. As mentioned before, E_T was generally higher under long-term adaptation and hence drainage was lower, compared with medium-term adaptation. This was largely caused by a stronger

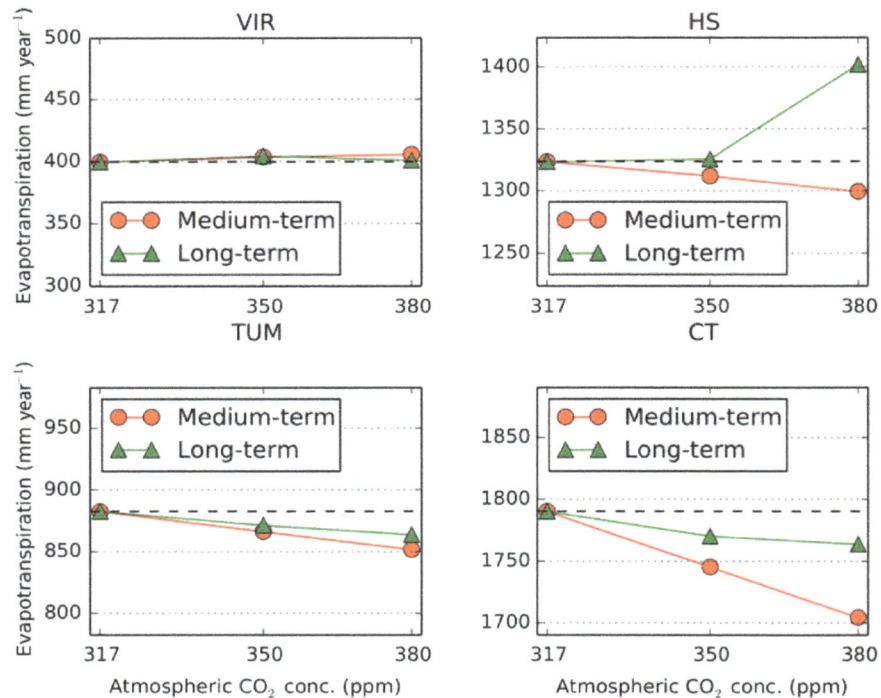

Figure 2. Simulated mean annual evapotranspiration rates for different atmospheric CO_2 concentrations (C_a). 'Medium-term' refers to simulations where constant vegetation properties (see Table 1) were optimized for $C_a = 317$ ppm, while dynamic vegetation properties were optimized for the respective C_a. 'Long-term' refers to simulations where all vegetation properties were optimized for the respective C_a. The horizontal black dashed lines are a visual guide to see the change relative to the E_T rates at 317 ppm C_a.

increase in FPC of perennial plants under long-term adaptation. CO_2 assimilation rates (A_g) increased stronger than FPC, and generally stronger under long-term adaptation. Root area index was generally lower under long-term adaptation at the two water-limited sites (VIR and HS).

Table 6 also reveals that, at the drier sites, simulated vegetation used more water under eCO_2, which was partly compensated for by decreases in soil evaporation, while at the wetter sites, the changes were reversed. CO_2 assimilation (A_g) at the drier sites also benefited more from eCO_2 than at the wetter sites, translating into an almost proportional scaling of WUE with atmospheric CO_2 in the simulations (relative sensitivities ranging between 0.6 for the wettest site and 1.2 for the driest site).

Discussion

The model presented here and its components have been tested at the HS site in previous publications (Schymanski *et al.* 2007, 2008*a*, *b*, 2009*b*) and further detailed tests are required before projections into the future are attempted. The simulations presented in this paper are intended as answers to the question what might be the response of vegetation to eCO_2 at different temporal and spatial scales *IF* vegetation were to adapt in a way to maximize its NCP in the long term. In other words, the model is used

as a tool to understand the implications of an optimality hypothesis on possible responses of vegetation to eCO_2 in different climates. Therefore, no attempts were made to improve any of the model results by parameter tuning or changes to the model structure. In that respect, one advantage of optimality-based models is that they simulate the adaptation of plants to their environment based on a principle that is not expected to change as the environment changes and hence their performance for predictions into the future is not expected to be fundamentally worse than that for predicting the past.

Ecological relevance of community-scale optimality

The origins of the community-scale optimality hypothesis adopted in the VOM can be traced back to Lotka (1922), who proposed that natural selection would yield communities that maximize their throughput of energy, Odum and Pinkerton (1955) who postulated that climax communities balance their primary productivity and maintenance cost while maximizing their biomass and Odum (1969), who extended the hypothesized goal of succession to include preservation of nutrients and protection from external perturbations by complex interactions including mutualism, commensalism and others. Maximization of the NCP can be seen as an approach to quantify the maximum amount of energy (in the form

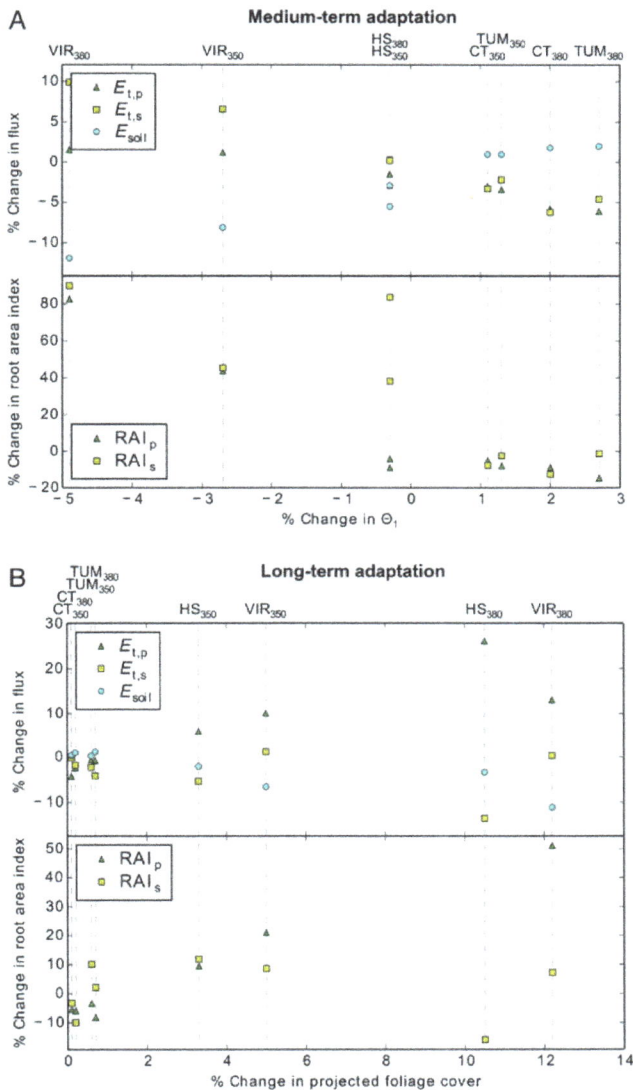

Figure 3. Relative changes in evaporative fluxes and RAI vs. relative changes in (A) surface soil moisture for medium-term and (B) FPC in long-term adaptation. E_t, transpiration; E_s, soil evaporation; Θ_1, relative saturation in the top 0.5 m of soil; RAI, root area index. Subscripts p and s refer to perennial and seasonal vegetation, respectively. Dashed lines link points belonging to a given site (codes following Table 2) and atmospheric CO_2 concentration (subscripts to side codes).

debate about the effects of natural selection at higher organizational levels is ongoing (Fussmann *et al.* 2007; Frank 2013; Pruitt and Goodnight 2014). A discussion of the consistency of optimality hypotheses with ecological theories is beyond the scope of this paper, but an overview of different optimality approaches relevant to ecohydrology can be found in Schymanski *et al.* (2009a).

Some consequences of simplifying assumptions

The results presented here revealed that the model does not reproduce the correct partitioning between perennial and seasonal vegetation at the two energy-limited sites (CT, TUM) examined. This is likely due to the fact that the only advantage of perennial vegetation in the present model is the ability to develop root systems deeper than 1 m. When rainfall is abundant throughout the year, and deep roots are not useful for increasing the long-term NCP, seasonal vegetation would have an additional advantage in the model of being able to reduce their FPC and associated maintenance costs on the rare and short occasions of insufficient water availability and be favoured over perennial vegetation. More realistic partitioning could likely be achieved if the advantage of being tall for light capture was considered in the model. Given that total FPC fell only very rarely below 0.99 at the wet sites (TUM and CT, Fig. 1) we expect that any such modifications would just shift the dominance from seasonal to perennial vegetation in the simulations, but not affect the overall fluxes very much. As discussed below, the big-leaf simplification complicates comparisons between simulated and observed leaf-scale properties. These may be further complicated by the neglect of carboxylation-limited photosynthesis in the VOM. Both simplifications were adopted to reduce computational burden in a model where optimal adaptation is computed using a large number of model runs. Given the prior performance of the model and its components (Schymanski *et al.* 2007, 2008a, b, 2009b; Lei *et al.* 2008), we assume that the structure of the costs and benefits of the optimized vegetation properties is captured adequately despite the simplifications.

Effect of spatial scale on eCO_2 responses

The effect of eCO_2 on evapotranspiration (E_T) has been known to decrease with increasing scale from leaf to canopy to catchment (Leuzinger *et al.* 2011). Our modelling results identify several mechanisms that may be responsible for this scale effect by buffering the leaf-scale reduction in transpiration at larger scales. First and most importantly, the reduction in leaf-scale transpiration with increasing atmospheric CO_2 causes more water to remain in the soil, which can either drain away and contribute to stream flow (as assumed in most papers

of assimilated carbon) that can be available within the ecosystem to support any such processes. It can been argued that natural selection does not act at the ecosystem level but instead acts on individuals and it has been shown on theoretical grounds that optimal resource use by competing plants is not equivalent to optimal resource use by a community of plants (Cowan 1982). However, consideration of more complex interactions in ecosystems resulted in the emergence of system-wide extrema in productivity or resource use (Loreau 1998) while the

Table 6. Relative CO_2 sensitivities in medium and long-term response scenarios derived from Table 5. Values indicate relative change per relative change in C_a as C_a was increased from 317 to 380 ppm. A value of, for example, 0.7 indicates that the relative response of this variable was 70 % of the relative change in C_a, i.e. 14 % increase for a 20 % increase in C_a. Negative values marked in red font. 'med.', constant vegetation properties (see Table 1) were optimized for $C_a = 317$ ppm; 'long', all vegetation properties were optimized for $C_a = 380$ ppm.

	VIR		HS		TUM		CT	
	Medium	Long	Medium	Long	Medium	Long	Medium	Long
Total Q	−0.2	0.0	0.4	−1.0	0.7	0.4	0.3	0.1
Total E_T	0.1	0.0	−0.1	0.3	−0.2	−0.1	−0.2	−0.1
$E_{t,p}$	0.1	0.7	−0.1	1.3	−0.3	−0.0	−0.3	−0.1
$E_{t,s}$	0.5	0.0	0.0	−0.7	−0.2	−0.2	−0.3	−0.1
E_s	−0.6	−0.6	−0.3	−0.2	0.1	0.1	0.1	0.1
$G_{s,p}$	−0.2	−0.5	−0.2	−0.4	−0.5	−0.4	−0.4	−0.2
$G_{s,s}$	−0.2	−0.3	−0.4	−0.4	−0.5	−0.3	−0.4	−0.1
FPC_p	0.0	1.3	0.0	1.6	0.0	0.2	0.0	0.1
FPC_s	0.3	0.3	0.3	−0.2	0.1	−0.0	0.0	−0.0
$A_{g,p}$	1.0	2.0	0.7	2.6	0.4	0.6	0.4	0.5
$A_{g,s}$	1.4	1.1	1.0	0.3	0.5	0.4	0.5	0.5
WUE_p	0.9	1.2	0.9	1.0	0.7	0.7	0.8	0.6
WUE_s	0.8	1.1	1.0	1.1	0.7	0.6	0.9	0.6
$iWUE_p$	1.1	1.2	0.9	1.0	0.8	0.8	0.7	0.7
$iWUE_s$	1.1	1.1	1.2	1.2	0.8	0.8	0.8	0.7
$J_{max25,p}$	0.4	0.2	0.2	0.3	0.1	0.1	0.1	0.1
$J_{max25,s}$	0.4	0.4	0.2	0.2	0.1	0.1	0.1	0.1
λ_p	−0.4	−0.9	0.0	−0.2	0.3	0.4	0.5	0.6
λ_s	−0.3	−0.7	−0.1	−0.4	0.3	0.7	0.2	0.7
RAI_p	4.2	2.6	−0.4	−0.8	−0.7	−0.4	−0.4	−0.3
RAI_s	4.5	0.4	4.2	−0.8	−0.1	0.1	−0.6	−0.5
Θ_1	−0.2	0.2	−0.0	0.2	0.1	0.1	0.1	0.1
$Av(\Theta)$	−0.2	−0.0	0.1	0.1	0.1	0.1	0.1	0.0

mentioned in the introduction) or be utilized by additional leaves or plants, especially in water-limited environments. The latter is supported by the general increase in simulated average FPC (Table 5), consistent with observational data compiled by Norby and Zak (2011), expressing the strongest increase in leaf area index for sites with initially low leaf area index and in line with conclusions drawn from remote sensing data by Donohue et al. (2013). In addition to increased drainage and/or FPC, the increase in soil moisture resulting from a decrease in transpiration may result in increased soil evaporation (E_s). This is indicated in the simulations for the energy-limited sites TUM and CT in Table 5, where both simulated soil moisture and soil evaporation rates increased with increasing C_a. However, in the simulations for the water-limited sites (VIR and HS), increased transpiration and associated decreases

in soil moisture and/or increases in ground shading due to increased FPC led to an overall reduction in soil evaporation in response to eCO2. In fact, the control on soil evaporation at the two water-limited sites shifted from soil moisture feedback in the medium-term simulations (Fig. 3A) to foliage cover feedback in the long-term simulations (Fig. 3B), whereas soil evaporation remained soil moisture controlled at the energy-limited sites (Table 5), where total FPC was close to 1 in all simulations (Fig. 1).

Long-term vs. medium-term responses

In addition to the dampening of leaf-scale reductions in transpiration at larger spatial scales, our modelling results also suggest a clear time-scale dependency at some of the sites and for some variables. Except for the driest site (VIR), the simulations representing medium-term adaptation, i.e. where slowly varying vegetation properties (years-decades) were kept constant, showed a stronger reduction in evapotranspiration (E_T) at eCO$_2$ than simulations representing long-term adaptation, i.e. where all vegetation properties were optimized to the respective C_a. In fact, at one of the sites (HS), long-term adaptation was predicted to lead to a dramatic reversal from an initial decrease in E_T by 30 mm year^{-1} in the medium-term to an increase by 100 mm year^{-1} in the long term, mainly caused by an increase in perennial vegetation cover and maximum rooting depth (Fig. 2). Except for the wettest site, there was a general shift in water use from seasonal to perennial vegetation between the medium-term and long-term adaptation simulations. Interestingly, the simulations showed a stronger increase in CO$_2$ assimilation in response to eCO$_2$ for seasonal plants compared with perennial plants in the medium-term, but a dramatic reversal, i.e. much stronger increases for perennial plants when long-term adaptation was considered. This was mainly due to increases in FPC for perennial plants in the long-term adaptation simulations, which were not permitted by design in the medium-term adaptation simulations. This implicates eCO$_2$ as a direct contributor to 'woody thickening'. In the present model, establishment of saplings and the effect of fires were not considered, so the reasons for the woody encroachment simulated here must be different from those suggested by Bond and Midgley (2000). Since all photosynthesis was modelled as C3 photosynthesis, the reasons for the woody thickening emerging from the model runs are also not related to physiological differences between woody C3 and herbaceous C4 plants, as suggested by Higgins and Scheiter (2012). In the VOM, perennial vegetation has the advantage of potential access to deeper soil water and can freely expand its FPC, whereas seasonal vegetation has a fixed rooting depth of 1 m and cannot exceed an FPC fraction of 1 minus the fractional cover of perennial vegetation. By its effect on WUE, eCO$_2$ acts to shift water-limited environments more towards energy-limited conditions, where the usually taller perennial plants have a selective advantage.

Effect of eCO$_2$ on stomatal control

An apparently paradox result is the negative correlation between trends in $\lambda = \partial E/\partial A$ and RAI. At the water-limited sites, λ decreases with increasing C_a (indicating more conservative water use), but RAI increases, while at the energy-limited sites, λ increases under eCO$_2$ (indicating less conservative water use) while RAI decreases. This can be better understood when looking at the combined effects of changes in C_a and λ on transpiration rates. If everything else stays constant, increasing values of λ increase transpiration rates. However, at

high constant values of λ, eCO$_2$ would commonly reduce transpiration rates (Fig. 4A) while at low constant values of λ, eCO$_2$ could also lead to an increase in transpiration rates (Fig. 4B). This is due to the non-linear effect of C_a on the shape of the $A_g(g_s)$ curve and hence on the slope $\lambda = \partial E/\partial A$. In the medium-term adaptation simulations, λ only responds to changes in soil moisture, i.e. reduction in transpiration under eCO$_2$ leads to increased soil moisture and hence increased λ and transpiration rates, representing a negative feedback loop. Conversely, increase in transpiration under eCO$_2$ at low values of λ would decrease soil moisture and hence decrease λ, resulting again in a negative feedback loop. In the long-term

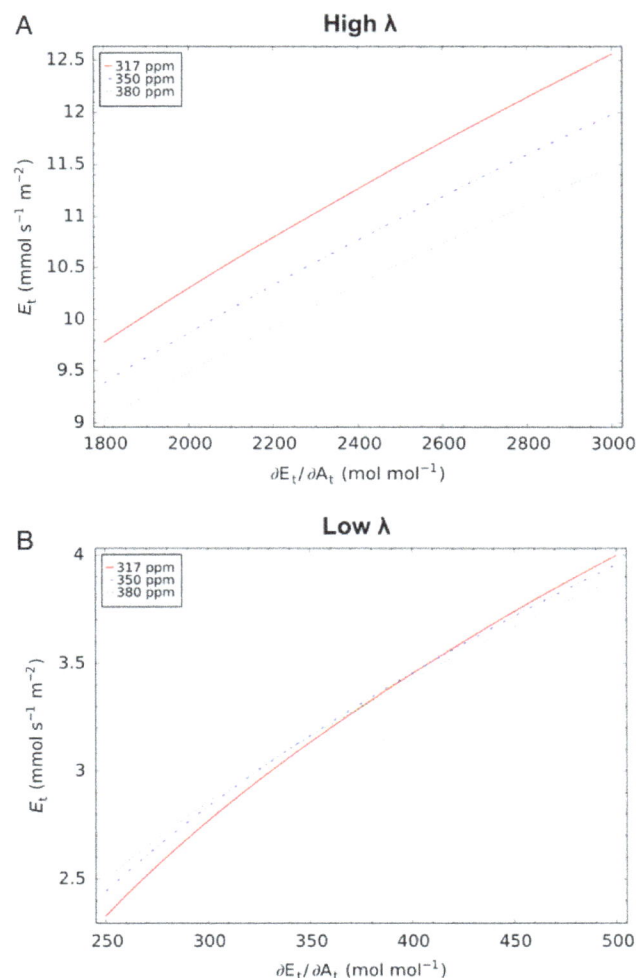

Figure 4. Sensitivity of transpiration rate (E_t, per unit leaf area) to $\lambda = \partial E_t/\partial A_g$ for different atmospheric CO$_2$ concentrations (see keys) at high λ (A) and low λ (B). Simulation conditions: 1000 μmol m^{-2} s^{-1} PPFD, 0.02 mol H$_2$O mol^{-1} air vapour deficit (equivalent to 2 kPa VPD), 40 ppm Γ*. Ranges of λ and J_{max} in (A) and (B) represent simulated values at the wettest site (CT, J_{max} = 485 μmol m^{-2} s^{-1}) and the driest site (VIR, J_{max} = 250 μmol m^{-2} s^{-1}), respectively (see Table 5).

adaptation simulations, the response of λ to soil moisture is optimized for maximum NCP, along with all other vegetation properties. The results suggest that λ should not be expected to scale with C_a in any straightforward way, as assumed by, for example, Katul et al. (2010), but is only part of the whole plant adaptation to its environment (Buckley and Schymanski 2014).

Effect of eCO_2 on photosynthesis and WUE

Another apparently paradoxical result is that, contrary to previously employed approaches (e.g. Prentice et al. 2011), where the assumption of constant c_i/c_a ratios resulted in decreased biochemical capacity (expressed as V_{cmax}) in response to eCO_2, the simulations here all resulted in increasing biochemical capacity (J_{max25}) with increasing c_a. This is in contrast to general findings in FACE studies, where leaf-scale biochemical parameters, J_{max} and V_{cmax}, have been found to decrease under eCO_2 (Ainsworth and Long 2005). This observed down-regulation of photosynthetic capacity is generally attributed to nutrient limitation or insufficient carbon sink capacity (Ainsworth and Long 2005; Leakey et al. 2009), neither of which are considered in the VOM. In contrast, the VOM results suggest that higher net carbon uptake rates could be achieved with higher J_{max25} under eCO_2. In this context, it is important to consider that the VOM simulates the properties of a 'big leaf', representing aggregated canopy properties rather than leaf-scale properties, whereas decreases in J_{max} under eCO_2 have been reported at the leaf scale (see above), with a simultaneous increase in leaf area index, even for relatively closed canopies (Norby and Zak 2011). These simultaneous responses may well have led to an overall increase in canopy-scale photosynthetic capacity (as predicted in

the present study) despite decreases at the leaf scale. Note that the big-leaf representation of the canopy and the neglect of nutrient limitation are also features of many global scale models (e.g. Prentice et al. 2011), so their apparent consistency with leaf-scale observations of decreasing V_{cmax} and/or J_{max} under eCO_2 may not be a good indication for the correct representation of acclimation to eCO_2.

The most coherent response to eCO_2 across all simulations is an increase in WUE and iWUE (assimilation rate divided by stomatal conductance), with relative responses varying between 0.6 and 1.2 (Table 6), i.e. roughly doubling WUE for a doubling in atmospheric CO_2 concentration. This coincides with the range observed using FACE (Table 7). In a study focussing on two FACE sites (Duke and Oak Ridge), De Kauwe et al. (2013) found that 11 state of the art process-based ecosystem models produced relative responses of WUE to eCO_2 between 0.24 and 0.88 while the observed relative responses were 0.65 and 0.93 at the two sites. It is remarkable that the unmodified optimality model employed here produced such a robust eCO_2 sensitivity across four very contrasting catchments, that was in close agreement with general trends in FACE results, while more empirically based models with direct parametrizations of stomatal sensitivity to eCO_2, partly based on other FACE experiments, produced much more scatter with a tendency to under-estimate the response of WUE to eCO_2.

Synthesis and comparison with FACE results

In summary, the results presented in this study suggest that the primary effects of eCO_2 are a reduction of stomatal conductance and an enhancement of CO_2 assimilation. The former leads to reduced transpiration

Table 7. Documented vegetation responses to eCO_2 vs. model predictions. Relative responses were deduced from reported relative change in vegetation property divided by relative change in C_a (e.g. for FACE experiments running at 580 ppm, relative change in C_a would be $580/380 - 1 = 0.5$. FACE, free-air CO_2 enrichment; WTC, whole tree chamber. Sources: [1]Ainsworth and Long (2005), [2]Norby and Zak (2011), [3]Franks et al. (2013), [4]Ainsworth and Rogers (2007), [5]Iversen (2010), [6]Ferguson and Nowak (2011), [7]Barton et al. (2012), [8]De Kauwe et al. (2013), [9]Tausz-Posch et al. (2013), [10]Battipaglia et al. (2013).

Property	Observed relative response	Source	Predicted relative response	
			Medium	Long
Stomatal conductance	−0.2 to −0.7	FACE[1,2,3,4]	−0.2 to −0.5	−0.1 to −0.5
LAI	0 to +1	FACE[2]	0 to +0.3	0 to +1.6
Tree rooting depth	0/+	FACE[2,5,6]	N/A	0 to 0.6
Fine roots	+/−	FACE[2,6]	−0.7 to +4.5	−0.8 to +2.6
Soil moisture	+	FACE[2]	−0.2 to +0.1	0 to +0.1
WUE	+0.7 to +1.4	FACE and WTC[7,8,9]	+0.7 to +1.0	+0.6 to +1.2
iWUE	+1 to +1.8	FACE[1,9,10]	+0.7 to +1.2	+0.6 to +1.2

per leaf area and an initially elevated soil moisture (Fig. 5), leading to increased drainage in energy-limited catchments. However, in water-limited catchments, elevated soil moisture is likely to result in increasing leaf area, while enhanced assimilation allows for the production of more and deeper roots, all of which would act to allow the vegetation to increase the light absorption by the canopy. The net effect is to either maintain, or even enhance, transpiration per unit ground area and to reduce soil moisture and drainage. Note that the relative increase in assimilation per unit ground area is very similar to relative increase in atmospheric CO_2 concentrations at the dry sites (VIR, HS) but at the wet sites, the response in assimilation to a 20 % increase in atmospheric CO_2 is more than halved, at around 10 % (Fig. 6).

The recent review of the FACE literature by Norby and Zak (2011) reveals that many of the long-term effects of eCO_2 predicted in the present study have already been observed experimentally, most notably increases in leaf area index, tree rooting depths and soil moisture. The observed responses are summarized in Table 7, which also indicates that the simulations presented here, both the medium and long-term adaptation scenarios, correspond very closely to general trends observed in FACE experiments. This is particularly remarkable given that most of the FACE data stems from experiments in temperate climates and managed ecosystems, whereas our model simulations refer to natural vegetation in semi-arid to tropical ecosystems. Both model simulations and FACE results illustrate a remarkable convergence in

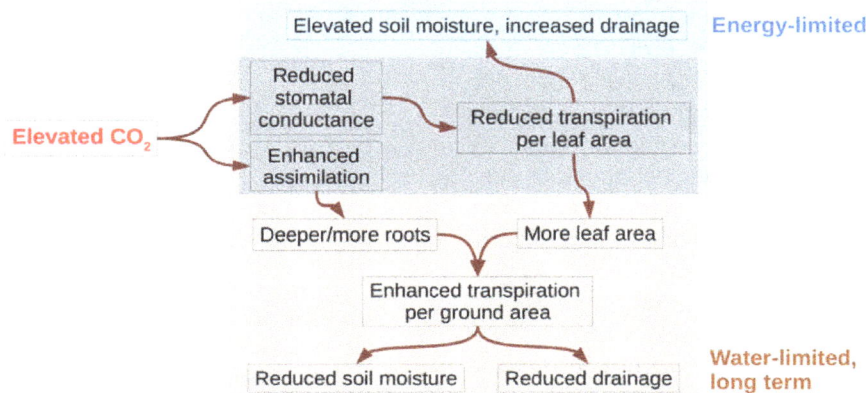

Figure 5. Summary of effects of eCO_2 on vegetation and water resources for constant climate. Effects specific to either water-limited or energy-limited catchments are in the respective coloured boxes. Note that decrease in transpiration per unit leaf area has an initial effect on increasing soil moisture in all catchments, whereas initially increased soil moisture and enhanced assimilation results in increasing leaf area and increased transpiration per ground area at the water-limited sites, reversing the initial effect on soil moisture.

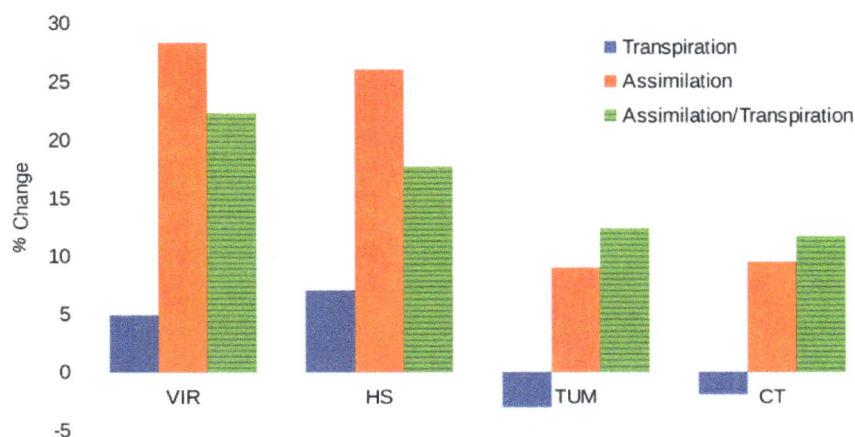

Figure 6. Relative response of transpiration, CO_2 assimilation and their ratio to a 20 % increase in atmospheric CO_2 concentrations at constant climate, assuming long-term adaptation.

some vegetation responses to eCO_2 across biomes and time scales, such as decreasing stomatal conductance, increasing WUE and in dry regions at least, an increase in the leaf area index.

Conclusions

The present analysis of the effects of eCO_2 on the economics of vegetation water use and carbon gain suggests that the assumption of optimal vegetation leads to results that are similar to observed patterns. From those results we also conclude that eCO_2 may be responsible for a large part of the globally observed shift towards more perennial vegetation ('woody thickening'). The study provides theoretical support for an eCO_2-vegetation feedback that has the capacity to dampen reductions in vegetation water use as a result of stomatal down-regulation by allowing more leaf area and/or plants to thrive in water-limited environments. Considering different time scales of adaptation for different vegetation properties, the study separates responses to eCO_2 likely occurring at different temporal scales and suggests that reductions in water use due to stomatal down-regulation should be expected in the shorter term, while unchanged or even increased water use due to an increase in leaf area, plant abundance and potentially rooting depths may result in the longer term in water-limited systems. This suggests that the still common assumption that eCO_2 will generally reduce vegetation water use due to reductions in leaf-level stomatal conductance is not justified in water-limited catchments.

Sources of Funding

This study was supported by Swiss National Science Foundation (S.J.S.: 200021-113442) and Australian Research Council (S.J.S.: DP120101676, M.L.R.: CE11E0098).

Contributions by the Authors

S.J.S. conceived and carried out the research, analysed the data and wrote the text in frequent collaboration with M.L.R. M.S. contributed thoughts and ideas, and helped develop and focus the text at various stages.

Acknowledgements

The authors acknowledge the assistance by Andreas Ostrowski and Steffen Richter (MPI Biogeochemistry, Jena, Germany) in cleaning up the model code and running numerical simulations. Martin de Kauwe and Randall Donohue kindly shared their helpful thoughts on different aspects of the work contained in this paper.

Literature Cited

Ainsworth EA, Long SP. 2005. What have we learned from 15 years of free-air CO_2 enrichment (FACE)? A meta-analytic review of the responses of photosynthesis, canopy properties and plant production to rising CO_2. *New Phytologist* **165**:351–371.

Ainsworth EA, Rogers A. 2007. The response of photosynthesis and stomatal conductance to rising [CO_2]: mechanisms and environmental interactions. *Plant, Cell and Environment* **30**:258–270.

Barton CVM, Duursma RA, Medlyn BE, Ellsworth DS, Eamus D, Tissue DT, Adams MA, Conroy J, Crous KY, Liberloo M, Löw M, Linder S, McMurtrie RE. 2012. Effects of elevated atmospheric [CO_2] on instantaneous transpiration efficiency at leaf and canopy scales in *Eucalyptus saligna*. *Global Change Biology* **18**:585–595.

Battipaglia G, Saurer M, Cherubini P, Calfapietra C, McCarthy HR, Norby RJ, Francesca Cotrufo M. 2013. Elevated CO_2 increases tree-level intrinsic water use efficiency: insights from carbon and oxygen isotope analyses in tree rings across three forest FACE sites. *New Phytologist* **197**:544–554.

Berry SL, Roderick ML. 2002. CO_2 and land-use effects on Australian vegetation over the last two centuries. *Australian Journal of Botany* **50**:511–531.

Betts RA, Boucher O, Collins M, Cox PM, Falloon PD, Gedney N, Hemming DL, Huntingford C, Jones CD, Sexton DMH, Webb MJ. 2007. Projected increase in continental runoff due to plant responses to increasing carbon dioxide. *Nature* **448**:1037–1041.

Bond WJ, Midgley GF. 2000. A proposed CO_2-controlled mechanism of woody plant invasion in grasslands and savannas. *Global Change Biology* **6**:865–869.

Bond WJ, Midgley GF. 2012. Carbon dioxide and the uneasy interactions of trees and savannah grasses. *Philosophical Transactions of the Royal Society B: Biological Sciences* **367**:601–612.

Buckley TN, Schymanski SJ. 2014. Stomatal optimisation in relation to atmospheric CO_2. *New Phytologist* **201**:372–377.

Cao L, Bala G, Caldeira K, Nemani R, Ban-Weiss G. 2010. Importance of carbon dioxide physiological forcing to future climate change. *Proceedings of the National Academy of Sciences of the USA* **107**:9513–9518.

Carsel RF, Parrish RS. 1988. Developing joint probability distributions of soil water retention characteristics. *Water Resources Research* **24**:755–769.

Cowan IR. 1982. Regulation of water use in relation to carbon gain in higher plants. In: Lange OL, Nobel PS, Osmond CB, Ziegler H, eds. *Physical plant ecology II. Encyclopedia of plant physiology*, Vol. 12B. Berlin: Springer, 589–613.

Cowan IR. 1986. Economics of carbon fixation in higher plants. In: Givnish TJ, ed. *On the economy of plant form and function*. Cambridge: Cambridge University Press, 133–171.

Cowan IR, Farquhar GD. 1977. Stomatal function in relation to leaf metabolism and environment. In: Jennings DH, ed. *Integration of activity in the higher plant*. Cambridge: Cambridge University Press, 471–505.

De Kauwe MG, Medlyn BE, Zaehle S, Walker AP, Dietze MC, Hickler T, Jain AK, Luo Y, Parton WJ, Prentice IC, Smith B, Thornton PE, Wang S, Wang YP, Wårlind D, Weng E, Crous KY, Ellsworth DS, Hanson PJ, Seok Kim H, Warren JM, Oren R, Norby RJ. 2013. Forest water use and water use efficiency at elevated CO_2: a model-data intercomparison at two contrasting temperate forest FACE sites. *Global Change Biology* **19**:1759–1779.

Donohue RJ, Roderick ML, McVicar TR. 2008. Deriving consistent long-term vegetation information from AVHRR reflectance data using a cover-triangle-based framework. *Remote Sensing of Environment* **112**:2938–2949.

Donohue RJ, Roderick ML, McVicar TR, Farquhar GD. 2013. Impact of CO_2 fertilization on maximum foliage cover across the globe's warm, arid environments. *Geophysical Research Letters* **40**: 3031–3035.

Drake BG, Gonzàlez-Meler MA, Long SP. 1997. More efficient plants: a consequence of rising atmospheric CO_2? *Annual Review of Plant Physiology and Plant Molecular Biology* **48**:609–639.

Duan QY, Gupta VK, Sorooshian S. 1993. Shuffled complex evolution approach for effective and efficient global minimization. *Journal of Optimization Theory and Applications* **76**:501–521.

Duan Q, Sorooshian S, Gupta VK. 1994. Optimal use of the SCE-UA global optimization method for calibrating watershed models. *Journal of Hydrology* **158**:265–284.

Eamus D, Palmer AR. 2008. Is climate change a possible explanation for woody thickening in arid and semi-arid regions? *International Journal of Ecology* **2007**; doi:10.1155/2007/37364.

Ferguson SD, Nowak RS. 2011. Transitory effects of elevated atmospheric CO_2 on fine root dynamics in an arid ecosystem do not increase long-term soil carbon input from fine root litter. *New Phytologist* **190**:953–967.

Frank SA. 2013. Natural selection. VII. History and interpretation of kin selection theory. *Journal of Evolutionary Biology* **26**: 1151–1184.

Franks PJ, Adams MA, Amthor JS, Barbour MM, Berry JA, Ellsworth DS, Farquhar GD, Ghannoum O, Lloyd J, McDowell N, Norby RJ, Tissue DT, von Caemmerer S. 2013. Sensitivity of plants to changing atmospheric CO_2 concentration: from the geological past to the next century. *New Phytologist* **197**:1077–1094.

Fussmann GF, Loreau M, Abrams PA. 2007. Eco-evolutionary dynamics of communities and ecosystems. *Functional Ecology* **21**:465–477.

Gedney N, Cox PM, Betts RA, Boucher O, Huntingford C, Stott PA. 2006. Detection of a direct carbon dioxide effect in continental river runoff records. *Nature* **439**:835–838.

Gopalakrishnan R, Bala G, Jayaraman M, Cao L, Nemani R, Ravindranath NH. 2011. Sensitivity of terrestrial water and energy budgets to CO_2 -physiological forcing: an investigation using an offline land model. *Environmental Research Letters* **6**: 044013.

Higgins SI, Scheiter S. 2012. Atmospheric CO_2 forces abrupt vegetation shifts locally, but not globally. *Nature* **488**:209–212.

Iversen CM. 2010. Digging deeper: fine-root responses to rising atmospheric CO_2 concentration in forested ecosystems. *New Phytologist* **186**:346–357.

Jeffrey SJ, Carter JO, Moodie KB, Beswick AR. 2001. Using spatial interpolation to construct a comprehensive archive of Australian climate data. *Environmental Modelling and Software* **16**: 309–330.

Katul G, Manzoni S, Palmroth S, Oren R. 2010. A stomatal optimization theory to describe the effects of atmospheric CO_2 on leaf photosynthesis and transpiration. *Annals of Botany* **105**: 431–442.

Leakey ADB, Ainsworth EA, Bernacchi CJ, Rogers A, Long SP, Ort DR. 2009. Elevated CO_2 effects on plant carbon, nitrogen, and water relations: six important lessons from FACE. *Journal of Experimental Botany* **60**:2859–2876.

Lei H, Yang D, Schymanski SJ, Sivapalan M. 2008. Modeling the crop transpiration using an optimality-based approach. *Science in China Series E: Technological Sciences* **51**:60–75.

Leuzinger S, Luo Y, Beier C, Dieleman W, Vicca S, Körner C. 2011. Do global change experiments overestimate impacts on terrestrial ecosystems? *Trends in Ecology and Evolution* **26**:236–241.

Loreau M. 1998. Ecosystem development explained by competition within and between material cycles. *Proceedings of the Royal Society B: Biological Sciences* **265**:33–38.

Lotka AJ. 1922. Contribution to the energetics of evolution. *Proceedings of the National Academy of Sciences of the USA* **8**: 147–151.

Muttil N, Liong SY. 2004. Superior exploration-exploitation balance in shuffled complex evolution. *Journal of Hydraulic Engineering-Asce* **130**:1202–1205.

Niu J, Sivakumar B, Chen J. 2013. Impacts of Increased CO_2 on the Hydrologic Response over the Xijiang (West River) Basin, South China. *Journal of Hydrology* **505**:218–227.

Norby RJ, Zak DR. 2011. Ecological lessons from free-air CO_2 enrichment (FACE) experiments. *Annual Review of Ecology, Evolution, and Systematics* **42**:181–203.

Odum EP. 1969. The strategy of ecosystem development. *Science* **164**:262–270.

Odum HT, Pinkerton RC. 1955. Time's speed regulator: the optimum efficiency for maximum power output in physical and biological systems. *American Scientist* **43**:331–343.

Oliveira PJC, Davin EL, Levis S, Seneviratne SI. 2011. Vegetation-mediated impacts of trends in global radiation on land hydrology: a global sensitivity study. *Global Change Biology* **17**: 3453–3467.

Piao S, Friedlingstein P, Ciais P, de Noblet-Ducoudré N, Labat D, Zaehle S. 2007. Changes in climate and land use have a larger direct impact than rising CO_2 on global river runoff trends. *Proceedings of the National Academy of Sciences of the USA* **104**:15242–15247.

Prentice IC, Harrison SP, Bartlein PJ. 2011. Global vegetation and terrestrial carbon cycle changes after the last ice age. *New Phytologist* **189**:988–998.

Pruitt JN, Goodnight CJ. 2014. Site-specific group selection drives locally adapted group compositions. *Nature* **514**:359–362.

Schymanski SJ. 2007. Transpiration as the leak in the carbon factory: a model of self-optimising vegetation. PhD Thesis, University of Western Australia, Perth, Australia.

Schymanski SJ, Roderick ML, Sivapalan M, Hutley LB, Beringer J. 2007. A test of the optimality approach to modelling canopy properties and CO_2 uptake by natural vegetation. *Plant, Cell and Environment* **30**:1586–1598.

Schymanski SJ, Roderick ML, Sivapalan M, Hutley LB, Beringer J. 2008a. A canopy-scale test of the optimal water-use hypothesis. *Plant, Cell and Environment* **31**:97–111.

Schymanski SJ, Sivapalan M, Roderick ML, Beringer J, Hutley LB. 2008b. An optimality-based model of the coupled soil moisture and root dynamics. *Hydrology and Earth System Sciences* **12**: 913–932.

Schymanski SJ, Kleidon A, Roderick ML. 2009a. Ecohydrological optimality. In: Anderson MG, McDonnell JJ, eds. *Encyclopedia of hydrological sciences. Theory organization and scale*. Chichester: John Wiley & Sons Ltd.

Schymanski SJ, Sivapalan M, Roderick ML, Hutley LB, Beringer J. 2009b. An optimality-based model of the dynamic feedbacks between natural vegetation and the water balance. *Water Resources Research* **45**:W01412.

Sellers PJ, Bounoua L, Collatz GJ, Randall DA, Dazlich DA, Los SO, Berry JA, Fung I, Tucker CJ, Field CB, Jensen TG. 1996. Comparison of radiative and physiological effects of doubled atmospheric CO_2 on climate. *Science* **271**:1402–1406.

Tausz-Posch S, Norton RM, Seneweera S, Fitzgerald GJ, Tausz M. 2013. Will intra-specific differences in transpiration efficiency in wheat be maintained in a high CO_2 world? A FACE study. *Physiologia Plantarum* **148**:232–245.

Thimijan RW, Heins RD. 1983. Photometric, radiometric, and quantum light units of measure—a review of procedures for interconversion. *HortScience* **18**:818–822.

von Caemmerer S. 2000. *Biochemical models of leaf photosynthesis, techniques in plant sciences*, Vol. 2. Collingwood: CSIRO Publishing.

Wong SC, Cowan IR, Farquhar GD. 1979. Stomatal conductance correlates with photosynthetic capacity. *Nature* **282**:424–426.

Wu Y, Liu S, Abdul-Aziz OI. 2012. Hydrological effects of the increased CO_2 and climate change in the Upper Mississippi River Basin using a modified SWAT. *Climatic Change* **110**:977–1003.

14

Fishing for nutrients in heterogeneous landscapes: modelling plant growth trade-offs in monocultures and mixed communities

Simon Antony Croft[1,2]*, Jonathan W. Pitchford[1,2] and Angela Hodge[1]

[1] Department of Biology, University of York, Wentworth Way, York YO10 5DD, UK
[2] York Centre for Complex Systems Analysis (YCCSA), The Ron Cooke Hub, University of York, Heslington, York YO10 5GE, UK

Associate Editor: James F. Cahill

Abstract. The problem of how best to find and exploit essential resources, the quality and locations of which are unknown, is common throughout biology. For plants, the need to grow an efficient root system so as to acquire patchily distributed soil nutrients is typically complicated by competition between plants, and by the costs of maintaining the root system. Simple mechanistic models for root growth can help elucidate these complications, and here we argue that these models can be usefully informed by models initially developed for foraging fish larvae. Both plant and fish need to efficiently search a spatio-temporally variable environment using simple algorithms involving only local information, and both must perform this task against a backdrop of intra- and inter-specific competition and background mortality. Here we develop these parallels by using simple stochastic models describing the growth and efficiency of four contrasting idealized root growth strategies. We show that plants which grow identically in isolation in homogeneous substrates will typically perform very differently when grown in monocultures, in heterogeneous nutrient landscapes and in mixed-species competition. In particular, our simulations show a consistent result that plants which trade-off rapid growth in favour of a more efficient and durable root system perform better, both on average and in terms of the best performing individuals, than more rapidly growing ephemeral root systems. Moreover, when such slower growing but more efficient plants are grown in competition, the overall community productivity can exceed that of the constituent monocultures. These findings help to disentangle many of the context-dependent behaviours seen in the experimental literature, and may form a basis for future studies at the level of complex population dynamics and life history evolution.

Keywords: Complexity; individual-based simulation; patchy environment; productivity; recruitment; stochastic model.

* Corresponding author's e-mail address: simon.croft@york.ac.uk

Introduction

The distribution of nutrients in soil is both spatially and temporally heterogeneous or 'patchy'. Plants must explore this heterogeneous environment and exploit the nutrient patches they encounter to obtain the resources needed for their growth and reproduction. This exploitation is achieved via the growth of a system of roots. These roots also play important roles in anchorage and water uptake, but the uptake of nutrients is the focus of this study. In what follows, we aim to: (i) summarize the key empirical features of root growth in patchy environments; (ii) draw parallels with, and identify contrasts between, root growth and the ecological and evolutionary processes driving a seemingly rather different system, namely the foraging and growth of fish larvae, (iii) show how these similarities and contrasts can be encapsulated within mathematical, computational and statistical models. This synthesis between biological disciplines allows us to develop a modelling framework that can help to answer some important strategic questions.

Growing root systems rely on integrating local environmental information in order to efficiently exploit available resources (Robinson et al. 2003). Because root systems are effectively modular, and the number of modules (roots) is not fixed, growing root systems can show a high degree of flexibility or 'plasticity' (Hodge 2004, 2006). Moreover, roots of different plant species do not always respond in the same way to nutrient patches (Campbell et al. 1991; Hodge et al. 1998), and the same plant species grown under the same experimental conditions can show differing responses depending on the type of nutrient patch encountered (Hodge et al. 1999a, 2000a). This response may be further modified by the presence of competitors (Cahill et al. 2010; Mommer et al. 2012; Hodge and Fitter 2013). Consequently, general 'rules' of how plants will respond to their heterogeneous environment have proved hard to predict.

There is experimental evidence that individual plants respond to small-scale resource heterogeneity (defined here as heterogeneity at scales comparable to individual plant roots) through a range of mechanisms. These include increased root proliferation (Drew 1975), root production (Pregitzer et al. 1993; Hodge et al. 1999a, b), altered lateral branching (Farley and Fitter 1999; Malamy 2005) and increased ion uptake (Jackson et al. 1990; Robinson et al. 1994). Such responses vary between species and may be context-specific, for example, root growth may depend on the attributes of the nutrient patch present (i.e. size, concentration and duration; Hodge et al. 2000a, b, c).

At larger scales, and in a more ecological context, plants have evolved to grow in competition. Resource availability is known to influence plant interactions (Hodge 2004; Cahill and McNickle 2011; Hodge and Fitter 2013). It is known that heterogeneity in physical or chemical properties of soils can influence both plant diversity (Fitter 1982) and vegetation patterns (Tilman 1982) and can promote species coexistence (Berendse 1981; Fitter 1982).

Although there are clear differences between the two systems, here we argue that some of the key elements of plant root growth and nutrient acquisition have fundamental commonalities with foraging and growth of fish larvae, and that therefore there is scope for cross-fertilization between the sub-disciplines of mathematical modelling.

Two similarities are especially germane. First, like plant roots, fish larvae typically have only very temporally and spatially local information about their environment. Nor are they renowned for their intellectual capacities. While factors such as turbulence, detailed fluid mechanics, environmental heterogeneity and predator–prey interactions may all play a role (Pitchford and Brindley 2001; Pitchford et al. 2003), the paradigm of an essentially agnostic and unintelligent biological entity (plant root or fish larva) foraging for heterogeneous resources using only local information is identical.

The second, less immediately obvious, commonality concerns the interplay between the roles of populations (of roots from a single plant, or of offspring from a single parent fish), evolution and 'luck'. An adult female fish will typically produce millions of eggs. Assuming equal sex ratios and constant population size and structure she needs two of these to hatch and grow to maturity over her lifetime; only a tiny minority of larvae, the 'luckiest', successfully reach adulthood (Pitchford et al. 2005). Evolution would therefore favour behaviours that increase the probability of an individual being 'lucky' (e.g. the ability to find, remain within and exploit an ephemeral food patch) rather than those which confer an advantage on average (e.g. faster swimming) (Pitchford et al. 2003). The success of a plant at below-ground resource capture, in contrast, depends on the integrated performance (and cost) of all of its constituent population of roots. However, each growing root could be thought of as an essentially independent forager seeking to exploit nutrients while subject to the possibility of mortality (root 'turnover'). It is not immediately clear whether investing in a population of fewer more resilient roots may confer more of a benefit to the plant than a larger number of faster growing, more ephemeral, roots. Plants generally have both root 'types', but the balance between the two differs among species.

The mathematics of stochastic ('random') processes provides the unifying tool to quantify these ideas. First, stochastic models of individuals foraging in patchy environments developed for fish larvae, can be transferred to the analogous plant root system. Secondly, the impact of individual-level variability at the population scale can be addressed: the crucial ingredient here is that in non-linear stochastic systems one cannot simply multiply the average success of an individual by the population size to estimate population-level performance. Jensen's well known (to statisticians) inequality states that 'the function of the average is not the same as the average of the function' (see, for example, Pitchford et al. 2005), and therefore more mathematically rigorous methods are required.

The preceding comments allow a logical framework to be developed which applies expertise and methodologies from models of larval growth to be transferred to plants. Several authors have applied models of animal behaviour to plants (Gersani et al. 2001; Maina et al. 2002; McNickle and Brown 2014) with varying degrees of success (see Hess and de Kroon 2007; Hodge 2009; Dudley et al. 2013). Nevertheless, the application of animal-inspired models to plant foraging offers a useful way forward, particularly given the difficulty in studying individual root systems in the first place, let alone the more realistic case when these have evolved to grow in a complex plant community.

This study uses methods motivated by foraging fish larvae to explore the growth of plant roots in an unpredictable and heterogeneous environment at the root system scale, and to account for intra- and inter-specific competition between plants with contrasting growth strategies. Growth models employing stochastic differential equations (SDEs) provide general results about the role of randomness (Pitchford et al. 2005). For animal foraging, extending these to so-called non-diffusive systems (allowing for more realistic movement patterns) has been particularly useful (Sims et al. 2008; Preston et al. 2010), but there are still open problems (Pitchford 2013). Perhaps more notably in this context, SDE results derived for fish (Lv and Pitchford 2007) have been applied to plant monoculture data using Bayesian methods to identify and quantify plant root competition at a phenomenological rather than at a mechanistic level (Lv et al. 2008).

In Croft et al. (2012), an idealized 1D model of plant growth, root proliferation, resource capture and inter-plant competition was developed and shown to match SDE representations; this model was used to study the effects of spatial heterogeneity in resource distribution on the evolutionarily optimal root proliferation strategy in monocultures. Details of the model implementation, and of its practical equivalence to SDE models, are provided in Croft et al. (2012). A hierarchy of factors emerged, with the 'optimal' (in an evolutionary context) root proliferation strategy depending on resource levels and their distribution, and on the presence or absence of competition.

In the present work, the model from Croft et al. (2012) is firstly adapted and expanded into two spatial dimensions, and secondly extended to allow competition between

several plant species. These developments, although necessarily 'strategic' in that they describe idealized growth and competition scenarios rather than particular species and environments, allow the trade-off between different root system growth strategies to be modelled explicitly. This allows the model to capture spatial and temporal crowding effects and plant–plant interactions, as well as more realistic resource distributions. It also allows results relating to growth in monocultures to be distinguished from the behaviour of mixed competitive communities.

Plants are modelled with different growth properties, some growing quickly at the sacrifice of the effectiveness of the root system to capture and uptake available nutrients, and others trading off speed and initial size for a root system better at capturing local resources (cf. the fish larvae modelled in Pitchford et al. (2003) and Preston et al. (2010), wherein the trade-off is between swimming faster to incur a deterministic cost in the hope of a stochastic gain in prey encounters). Spatially averaged resource densities are the same between different environmental types, but the relative levels of resource heterogeneity differ (again following the analogy of Pitchford and Brindley (2001), Pitchford et al. (2003), Preston et al. (2010)). These ecological extensions to the established idealized model provide a theoretical framework within which to ask three important strategic questions:

(1) How does the growth strategy adopted by a single plant impact upon its performance in a monoculture?
(2) When plant species are grown in mixed competition for resources, what is the impact on individual, population and community productivity?
(3) What is the role of resource heterogeneity in the above questions?

Answers to these questions are of importance to food security and the development of efficient agriculture, and are also relevant to more general issues of ecological diversity and productivity. The methods used are necessarily idealized but can offer useful general insights and to provide focus for future theoretical and experimental work.

Methods

Overview

A new computational model was created within the Matlab-coding environment, building upon methodologies developed, tested and described in detail in Croft et al. (2012) and Croft (2013). At its core, the model allows the root systems of individual plants to grow and compete for finite resources, using probabilistic methods to allow the broad-scale properties of root system growth, and the stochastic interactions between roots and environment, to be described with a small number of parameters.

The model is summarized conceptually, below, with emphasis on the strategic modelling approach and the key biological factors: different root system growth strategies; descriptions of environmental heterogeneity and the contrasts between isolated growth, monocultures and mixed communities. Technical details of mathematical and computational implementation are available in the Technical Methodology and in Croft (2013) **[see Supporting Information—File S1]**.

The parameter values chosen in this study are given in the Technical Methodology **[see Supporting Information—File S1]**. These should be considered only relative to one another, rather than as pertaining to any particular biological system. In this sense, the total time ($T = 1$) for each simulation is arbitrary. It may be helpful to think of this time scale as referring to a single growing season. The plants have intrinsic growth rates (g) allowing them to approach some upper size limit (L_{max}) on this time scale, but this size limit also depends on the success of their root system in finding resources. These resources are distributed throughout the environment as a set of n individual point resources, which may be encountered by a growing root system. The efficiency of the root system in finding these resources is described by the root system efficiency (SDE) measured on a scale of 0 (no utilization of encountered resources) to 1 (perfect utilization). In this way, the trade-off between growing fast but potentially unreliable root systems can be contrasted with more efficient but slower growing roots. The model updates on a time scale ($dt = 10^{-4}$) in the order of 1 h.

Modelling the resource environment

The environment is defined as a square of continuous space with periodic boundaries (i.e. one edge connects to the opposite edge). The environments are sized sufficiently large so as to not inhibit growth of an isolated individual due to space limitation, and are scaled according to the number of plants being grown within a numerical simulation so that plant density (in terms of number of plants per unit area) is constant. These two measures ensure that space is not a limited resource at the population level, and facilitate comparison across all simulation scenarios.

Resources occur in the environment in a finite number of discrete locations. Each of these discrete resources is of the same quality, i.e. it confers the same relative growth benefit to a plant able to acquire it.

Across all environments, the mean resource density is kept constant. Combined with the spatial scaling detailed above, this ensures that the total quantities of resources per plant, as well as the total resource density across the entire environment, are consistent across all scenarios. This allows the role of resource heterogeneity to be addressed without ambiguity.

Two types of probabilistic environmental heterogeneity are considered: 'uniformly random' and 'patchy'. The uniformly random environments (Fig. 1A) are created by placing each discrete resource within the environment independently according to a 2D uniform random distribution. This creates a statistically homogeneous environment, with a given resource point providing no information about the relative location of any other. In contrast, the patchy environments (Fig. 1B) are created by a random walk

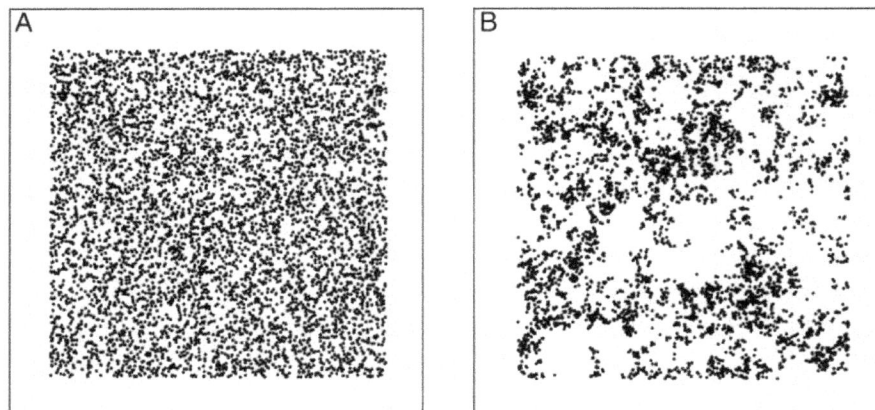

Figure 1. Visual representations of a 'uniformly random' environment (A) and 'patchy heterogeneous' environments (B). Each environment is comprised of 6400 individual point resources. In uniformly random environments, these are distributed according to a 2D independent uniformly random distribution. There is no structure to the patch distribution, with each patch independent from all others. In patchy heterogeneous environments, resources are distributed according to a 2D random walk. The random walk step lengths are sampled from a long-tailed Pareto distribution, and rotations sampled from a uniformly random distribution. Individual resource positions are not independent, with a patchy heterogeneous structure emerging.

Fishing for nutrients in heterogeneous landscapes: modelling plant growth trade-offs in monocultures...

155

process sampling rotations from a uniformly random distribution, and step lengths from a long-tailed Pareto distribution (Preston *et al.* 2010). This results in statistically 'patchy' environments, where individual resource points are likely to aggregate to form a structured distribution.

This approach assumes that nutrient resources occur as a finite number of discrete points. However, because the simulations use a large number of resource points, the overall resource distribution is essentially continuous at the scale of a plant (see Figs 1, 6, and 7) while maintaining a computationally tractable model.

Randomizing initial plant locations

When grown in isolation, an individual plant is placed in the centre of the environment. Since the boundaries are periodic and the environments randomly generated, it does not actually matter where in the environment an isolated plant is placed (i.e. there will be no boundary effects or environmental bias by being placed centrally); the centre is chosen merely for convenience.

When a simulation is to comprise of multiple plants growing and competing simultaneously (either as a monoculture or in mixed competition), each individual is placed independently according to a 2D uniform distribution within the environment. This means that the placement of each individual is random within the environment, and that the presence/absence of competitors within an area, or the type of plant, does not affect this placement. The resulting distributions of competitors within the neighbourhood, which are statistically uniform on average, may lead to varying levels of localized grouping and competition within and across each realization of the simulations.

Implementing root system growth

Each individual plant's root system starts as a point and expands radially (i.e. as an expanding circle) with growth at a constant rate (by area). Each individual has its own initial upper size limit, and growth ceases when the plant reaches this size. This initial upper limit can be thought of as representing possible growth due resources in the seed and/or background resource concentration, and is necessary to 'kickstart' the growth/resource acquisition. This initial size limit is parameterized to be equal to one-tenth of the expected final size of an individual growing in isolation with available resources.

Whenever a plant's root system expands to overlap a resource point, the plant has a chance (detailed in Growth strategies and competition section) to acquire this resource and allocate it to growth.

With the successful acquisition of each resource point, the plant experiences an instantaneous growth (i.e. a jump

in size), and the upper size limit increases by an amount equal to the growth jump (i.e. growth is resource limited, and by acquiring resources this ceiling limit on size increases). The size of this jump is equal to the quality of the patch, *p*, and the individual plant's relative marginal benefit factor parameter, *mbf*. Individuals are not directly affected by competing neighbours, so root systems can overlap. Indirectly, plants growing in crowded areas, and whose root systems overlap with neighbours, risk finding themselves growing into areas depleted of resources by their competitors.

This method has been shown to successfully replicate the non-linear growth of an individual growing according to Gompertz growth functions (Purves and Law 2002; Schneider *et al.* 2006; Lv *et al.* 2008), where resource acquisition results in an increase in asymptotic limit and current growth rate (Croft *et al.* 2012; Croft 2013), as well as preserving results of competition between multiple plants (Croft 2013). It is noted that the Gompertz equations arise naturally via the Von Bertalanffy fish growth models (Lv and Pitchford 2007) which motivated this work. Simulating with linear growth and instantaneous resource depending growth as described here is significantly computationally quicker than direct implementation of Gompertz models (Croft 2013).

Note that, because this model concerns below-ground interactions, plant growth and root system growth are synonymous; one can consider above-ground growth to be reflected by below-ground growth, with above-ground effects such as shading and carbon limiting neglected (i.e. growth is purely below-ground resource limited). Root systems appear as circles representing their size, but this does not prevent the model from probabilistically accounting for finer scale structure, as detailed below.

Growth strategies and competition

The model allows root system growth strategies involving rapid growth of ephemeral and/or sparse root systems to be distinguished from those involving slower growth and possibly more exhaustive exploitation of local surroundings. Explicitly, at any time each plant's root system has a size (area) *A* and a probability determined by its 'RDE' of acquiring available resources which its root system overlaps.

Figure 2 summarizes, schematically, the way in which these properties change with time for four contrasting idealized plant growth strategies (labelled 'species' for conciseness). Plants of type 1 are represented by red, type 2 by blue, type 3 by magenta and type 4 by green. For clarity, this colour scheme is maintained throughout all subsequent figures, with darker shading to indicate

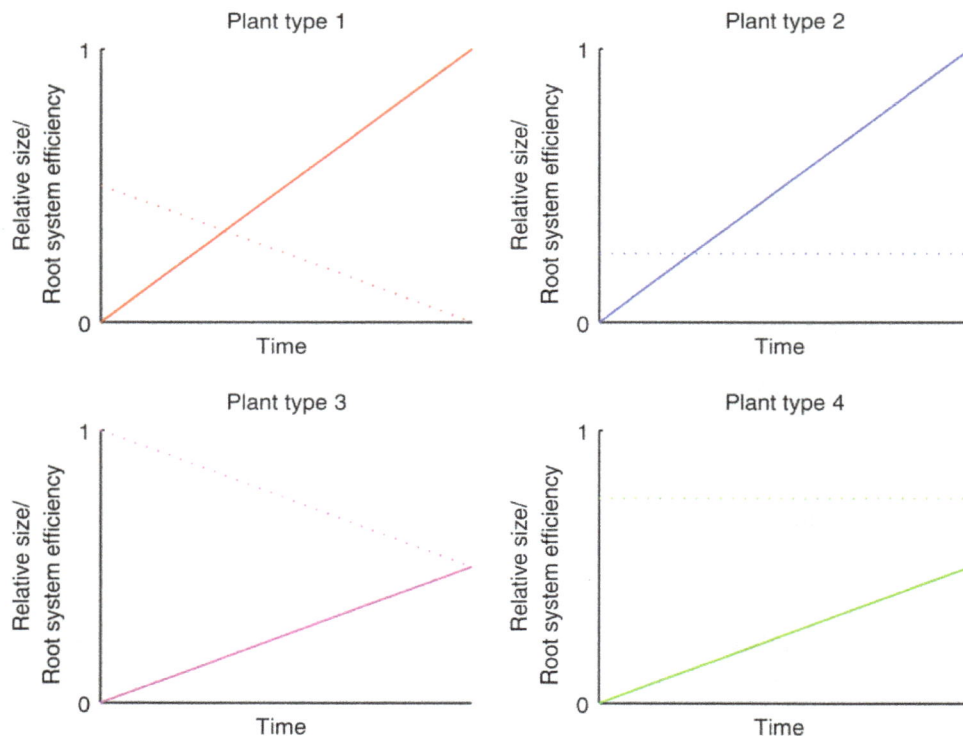

Figure 2. Visual representations of different growth strategies. The solid lines denote relative growth, and the dotted lines relative *RSE*. Plant types 1 and 2 experience faster growth rates at the expense of lower *RSE*; Plant types 3 and 4 instead have slower growth rates but higher *RSE*. Plant types 1 and 3 have declining *RSE*, while Plant types 2 and 4 have constant *RSE*, equal to the average *RSE* of types 1 and 3, respectively. Plant type 1 is represented by red lines, type 2 by blue lines, type 3 by magenta lines and type 4 by green lines. This colour coding will remain consistent throughout subsequent figures.

plants grown in uniformly random environments and lighter shading to indicate growth in patchy environments.

Strategies are defined by relative growth rates and relative abilities to acquire available resources. As well as relative levels, some plants exhibit a constant ability to obtain available resources, while others see this ability decline with time. Type 1 and 2 both grow equally quickly, and have the same average *RSE* throughout the period of simulation. However, type 2 has a root system whose *RSE* starts relatively high and then declines with time (reflecting an ephemeral root system where the ability to forage effectively diminishes as the root system becomes more diffuse) whereas type 2 has a constant *RSE* (reflecting more investment in maintenance of the root system at the expense of initial efficiency). Type 3 and 4 grow slowly (relative to type 1 and 2), but they benefit from investing in a more efficient root system (i.e. one which will statistically capture more available resource per unit area occupied) which better exploits available resources in a way which either starts high and declines with time (type 3) or remains constant with time (type 4).

Parameter values for resource quantity/quality are chosen such that, when grown in isolation in uniformly random

environments, all four plant species perform equally well on average. This provides a normalized level of performance against which to measure the relative performance of the different plant species in varying conditions. By accounting for trade-offs in this way and normalizing behaviour in idealized conditions, the study retains its focus on the role of intra- and inter-specific competition, and its modulation by resource heterogeneity.

Results

The numerical implementation of the model is carried out as follows. First, in a series of 'control' tests, a single individual is placed in an environment and allowed to grow in the absence of competition. The results from this (Fig. 3) not only confirm that, on average each of the 'types' of plant under consideration performs equally well, but also illustrate where environmental heterogeneity can cause substantial variability about that average. Having established a level playing field for plants in isolation, the simulations are then extended to model the growth of several plants competing within a monoculture (Fig. 4), and finally to investigate competition and growth within a mixed community (Figs 5–7).

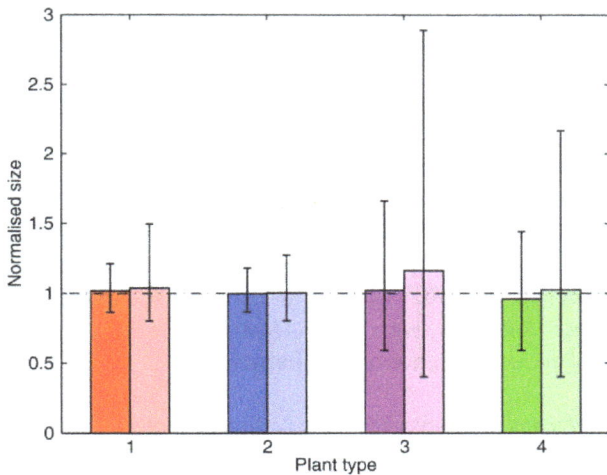

Figure 3. Normalized mean size of individuals grown in control conditions (i.e. in isolation) in uniformly random (darker bars) and patchy heterogeneous (lighter bars) environments. The mean size across all four plant types in the uniformly random environments is taken as the base level to which results are normalized. Results for each plant type/environment type combination show mean size for 10 000 repetitions, with vertical bars denoting 5th and 95th percentiles. Plant type 1 is represented by red bars, type 2 by blue bars, type 3 by magenta bars and type 4 by green bars.

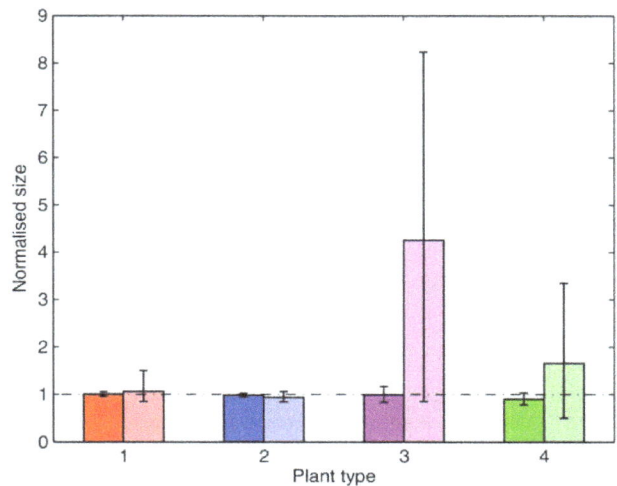

Figure 5. Normalized mean size of individuals when grown in mixed competition with all plants types. Results normalized against baseline (control tests in uniformly random environments) results. Darker bars show results in uniformly random environments, with lighter bars showing results in patchy heterogeneous environments. Vertical bars denote 5th and 95th percentiles for normalized plant type population level results across 1000 repetitions of 64 plants (16 of each type). Plant type 1 is represented by red bars, type 2 by blue bars, type 3 by magenta bars and type 4 by green bars.

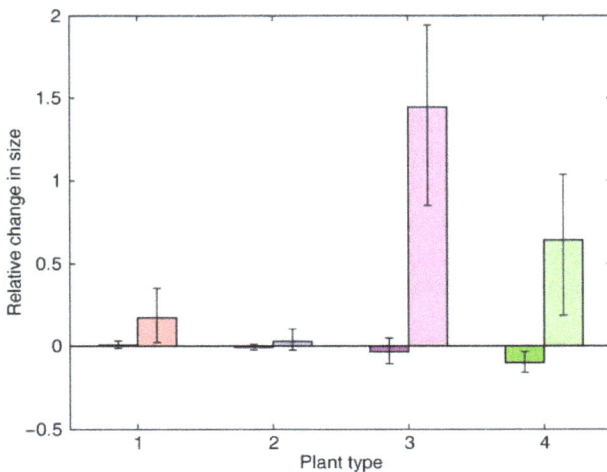

Figure 4. Relative change in size for each plant type when grown as a monoculture (in competition with its own kind) compared with baseline (control tests in uniformly random environments) results. Darker bars show results in uniformly random environments, with lighter bars showing results in patchy heterogeneous environments. Vertical bars denote 5th and 95th percentiles for normalized population level results across 100 repetitions of 64 plants. Plant type 1 is represented by red bars, type 2 by blue bars, type 3 by magenta bars and type 4 by green bars.

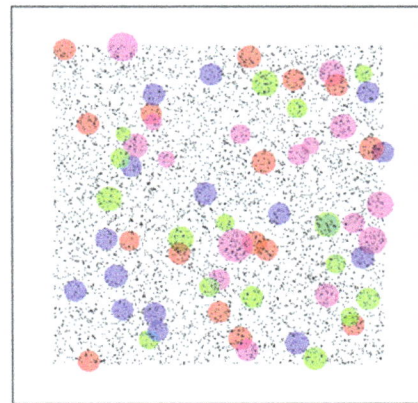

Figure 6. Visual representation of mixed competition experiments in a uniformly random environment. Population comprised of 64 plants (16 of each plant type) placed uniformly randomly within the environment. Environment has periodic boundaries which are not shown in this figure for clarity of distribution of individuals and their sizes. Plants of type 1 are represented by red circles, type 2 by blue circles, type 3 by magenta circles and type 4 by green circles.

The different plant types were tested in control, monoculture competition and mixed competition conditions within the uniformly random and patchy environments.

Throughout the results (Figs 3–5), each pair of grouped bars represent an individual species, with the darker (left hand) bars signifying growth in uniformly random environments, and the lighter (right hand) bars growth in patchy environments. The different plant types continue to be represented in figures by the same colours as in Fig. 2.

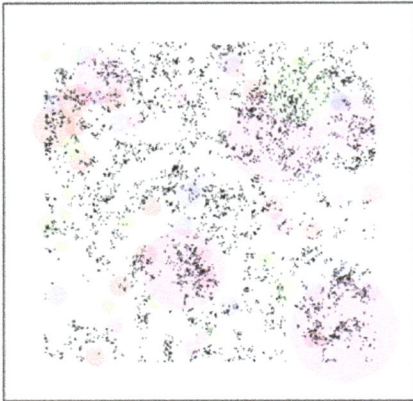

Figure 7. Visual representation of mixed competition experiments in a patchy heterogeneous environment. Population comprised of 64 plants (16 of each plant type) placed uniformly randomly within the environment. Environment has periodic boundaries which are not shown in this figure for clarity of distribution of individuals and their sizes. Plants of type 1 are represented by red circles, type 2 by blue circles, type 3 by magenta circles and type 4 by green circles.

Control tests: one plant in isolation

The results for the control tests (individuals grown in isolation) are summarized in Fig. 3. Bars show average performance across 10 000 individuals, with the 5th and 95th percentiles shown to demonstrate relative variability.

Uniformly random environments. As discussed in Methods, the quality of the individual resource patches (in terms of the marginal benefit to the acquiring plant) was chosen so as to best normalize performance across the four different plant types in control conditions within uniformly random environments. As such, when grown in isolation within the uniformly random environments, average performance is relatively even among the different plant types and their different growth strategies, with Plant types 3 and 4 (the slower growing plants types with higher *RSE*) showing higher relative variability.

Patchy environments. When the control tests are repeated within the patchy environments, average performance remains largely unchanged from the comparative results for uniformly random environments. All plant types exhibit little change in average performance, but Plant types 3 and 4 experience a significant increase in variability.

Monoculture tests: intraspecific competition

Figure 4 shows the relative change in performance for each of the four plant types when grown as a monoculture in competition. Relative performance is gauged against the baseline normalized performance for individuals grown in control conditions in uniformly random environments (Fig. 3). Bars show average performance across 100 populations of 64 plants, with the 5th and 95th percentiles shown to demonstrate relative variability.

Uniformly random environments. None of the plant species exhibit any important change in average performance when grown within competitive monocultures in the uniformly random environments. Plant types 1 and 2 (the faster growing plants types with lower *RSE*) also exhibit very little variability, but Plant types 3 and 4 (the slower growing plants types with higher *RSE*) demonstrate slightly higher variability, with type 4 (constant *RSE*) showing a small reduction in average performance.

Patchy environments. The introduction of competition within monocultures in patchy heterogeneous environments sees a significant shift in the relative performance across the different plant types. Of the two faster growing plant types with lower *RSE*, Plant type 1 (decreasing *RSE*) demonstrated a small increase in performance while Plant type 2 (constant *RSE*) experienced little difference compared with the control tests. In contrast, the slower growing plant types with higher *RSE* exhibited significant gains in performance when grown as monocultures compared with in control conditions. Plant type 3 (declining *RSE*) experienced a much larger gain than type 4 (constant *RSE*). Variability of results also increased markedly, especially in Plant types 3 and 4.

Mixed competition: community-level productivity

The results for when all four plant types are grown simultaneously in mixed competition are summarized in Fig. 5. Bars show the average performance across 1000 sub-populations of 16 plants, with the 5th and 95th percentiles shown to demonstrate relative variability.

Uniformly random environments. When grown in mixed competition within uniformly random environments, none of the four plant species demonstrated any significant difference in performance from the control baseline result. Figure 6 shows a visualization of one of the simulation runs.

Patchy environments. Growing all four plant types together in mixed competition within patchy environments resulted in significant gains for Plant type 3. Plant type 4 had little change from performance as a monoculture, but still outperformed the faster growing plants types with lower *RSE* (types 1 and 2) which demonstrated little difference in performance from previous numerical simulations. Plant types 3 and 4 (the slower growing plants types with higher *RSE*) demonstrate large variability. Figure 7 shows a visualization of one of the simulation runs.

Discussion

There is increasing interest in applying the more developed models of animal behaviour to plants to explain foraging behaviour. However, important differences between plants and animals exist. For example, animals will often only be able to exploit one 'patch' at a time, and thus must decide to exploit that patch or try to find a potentially more rewarding patch. Conversely, roots may simultaneously exploit several patches of varying quality. However, 'decisions' are still required by the plant in determining which of these patches to fully exploit (Duke and Caldwell 2000; Hodge 2009). In this work, four different plant types with similar behaviour in isolation were tested under a number of different combinations of conditions of competition and resource distribution. The different plant parameterizations trade growth rate and initial size constraints against the root systems' effectiveness (*RSE*) in acquiring resources. Even without explicit plastic root responses such as altered root length, root demography etc. (see Hodge 2004, 2006), it is shown that resource distribution could have significant effects on the outcomes of different growth scenarios, with competitive growth being significantly influenced by resource heterogeneity.

The variability in the simulation results arises principally through the environment (resources) and probabilistic nutrient acquisition, and indirectly by the neighbourhood (competitors). The plants possessed no ability to respond directly to their environmental conditions, and therefore 'grew' in a purely passive manner. For two of these plant types (types 1 and 2; the faster growing plant types with lower overall relative root system effectiveness), there was no significant difference in final size irrespective of the presence (or nature) of competition or the resource distribution. In contrast, for the other two plant types (types 3 and 4; the slower growing plant types with higher root system effectiveness) there was a markedly different performance depending on growing conditions. The notion of plants of differing growth strategies trading scale against precision of response to a patchy environment is not new (Campbell et al. 1991), though is also far from being universally accepted as being the norm across all plant species (see Kembel and Cahill 2005; Kembel et al. 2008). However, precision of foraging is not a fixed trait (Wijesinghe et al. 2001) and the response by the plant can vary depending on the way nutrient patches are presented to the plant, again highlighting the importance of the attributes of the patch to the response observed.

The presence of competitors can influence root placement and foraging capability (see Jumpponen et al. 2002; Cahill et al. 2010; Mommer et al. 2012), and the outcomes of competitive interactions are not always predictable from extrapolations from growth as monocultures

(see Hodge 2003; Cahill et al. 2010; Padilla et al. 2013), nor in different ecosystems (cf. Jacob et al. 2013 with Mommer et al. 2010). However, in the model presented here, space and resources per plant were consistent between the different numerical simulations. Thus, the introduction of competition within this framework does not lead to a decrease in available space or resources per plant (which is recognized as an important consideration, although the impact of 'space' can be highly variable among species; see McConnaughay and Bazzaz 1991; Murphy et al. 2013; McNickle and Brown 2014). It should be noted that the possibility of local overcrowding does result in direct competition between neighbours for locally available resources. Growth into an area of overlap with a competitor will statistically mean growth into an area with lower average resources, reducing the scope for subsequent growth. Plants have been observed to demonstrate root segregation (Schenk et al. 1999) which makes sense from this perspective; however in different contexts they have been found to actively proliferate into areas of competition (Hodge et al. 1999c; Robinson et al. 1999). In patchy heterogeneous conditions, the increase in average performance by Plant types 3 and 4 when grown as monocultures as opposed to in isolation highlights an increased ability to exploit available resources. At the population level, both plant types had better per plant performance than when grown in isolation, reflecting the acquisition of a higher proportion of the available resources on average.

Although Plant types 1 and 2 demonstrate a slight reduction in performance when grown in mixed competition in patchy heterogeneous conditions (Fig. 5) compared with when grown as a monoculture (Fig. 4), this reduction is less than the increase in performance experienced by type 3 (as mentioned, type 4 sees little change in performance). This means that while Plant type 3 enjoys an advantage when grown in mixed competition, that advantage is not wholly at the expense of its competitors.

When observing real plants and their performance, behaviour and response to different environmental conditions and stresses, the consistent (if perhaps unhelpful) message is that results are context sensitive (reviewed by Hodge 2004; Karst et al. 2012). A remarkable number of different root traits that have been demonstrated to be important for nutrient acquisition from a heterogeneous or 'patchy' nutrient environment under different experimental conditions (Hodge 2004; Cahill and McNickle 2011). It follows that any model hoping to capture and replicate all observed behaviour is necessarily going to require a level of complexity and parameterization which, even if it were possible and the required knowledge and understanding were available, would negate the need for such models in the first place. In this work, elements such as plastic responses to the environment are omitted in favour of isolating and investigating mechanistic and stochastic-driven impacts of environmental heterogeneity on growth and competition.

The strength of the work presented here is the use of foraging analogies developed elsewhere to condense a number of these complex traits into two essential mechanistic factors: root system 'growth' and 'effectiveness' (RSE). By categorizing these factors (fast/slow growth, high/low RSE) and normalizing so that isolated plants in homogeneous environments behave identically on average, it is possible to isolate the predicted influence of these factors at the individual, population and community level in both homogeneous and patchy environments.

Day et al. (2003) observed that populations, when grown under conditions of varying levels of scale and heterogeneity, demonstrated little change in population level yields providing the same total levels of nutrient supply were available. Similarly, Casper and Cahill (1996, 1998) observed soil nutrient heterogeneity had no impact upon productivity or population structure of Abutilon theophrasti Medik. monocultures. In contrast, Hutchings and Wijesinghe (2008) observed resource distribution having a distinct effect on overall population level yield. These contrasting results demonstrate the importance of context sensitivity, and the work presented here displays both of these types of behaviour depending on plant characteristics and community composition.

The modelling framework developed here is, to our knowledge, unique in its consideration of stochastic root system growth, maintenance and competition in heterogeneous environments. O'Brien et al. (2007) move beyond the traditional 'zone of interaction models' (where interaction and overlap between root systems are typically controlled by predefined rules; see for example, Berger et al. 2002) to use a game-theoretic spatially explicit model to predict root system distribution of two competing plants. By simplifying resource uptake and depletion, the authors are able to solve deterministic equations for optimal (in cost-benefit terms) growth in competition. This reveals information about root proliferation, overlap and below-ground resource foraging consistent with some empirical studies, such as the reduction of lateral root spread in the presence of a competitor, and an increase in lateral root spread with the introduction of resource heterogeneity. However, while the model can accommodate environmental heterogeneity, an essentially deterministic model such as this cannot capture the stochastic growth dynamics present in reality.

Useful comparisons can also be made with Craine et al. (2005) and Craine (2006), where continuous-time uptake and growth mechanisms are employed to model root growth and competition, using a spatially explicit set of 2D (horizontal and vertical) root growth simulations and grid-based diffusion at small (cm) scales. This allows inferences to be made about optimal resource allocation and competition, contingent upon these simplifying assumptions, but does not allow generalization to more than two competitors. The modelling framework developed in our work adds a spatially explicit account of stochastic interaction and depletion of patchy resources, and includes multiple individuals and growth strategies; future hybrids of these modelling approaches may prove fruitful in resolving the mechanisms behind the context-dependent empirical results highlighted above.

Conclusions

This work shows that ideas, and mathematical and computational methods, borrowed from animal growth and foraging can be used to help to disambiguate the many context-dependent results observed in studies of plant root growth and plasticity. Combining complex processes into idealized properties of growth and efficiency allows the roles of resource heterogeneity and intra- and inter-specific competition to be disentangled. Returning to the questions presaged in the Introduction section:

(1) How does the growth strategy adopted by a single plant impact upon its performance in a monoculture?

In homogeneous environments, intra-specific competition has little impact on plant performance regardless of growth strategy. However, different growth strategies can lead to greatly different performance when grown in intra-specific competition in heterogeneous environments. In these conditions, sacrificing growth rate for RSE conveys a clear advantage to the population, and at the individual level provides a better chance of being 'lucky'.

(2) When plant species are grown in mixed competition for resources, what is the impact on individual, population and community productivity?

During inter-specific competition, there is very little difference in performance at the individual, population or community scale in homogeneous environments. However, in heterogeneous environments, the slower growing plants with higher RSE perform significantly better (on average and in terms of best performing individuals) than the other species, and also better than when grown in monocultures. Only a small part of this increase is at the direct expense of the other species, resulting in a community-level increase in productivity.

(3) What is the role of resource heterogeneity in the above questions?

In answering the first two questions, it is impossible to avoid the effect of resource heterogeneity. The results underline the fact that the effects of growth strategy, competition and resource distribution on individuals, populations and communities are intrinsically interlinked.

This work highlights the utility of mathematical and computational models to frame complex problems in a relatively simple and tractable form. Within this work, all four plant types operate within the same framework; they differ only in the parameterization of growth and *RSE*. Yet they are shown to display near identical or markedly different behaviour depending on the context. Different aspects of this context can be individually and independently adjusted to isolate the effects of one factor or another. The story which emerges is consistent with the empirical literature; individual factors generally do not have clear impacts on performance. It is only by considering all factors together that the impacts of different factors can be usefully assessed.

In the discussion of these and other experimental results, a large emphasis is placed on context. However, when talking about 'optimal' behaviour, and metrics of performance, one has to be mindful of exactly what it means to perform 'better' or 'optimally' (Currey *et al.* 2007; Preston *et al.* 2010). The results shown here complement experimental evidence in terms of performance and results under a given set of conditions, but a key strength of this approach is that such frameworks can be tested within an evolutionary context (Croft *et al.* 2012). The next step would be to compare the behaviour and performance of different strategies not just over multiple replications, but rather over a series of dependent iterations. It is arguable only when evolutionarily relevant metrics of performance and optimality are considered that a truly relevant context is considered.

Sources of Funding

This work was funded by the Biotechnology and Biological Sciences Research Council (BBSRC), UK.

Contributions by the Authors

S.A.C., J.W.P. and A.H. designed the research. S.A.C. performed the research and analysis. S.A.C., J.W.P. and A.H. wrote the paper.

Acknowledgements

We thank Professors Alastair Fitter, Richard Law and Calvin Dytham for their advice and guidance during the development of this work.

Literature Cited

Berendse F. 1981. Competition between plant populations with different rooting depths. II. Pot experiments. *Oecologia* **48**:334–341.

Berger U, Hildenbrandt H, Grimm V. 2002. Towards a standard for the individual-based modeling of plant populations: self-thinning and the field-of-neighborhood approach. *Natural Resource Modeling* **15**:39–54.

Cahill JF, McNickle GG. 2011. The behavioral ecology of nutrient foraging by plants. *Annual Review of Ecology, Evolution, and Systematics* **42**:289–311.

Cahill JF, McNickle GG, Haag JJ, Lamb EG, Nyanumba SM, St. Clair CC. 2010. Plants integrate information about nutrients and neighbors. *Science* **328**:1657.

Campbell BD, Grime JP, Mackey JML. 1991. A trade-off between scale and precision in resource foraging. *Oecologia* **87**:532–538.

Casper BB, Cahill JF. 1996. Limited effects of soil nutrient heterogeneity on populations of *Abutilon theophrasti* (Malvaceae). *American Journal of Botany* **83**:333–341.

Casper BB, Cahill JF. 1998. Population-level responses to nutrient heterogeneity and density by *Abutilon theophrasti* (Malvaceae): an experimental neighborhood approach. *American Journal of Botany* **85**:1680–1687.

Crain JM, Fargione J, Sugita S. 2005. Supply pre-emption, not concentration reduction, is the mechanism of competition for nutrients. *New Phytologist* **166**:933–940.

Craine JM. 2006. Competition for nutrients and optimal root allocation. *Plant and Soil* **285**:171–185.

Croft SA. 2013. Stochastic models of plant growth and competition. PhD Thesis, University of York, York, UK.

Croft SA, Hodge A, Pitchford JW. 2012. Optimal root proliferation strategies: the roles of nutrient heterogeneity, competition and mycorrhizal networks. *Plant and Soil* **351**:191–206.

Currey JD, Baxter PD, Pitchford JW. 2007. Variability of the mechanical properties of bone, and its evolutionary consequences. *Journal of the Royal Society Interface* **4**:127–135.

Day KJ, Hutchings MJ, John EA. 2003. The effects of spatial pattern of nutrient supply on yield, structure and mortality in plant populations. *Journal of Ecology* **91**:541–553.

Drew MC. 1975. Comparison of the effects of a localised supply of phosphate, nitrate, ammonium and potassium on the growth of the seminal root system, and the shoot, in barley. *New Phytologist* **75**:479–490.

Dudley SA, Murphy GP, File AL. 2013. Kin recognition and competition in plants. *Functional Ecology* **27**:898–906.

Duke SE, Caldwell MM. 2000. Phosphate uptake kinetics of *Artemisia tridentata* roots exposed to multiple soil enriched-nutrient patches. *Flora* **195**:154–164.

Farley RA, Fitter AH. 1999. The responses of seven co-occurring woodland herbaceous perennials to localized nutrient-rich patches. *Journal of Ecology* **87**:849–859.

Fitter AH. 1982. Influence of soil heterogeneity on the coexistence of grassland species. *Journal of Ecology* **70**:139–148.

Gersani M, Brown JS, O'Brien EE, Maina GM, Abramsky Z. 2001. Tragedy of the commons as a result of root competition. *Journal of Ecology* **89**:660–669.

Hess L, de Kroon H. 2007. Effects of rooting volume and nutrient availability as an alternative explanation for root self/non-self discrimination. *Journal of Ecology* **95**:241–251.

Hodge A. 2003. Plant nitrogen capture from organic matter as affected by spatial dispersion, interspecific competition and mycorrhizal colonization. *New Phytologist* **157**:303–314.

Hodge A. 2004. The plastic plant: root responses to heterogeneous supplies of nutrients. *New Phytologist* **162**:9–24.

Hodge A. 2006. Plastic plants and patchy soils. *Journal of Experimental Botany* **57**:401–411.

Hodge A. 2009. Root decisions. *Plant, Cell and Environment* **32**: 628–640.

Hodge A, Fitter AH. 2013. Microbial mediation of plant competition and community structure. *Functional Ecology* **27**:865–875.

Hodge A, Stewart J, Robinson D, Griffiths BS, Fitter AH. 1998. Root proliferation, soil fauna and plant nitrogen capture from nutrient-rich patches in soil. *New Phytologist* **139**:479–494.

Hodge A, Robinson D, Griffiths BS, Fitter AH. 1999a. Nitrogen capture by plants grown in N-rich organic patches of contrasting size and strength. *Journal of Experimental Botany* **50**:1243–1252.

Hodge A, Stewart J, Robinson D, Griffiths BS, Fitter AH. 1999b. Plant, soil fauna and microbial responses to N-rich organic patches of contrasting temporal availability. *Soil Biology and Biochemistry* **31**:1517–1530.

Hodge A, Robinson D, Griffiths BS, Fitter AH. 1999c. Why plants bother: root proliferation results in increased nitrogen capture from an organic patch when two grasses compete. *Plant, Cell and Environment* **22**:811–820.

Hodge A, Stewart J, Robinson D, Griffiths BS, Fitter AH. 2000a. Competition between roots and soil micro-organisms for nutrients from nitrogen-rich patches of varying complexity. *Journal of Ecology* **88**:150–164.

Hodge A, Stewart J, Robinson D, Griffiths BS, Fitter AH. 2000b. Spatial and physical heterogeneity of N supply from soil does not influence N capture by two grass species. *Functional Ecology* **14**: 645–653.

Hodge A, Stewart J, Robinson D, Griffiths BS, Fitter AH. 2000c. Plant N capture and microfaunal dynamics from decomposing grass and earthworm residues in soil. *Soil Biology and Biochemistry* **32**: 1763–1772.

Hutchings MJ, Wijesinghe DK. 2008. Performance of a clonal species in patchy environments: effects of environmental context on yield at local and whole-plant scales. *Evolutionary Ecology* **22**: 313–324.

Jackson RB, Manwaring JH, Caldwell MM. 1990. Rapid physiological adjustment of roots to localized soil enrichment. *Nature* **344**: 58–60.

Jacob A, Hertel D, Leuschner C. 2013. On the significance of belowground overyielding in temperate mixed forests: separating species identity and species diversity effects. *Oikos* **122**: 463–473.

Jumpponen A, Högberg P, Huss-Danell K, Mulder CPH. 2002. Interspecific and spatial differences in nitrogen uptake in monocultures and two-species mixtures in north European grasslands. *Functional Ecology* **16**:454–461.

Karst JD, Belter PR, Bennett JA, Cahill JF Jr. 2012. Context dependence in foraging behaviour of *Achillea millefolium*. *Oecologia* **170**:925–933.

Kembel SW, Cahill JF Jr. 2005. Plant phenotypic plasticity belowground: a phylogenetic perspective on root foraging trade-offs. *The American Naturalist* **166**:216–230.

Kembel SW, de Kroon H, Cahill JF Jr, Mommer L. 2008. Improving the scale and precision of hypotheses to explain root foraging ability. *Annals of Botany* **101**:1295–1301.

Lv Q, Pitchford JW. 2007. Stochastic Von Bertalanffy models, with applications to fish recruitment. *Journal of Theoretical Biology* **244**:640–655.

Lv Q, Pitchford JW, Schneider MK. 2008. Individualism in plant populations: using stochastic differential equations to model individual neighbourhood-dependent plant growth. *Theoretical Population Biology* **74**:74–83.

Maina GG, Brown JS, Gersani M. 2002. Intra-plant versus inter-plant root competition in beans: avoidance, resource matching or tragedy of the commons. *Plant Ecology* **160**:235–247.

Malamy JE. 2005. Intrinsic and environmental response pathways that regulate root system architecture. *Plant, Cell and Environment* **28**:67–77.

McConnaughay KDM, Bazzaz FA. 1991. Is physical space a soil resource? *Ecology* **72**:94–103.

McNickle GG, Brown JS. 2014. An ideal free distribution explains the root production of plants that do not engage in a tragedy of the commons game. *Journal of Ecology* **102**:963–971.

Mommer L, van Ruijven J, de Caluwe H, Smit-Tiekstra AE, Wagemaker CAM, Ouborg NJ, Bögemann GM, Van der Weerden GM, Berendse F, de Kroon H. 2010. Unveiling belowground species abundance in a biodiversity experiment: a test of vertical niche differentiation among grassland species. *Journal of Ecology* **98**:1117–1127.

Mommer L, van Ruijven J, Jansen C, van de Steeg HM, de Kroon H. 2012. Interactive effects of nutrient heterogeneity and competition: implications for root foraging theory? *Functional Ecology* **26**: 66–73.

Murphy GP, File AL, Dudley SA. 2013. Differentiating the effects of pot size and nutrient availability on plant biomass and allocation. *Botany* **91**:799–803.

O'Brien EE, Brown JS, Moll JD. 2007. Roots in space: a spatially explicit model for below-ground competition in plants. *Proceedings of the Royal Society B: Biological Sciences* **274**:929–934.

Padilla FM, Mommer L, de Caluwe H, Smit-Tiekstra AE, Wagemaker CA, Ouborg NJ, de Kroon H. 2013. Early root overproduction not triggered by nutrients decisive for competitive success belowground. *PLoS ONE* **8**:e55805.

Pitchford JW. 2013. Applications of search in biology: some open problems. In: *Search theory: a game theoretic perspective*. New York: Springer.

Pitchford JW, Brindley J. 2001. Prey patchiness, predator survival and fish recruitment. *Bulletin of Mathematical Biology* **63**: 527–546.

Pitchford JW, James A, Brindley J. 2003. Optimal foraging in patchy turbulent environments. *Marine Ecology Progress Series* **256**: 99–110.

Pitchford JW, James A, Brindley J. 2005. Quantifying the effects of individual and environmental variability in fish recruitment. *Fisheries Oceanography* **14**:156–160.

Pregitzer KS, Hendrick RL, Fogel R. 1993. The demography of fine roots in response to patches of water and nitrogen. *New Phytologist* **125**:575–580.

Preston MD, Pitchford JW, Wood AJ. 2010. Evolutionary optimality in stochastic search problems. *Journal of the Royal Society Interface* **7**:1301–1310.

Purves DW, Law R. 2002. Experimental derivation of functions relating growth of *Arabidopsis thaliana* to neighbour size and distance. *Journal of Ecology* **90**:882–894.

Robinson D, Linehan DJ, Gordon DC. 1994. Capture of nitrate from soil by wheat in relation to root length, nitrogen inflow and availability. *New Phytologist* **128**:297–305.

Robinson D, Hodge A, Griffiths BS, Fitter AH. 1999. Plant root proliferation in nitrogen-rich patches confers competitive advantage. *Proceedings of the Royal Society B: Biological Sciences* **266**: 431–435.

Robinson D, Hodge A, Fitter AH. 2003. Constraints on the form and function of root systems. In: de Kroon H, Visser EJW, eds. *Root ecology*. New York: Springer, 1–31.

Schenk HJ, Callaway RM, Mahall BE. 1999. Spatial root segregation: are plants territorial? *Advances in Ecological Research* **28**:145–180.

Schneider MK, Law R, Illian JB. 2006. Quantification of neighbourhood-dependent plant growth by Bayesian hierarchical modelling. *Journal of Ecology* **94**:310–321.

Sims DW, Southall EJ, Humphries NE, Hays GC, Bradshaw CJ, Pitchford JW, James A, Ahmed MZ, Brierley AS, Hindell MA, Morritt D, Musyl MK, Righton D, Shepard ELC, Wearmouth VJ, Wilson RP, Witt MJ, Metcalfe JD. 2008. Scaling laws of marine predator search behaviour. *Nature* **451**:1098–1102.

Tilman D. 1982. *Resource competition and community structure*. Princeton: Princeton University Press.

Wijesinghe DK, John EA, Beurskens S, Hutchings MJ. 2001. Root system size and precision in nutrient foraging: responses to spatial pattern of nutrient supply in six herbaceous species. *Journal of Ecology* **89**:972–983.

15

Experimental assessment of factors mediating the naturalization of a globally invasive tree on sandy coastal plains

Thalita G. Zimmermann[1]*, Antonio C. S. Andrade[1] and David M. Richardson[2]

[1] Laboratório De Sementes. Instituto De Pesquisas Jardim Botânico Do Rio De Janeiro. Rua Pacheco Leão, 915, Jardim Botânico, Rio De Janeiro, RJ 22460-030, Brazil
[2] Department of Botany and Zoology, Centre for Invasion Biology, Stellenbosch University, Matieland 7602, South Africa

Guest Editor: Heidi Hirsch

Abstract. As all naturalized species are potential invaders, it is important to better understand the determinants of naturalization of alien plants. This study sought to identify traits that enable the alien tree *Casuarina equisetifolia* to overcome barriers to survival and reproductive and to become naturalized on sandy coastal plains. Restinga vegetation in Brazil was used as a model system to conceptualize and quantify key stressors (high temperature, solar radiation, drought and salinity) which can limit the initial establishment of the plants. Experiments were conducted to evaluate the effects of these environmental factors on seed persistence in the soil (field), germination (laboratory), survival, growth, phenotypic plasticity and phenotypic integration (greenhouse). Results show that the expected viability of the seeds in the soil was 50 months. Seeds germinated in a similar way in constant and alternating temperatures (20–40 °C), except at 40 °C. Low light, and water and salt stresses reduced germination, but seeds recovered germination when stress diminished. Young plants did not tolerate water stress (<2 % of soil moisture) or deep shade. Growth was greater in sunny than in shady conditions. Although a low degree of phenotypic plasticity is important in habitats with multiple stress factors, this species exhibited high germination plasticity, although young plants showed low plasticity. The positive effect of phenotypic integration on plastic expression in the shade shows that in stressful environments traits that show greater phenotypic plasticity values may have significant phenotypic correlations with other characters, which is an important factor in the evolutionary ecology of this invasive species. Long-term seed persistence in the soil, broad germination requirements (temperature and light conditions) and the capacity to survive in a wide range of light intensity favours its naturalization. However, *C. equisetifolia* did not tolerate water stress and deep shade, which limit its potential to become naturalized on sandy coastal plain.

Keywords: Biological invasions; germination; growth; phenotypic integration; phenotypic plasticity; shade; survival; trait; tree invasions; water stress.

* Corresponding author's e-mail address: thalitagabriella@gmail.com

Introduction

Biological invasions are conceptualized as occurring along an introduction–naturalization–invasion continuum (Blackburn et al. 2011; Richardson and Pyšek 2012). As all naturalized species have the potential to become invasive, naturalization is a critical stage of the invasion process (Richardson and Pyšek 2012). For an introduced population to become naturalized, it must overcome biotic and abiotic barriers to survival and reproduction (Blackburn et al. 2011). Research on naturalized populations is important for elucidating the ecological factors and species traits that mediate the transition of a population from casual to naturalized, but it is surprising that this phase is rarely explored in studies of invasions (Pyšek et al. 2008; Richardson and Pyšek 2012). In general, reproductive traits, such as seed bank longevity, seed germination and seedling survival and growth (Pyšek and Richardson 2007), in addition to high phenotypic plasticity and high phenotypic integration (Pigliucci 2003; Hamilton et al. 2005; Richards et al. 2006) are considered to be important determinants of invasiveness. However, we know of no studies that evaluate the importance of all these factors together in mediating the transition of a population from casual to naturalize.

High levels of plasticity can increase the average fitness of a species, thereby expressing advantageous phenotypes that facilitate invasion across a wide range of new environments (Richards et al. 2006; Funk 2008; Molina-Montenegro et al. 2012). Nonetheless, plasticity is not necessarily a crucial factor in invasiveness (Peperkorn et al. 2005; Godoy et al. 2011; Palacio-López and Gianoli 2011). It seems to be less relevant in habitats that experience the effects of multiple stress factors, where convergence to a low degree of phenotypic plasticity and high canalization may be advantageous (Valladares et al. 2007). Considering that the phenotype expressed by plants is the result of the integration of their characters in each environmental condition (Pigliucci 2003), it has been suggested that phenotypic integration (i.e. the pattern and magnitude of functional correlation among different plant traits, Pigliucci 2003), may play a role in constraining phenotypic plasticity (Gianoli 2004; Valladares et al. 2007; Gianoli and Palacio-López 2009). An integrated phenotype may have an important advantage in the invasion process because it can respond to environmental variation more efficiently, producing a more adaptive response to the environment than less integrated phenotypes (Schlichting 1989; Gianoli 2004). Consequently, plants with a more integrated phenotype should be less plastic than plants that show lower number of correlations among their traits (Valladares et al. 2007; Gianoli and Palacio-López 2009).

However, phenotypic plasticity and phenotypic integration can both favour plant fitness (Godoy et al. 2012). Further research is thus necessary to elucidate the direction of phenotypic change in invasive species for a better understanding of how ecological traits are influenced by new environmental conditions (Flores-Moreno et al. 2015).

A genus of trees that has been widely planted outside its native range is *Casuarina* (Casuarinaceae) (Potgieter et al. 2014a). Casuarinas differ from other well-studied invasive trees (e.g. Australian acacias, *Eucalyptus* spp. and *Pinus* spp.; Kueffer et al. 2013) in that they invade a distinctive set of habitats (e.g. beach crests, rock coasts, young volcanic flows, riparian ecosystems) and their requirements for successful invasion differ from those of other tree taxa (Morton 1980; Potgieter et al. 2014a, c). This genus provides a useful model for understanding how interactions between ecological factors and species traits mediate naturalization and other stages along the introduction–naturalization–invasion continuum (Potgieter et al. 2014a). *Casuarina equisetifolia* L. is the most widely planted species in the genus and is one of the most invasive alien tree species in the world (Rejmánek and Richardson 2013, Potgieter et al. 2014a); it invades mainly coastal regions (Wheeler et al. 2011). In Brazil, the species was introduced along the entire coast, especially in sandy coastal plains (I3N Brazil 2015). The species is widely naturalized, but it is not yet invasive in this country (Zenni and Ziller 2011; Potgieter et al. 2014a). Given the large extent of climatically suitable areas for *C. equisetifolia* in Brazil, including many areas with substantial plantings (high propagule pressure), further naturalizations and invasions of this species are likely in the future (Potgieter et al. 2014a).

Sandy coastal plain ecosystems are characterized by multiple stressful conditions (e.g. high solar radiation, drought, nutrient-poor sandy substrate, high temperatures and salinity, Reinert et al. 1997; Hesp and Martínez 2007). These factors have the potential to limit germination, survival and growth of plants (Maun 1994; Scarano 2009). Communities of sandy coastal plains called 'restinga' (sensu Araújo 1992) occupy 79 % of the Brazilian coast (5.820 km), extending from the Equator to below the Tropic of Capricorn—a distance of ~3.900 km (67 % in the tropics; Lacerda et al. 1993). The restingas occur on sandy soils and have several formations which vary in species composition and vegetation structure, due to varying abiotic conditions (Lacerda et al. 1993). Some restingas have a patchy structure and are classified as open scrub vegetation. In many parts of the world, extensive areas of sandy coastal plains are covered by open scrub vegetation that may occur behind the coastal thicket or farther inland (Araújo and Pereira

2002). This vegetation provides a spatial heterogeneity of resources, resulting in two distinct microsites: vegetation patches and open areas (Araújo and Pereira 2002) [**see Supporting Information—Fig. S1**]. Woody species (up to 5 m high) dominate and vines are also common components of the vegetation patches (Araújo and Pereira 2002, Araújo et al. 2009). Inside the patches, environmental conditions may be less harsh than in open areas due to higher water supply and lower solar irradiation (Gómez-Aparicio et al. 2005). Nevertheless, shade beneath patch canopies can limit plant growth by reducing photosynthesis (Callaway and Walker 1997; Hastwell and Facelli 2003). The two distinct environmental conditions found in the restinga (high irradiance and low water (open area) versus low irradiance and high water (patches) (Matos 2014)) allow for the evaluation of the combined effects of shade and drought in the naturalization process.

The restinga ecosystems are associated with the Brazilian Atlantic Forest domain which is highly degraded; only 11.7 % of the original vegetation remains, which 0.5 % comprises remaining restingas and mangroves (Ribeiro et al. 2009).The restinga is highly degraded (Araújo and Pereira 2002; Rocha et al. 2007) mainly as a result of vegetation removal for housing development, the collection of plants for sale and the establishment of alien plant species such as C. equisetifolia (Rocha et al. 2007). Despite its high invasive potential and its increasing biological and economic impacts on sandy coastal plains in many parts of the world (Potgieter et al. 2014a), relatively little is known about the ecophysiological traits that favour C. equisetifolia invasiveness. Thus, analysis of seed persistence in the soil, germination behaviour and plant growth performance in response to different environmental factors could allow a better understanding of the factors that make C. equisetifolia one of the most widespread invasive trees in coastal regions of the world (Rejmánek and Richardson 2013; Potgieter et al. 2014a).

The main objective of the study was to identify the sets of traits that enable C. equisetifolia to overcome the survival and reproductive barriers (Blackburn et al. 2011) and to become naturalized in the restinga. The hypotheses were: (i) C. equisetifolia forms a persistent soil seed bank that favours invasion; (ii) given the wide climatic amplitude in its native range (Whistler and Elevitch 2006; Potgieter et al. 2014a), C. equisetifolia seeds can germinate across a broad range of temperatures; (iii) because the species is shade-sensitive and mostly found near water bodies (U.S. National Research Council 1984; Parrotta 1993), drought and shade should reduce its germination, survival and growth; (iv) C. equisetifolia should display a low trait plasticity and (v) phenotypic plasticity and

phenotypic integration of traits are inversely related in this species (Gianoli 2004; Gianoli and Palacio-López 2009). A better understanding of the traits and the environmental factors that facilitate its naturalization will help to elucidate the magnitude of the invasion debt (sensu Rouget et al. 2016) for this species in many parts of the world where it has been planted but where invasions have not yet manifested. This study will improve our knowledge about how key stressors (high temperature, solar radiation, drought and salinity) can limit the initial establishment of an alien species and the transition of a population from casual to naturalized. Further, understanding why and under which circumstances species become naturalized may facilitate the prediction of future invasions, determine the best ways to control invasive species, and elucidate the impact of invasive species on native communities (Pyšek and Richardson 2007; Richardson and Pyšek 2012).

Methods

Study species

Casuarina equisetifolia (Australian pine or coastal sheoak) is an evergreen, fast-growing tree that attains a height of 10–40 m. The species has the largest natural distribution in the genus and is native to the east coast of Australia and Southeast Asia (Parrotta 1993). Reproduction is mainly by seeds (Morton 1980; Apfelbaum et al. 1983), but it can also propagate vegetatively (Rentería 2007). Dispersal is mainly by wind (Morton 1980), but also by water (Rentería 2007) and birds (Ferriter et al. 2007). The species tolerates saline conditions and low soil fertility (Morton 1980). Symbiotic associations with N-fixing actinomycete in the genus Frankia as well as ecto-, endo- and arbuscular mycorrhizal fungi allow C. equisetifolia to grow on nutrient-poor substrates (Zhong et al. 1995, Diagne et al. 2013). It has been planted in coastal regions in many parts of the world, mainly to stabilize dunes and for windbreaks (Morton 1980; Parrotta 1993). Casuarina equisetifolia has the capacity to invade open areas in the dunes and replace the native vegetation, threatening biodiversity in coastal regions (Wheeler et al. 2011). Further, it produces large amounts of litter, which can limit the establishment of native plants (Hata et al. 2010). The species is naturalized in at least 32 countries and it has become invasive in 10 geographical regions, including North America (Florida), Central America, South America, Asia, the Middle East, southern Africa and on many islands (Pacific, Indian Ocean, Atlantic and Caribbean Islands) (Rejmánek and Richardson 2013; Potgieter et al. 2014a). In Brazil, it was introduced and disseminated mainly after 1950, especially in the restingas of southern, south-eastern and northeastern Brazil (I3N Brazil 2015). There are no records of the species being invasive in Brazil, although it is widely naturalized (Zenni and Ziller 2011; Potgieter *et al.* 2014*a*).

Study area

The study was conducted in a naturalized population of *C. equisetifolia* (sea level, 22° 58′S, 42° 01′W) in the restinga of the State Park of Costa do Sol, in the municipality of Arraial do Cabo, State of Rio de Janeiro, Brazil (Fig. 1). This is one of the largest *Casuarina* stands (2.2 ha) in the park, and has 0.31 individuals m^{-2} (3.048 ind ha^{-1}), average height of 7.27 ± 3.86 m and diameter at breast height of 5.77 ± 5.18 cm ($n = 450$). In the state of Rio de Janeiro, at least 42 % of restingas are degraded (Rocha *et al.* 2007), but this percentage is now probably substantially higher as disturbance in this ecosystem has increased markedly in recent years (Cosendey *et al.* 2016). The remaining restingas comprise fragments, mostly of small size, with few areas occurring within official Conservation Units (Rocha *et al.* 2007). One of the restingas with the most critical situations in terms of degradation is in the State Park of Costa do Sol (Rocha *et al.* 2007). This restinga is located between the Atlantic Ocean and the Araruama lagoon, the largest hypersaline lagoon in the world. This region is characterized by a hot, semiarid climate, with 800 mm of annual precipitation occurring predominantly during the summer (November to February) (Barbiére 1984). The mean annual

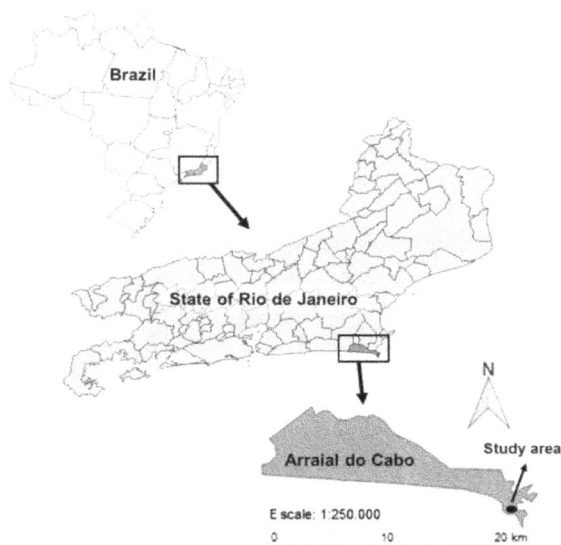

Figure 1. Study area (sea level, 22° 58′S, 42° 01′W) in the restinga of the State Park of Costa do Sol, in the municipality of Arraial do Cabo, State of Rio de Janeiro, Brazil.

temperature is 25 °C, with minimum and maximum temperatures of 12 and 36 °C, respectively (Scarano 2002).

Seed collection

Approximately 8000 seeds of *C. equisetifolia* were randomly collected from 20 trees, sampled with a minimal distance of 10 m from each other in August 2012. Mature seeds from opened dry dehiscent fruits were dried (18 °C; 18 % relative humidity) for 3–5 days, and hermetically stored in sealed plastic bags at -20 °C (Bonner 2008).

Seed longevity in the soil

To evaluate the longevity of *C. equisetifolia* seeds in the soil, the seeds were packed in nylon mesh bags with sterilized (autoclaved at 121 °C for 0.5 h) sandy soil collected in the restinga (open area). Seventy bags (40 seeds per bag) were buried at a depth of 5 cm in the same area as the seeds were collected. Groups of 10 bags were dug up after 1, 3, 6, 9, 12, 18 and 24 months and the viability of the seeds buried in the soil was evaluated in a laboratory by germination tests. To test the effect of the light in germination of buried seeds, germination tests were carried out under light (photoperiod of 8/16 h) and dark conditions. To compare the viability of the seeds ($n = 2800$) buried in the soil with optimal storage conditions, ~1500 seeds were stored at -18 °C (control group) over the same period that they were buried. Seed germination tests of the control group were carried out under light conditions. Seeds were germinated in Petri dishes (9 cm diameter), lined with two filter paper discs, moistened with 5 mL of distilled water. The germination tests had a randomized design, with five replicates of 40 seeds; the seeds in each bag constituted a replication.

Seed traits and germination tests

Dry weight and moisture content of the seeds (five replicates of five seeds) were determined according to the low-constant-temperature-oven method (103 °C/17 h; ISTA 1999). Length and width were measured with a digital calliper for 50 samaras (whole winged fruit, including the seed).

Germination tests were carried out to evaluate the effects of temperature, red/far-red light ratio (R:FR), water and salt stresses. The seeds were germinated in Petri dishes (9 cm diameter) lined with two filter paper discs, moistened with 5 mL of distilled water or specific osmotic solutions (sodium chloride (NaCl) or polyethylene glycol 8000 (PEG 8000)). The temperature of the germination chamber was determined by the temperature experiment. Unless light was an intended variable, a regime of 8 h light/16 h darkness was applied

(4×20 W white fluorescent lamps; total flux rate of 90 μmol/m²/s).

The temperature experiment was represented by constant temperatures of 15, 20, 25, 30, 35 and 40 °C (± 1.0 °C) and by alternating regimes of 25/20, 30/20, 35/20 and 40/20 °C (8/16 h, respectively; the alternating temperature treatment was 8 h in the light at the higher temperature and 16 h in the dark at the lower temperature). In the temperatures of 25, 30 and 30/20 °C the germination was also evaluated in the dark, and the Petri dishes were wrapped in two aluminium foils. The optimal germination temperature was used in light, water and saline stresses experiments.

The light experiment included six R:FR irradiance treatments: 0.0, 0.2, 0.4, 0.6, 0.8 and 1.0. Zero irradiance treatment was produced by wrapping the Petri dishes in two aluminium foils. The greatest R:FR treatment (1.0) was obtained by leaving the Petri dishes free of filters. Spectrum was provided by two fluorescent 22 W white lamps and one incandescent 15 W lamps, totalling 1.0 R:FR, which is close to the 1.19 R:FR of full sunlight (Smith 2000). The four remaining R:FR irradiance treatments were achieved by wrapping the Petri dishes with different colours of LEE filters. The R:FR irradiance was measured with sensors SKR 110 and SKP 215, coupled to SpectroSense (Skye Instruments Inc.).

The effect of water and salt stresses in the germination was tested with PEG 8000 and NaCl solutions, respectively. The osmotic potentials used were: 0.0, −0.25, −0.5, −0.75, −1.0, −1.25 and −1.5 MPa. These different potentials were found in the restinga (Martins et al. 2012). PEG 8000 and NaCl solutions were prepared according to Villela and Beckert (2001) and Salisbury and Ross (1992), respectively. To minimize water potential variation, seeds were transferred to a new Petri dish with the solution every 7 days. After 30 days, in a recovery treatment, the ungerminated seeds from PEG 8000 and NaCl solutions were washed with distilled water. The seeds were then transferred to Petri dishes with distilled water to evaluate the germination potential.

In all experiments, the positions of Petri dishes inside germination chambers were randomly changed every day. A seed was considered to have germinated when its radicle emerged to a length of 1 mm. Germination was recorded daily for 30 days, and germinated seeds were removed from Petri dishes. In the light experiment, the germination was evaluated in a dark and closed room, with a green safelight. Five replicates of 40 seeds were used in all experiments. Seeds that did not germinate were subjected to the application of pressure with tweezers, and were either empty or had been colonized by fungi.

Survival and growth

To minimize genetic variation, all seeds used in this experiment came from a single tree, so the seedlings were half-siblings. Seeds were germinated in germination chambers (30 °C; 8 h photoperiod) and after 2 months, seedlings were transplanted to individual plastic bags (2L) and transferred to the greenhouse of the Rio de Janeiro Botanic Garden. Soil substrate consisted of 1:1:1 volume homogenized mixture of soil of the area with C. equisetifolia invasion, sand collected inside the patches and bare sand. This mixture was used to provide a substrate with macro and micronutrients found in the restinga.

After 4 months, the height and stem diameter of the young plants of C. equisetifolia were measured. These plants were submitted to a factorial experiment to simulate the light intensity and water availability found in three microsites of the restinga (inside vegetation patches, edge and open area) and in the C. equisetifolia stands. This experiment had eight treatments, with four light levels and two watering regimes. The plants were separated in eight groups and there were no significant differences in initial height of the individuals between groups ($P < 0.05$). Distinct conditions of light were established with shade cages of wood (1 m×1 m×1 m), covered with cloth layers of different colours and thicknesses. The photosynthetic photon-flux density (PPFD%) and R:FR (mol mol^{-1}) inside each shade cage were: ∼2 %, 0.29 mol mol^{-1} (inside vegetation patches); ∼15 %, 0.48 mol mol^{-1} (edge); ∼70 %, 1.05 mol mol^{-1} (C. equisetifolia stand) and ∼100 %, 1.12 mol mol^{-1} (open area). At each light intensity, half of the young plants were grown under high water (>10 % of soil water content) and other half at low water conditions (<2 % of soil water content). Soil water content was monitored weekly from four soil samples per treatment, and was determined by gravimetric method (24 h/103 °C). The soil was irrigated once or twice a week by applying 30 (2 %, low water) to 150 ml (100 %, high water) of water.

The values of PPFD%, R:FR and watering regimes inside patches, edge and open area in the restinga were obtained by Matos (2014). Data of PPFD% and R:FR of C. equisetifolia stands were measured at 20 random points (68.5 ± 11.2 % PPFD%, 1.05 ± 0.10 μmol m^{-2} s^{-1}). The values of PPFD% were calculated taking as reference the mean full sunlight (100 % PPFD = 2305.3 μmol m^{-2} s^{-1}). All measurements were made at midday, on sunny cloud-free days, with a radiometer SKR-100 linked to a SpectroSense 2 SKL 904 (Skye Instruments, Llandrindod Wells, UK). To minimize experimental error due to light variability inside the shade cages, positions of the young plants were rotated once a week. For survival analysis, 15 individuals per treatment were monitored weekly, for 16 weeks. Plants that lost all their aerial structure and did not have any photosynthetic active leaf were recorded as dead.

At the end of the experiment, samples of all young plants that survived were harvested to measure stem length, main root length and collar diameter. Thereafter, they were separated into leaves stems and roots, and each fraction was dried (80 °C/48 h) and weighted. Total dry mass (TDM), Leaf mass fraction (LMF = leaf dry mass/plant dry mass), stem mass fraction (SMF = stem dry mass/plant dry mass), root mass fraction (RMF = root dry mass/plant dry mass), shoot: root ratio (RS = shoot dry mass/root dry mass), slenderness index (SI = stem height/collar diameter), specific stem length (SSL = stem length/stem dry mass), specific root length (SRL = root length/root dry mass), total leaf mass (TLM), total leaf area (TLA), specific leaf area (SLA = leaf area/total leaf mass) and leaf area ratio (LAR = leaf area/total plant dry mass). Leaf area and SLA were calculated following the protocol proposed by Gómez-Aparicio et al. (2006) for pines needles. Relative growth rates were calculated for total biomass (RGRb) and total leaf area (RGRa) using the pairing method (Evans 1972). RGR was calculated as RGR = $(\ln x_2 - \ln x_1)/(t_2 - t_1)$, where x_1 is the trait measured in time 1 (t_1) and x_2 is the trait measured in time 2 (t_2).

Phenotypic plasticity and phenotypic integration

Phenotypic plasticity in response to light for each trait was calculated as the relative distance plasticity index (RDPI = $\sum(\text{dij} \rightarrow i'j'/(xi'j' + xij))/n$), where n is the total number of distances, and j and j' are two individuals belonging to different treatments (i and i'). This index ranges from 0 (no plasticity) to 1 (maximal plasticity). Overall RDPI was calculated by summing all relative distances obtained and dividing by the total number of distances (Valladares et al. 2006). It was not possible to calculate RDPI in relation to water regime because almost all young plants died under low water conditions.

Phenotypic integration was estimated as the number of significant correlations ($P < 0.05$; Spearman's rank correlation coefficient) with the other traits (pairwise comparison) for 15 % of light (shady condition) and 100 % of light (sunny condition) (Gianoli and Palacio-López 2009). Phenotypic integration index in each light condition was calculated based on the variance of the eigenvalues of the correlation matrix between phenotypic traits (Wagner 1984).

Data analysis

In the experiments to determine seed longevity in the soil and the effect of temperature, PEG 8000 and NaCl solutions, germination was evaluated by germination percentage and germination rate ($v = \Sigma ni/(\Sigma ni \cdot ti)$); where 'ni' is the number of seeds germinated per day and 'ti' is the incubation time (days) (Labouriau and Pacheco 1978). In the light experiment only the final germination percentage was evaluated.

The longevity of C. equisetifolia seeds in the soil and cold conditions was analysed through germination percentage and germination rate parameters by linear regression. An analysis of covariance (ANCOVA) was used to compare the slopes of regression lines between the two storage conditions of the seeds (cold storage X soil storage) and the effect of the light conditions on germination of the buried seeds in the soil (light X dark). The ANCOVA was used with germination percentage and germination rate as dependent variables, storage and light conditions as factors and storage time (1, 3,…, 24 months) as covariate. The interaction between the conditions and time in the germination process was evaluated. Homogeneity of slopes was confirmed before conducting each ANCOVA. The differences in ANCOVA were in relation to the inclination.

The recovery germination percentage in the PEG 8000 and NaCl solutions was calculated by adding the germination values of each iso-osmotic solution and their respective germination value after transferal to distilled water. In the experiments of temperature, PEG 8000 and NaCl solutions data were analysed for normality using the Kolmogorov–Smirnov test and for homogeneity of variance using Levene's test. For data that did not show normality and/or variance homogeneity, germination percentage was arcsine $\sqrt{}$ transformed and germination rate transformed to $\log(x + 1)$ (Zar 1999). Germination percentage and germination rate were tested in a factorial ANOVA, followed by a post hoc Tukey's test ($P < 0.05$). In the experiment of light the relationship between germination percentage (y) and R:FR (x) was determined using a logistic function (Pearson et al. 2003) and described by the following equation: $y = a/\{1 + \exp[-((x-x_0)/b)]\}$, where a is a coefficient describing the maximum germination percentage, x_0 is a coefficient estimating the R:FR at 50 % of maximum germination and b is a coefficient of the slope of the germination response calculated from estimates of R:FR.

For survival analysis the Kaplan–Meier product limit method was used to estimate the survival function, and the log-rank test was used to assess for significant differences in survival curves among treatments. Cox regression was used to evaluate the effects of light, water and their interactions on probability of the death of young plants.

Growth analyses were performed only in treatments of 15, 70 and 100 % of light under high water conditions due to high mortality rates under low water conditions and in deep shade (2 %). To test the effect of light for all morphological and biomass allocation traits together Multivariate analysis of variance (MANOVA) was used. Traits that showed a significant effect in the MANOVA results were tested separately by one-way ANOVA, followed by a *post hoc* Tukey's test ($P < 0.05$). Before the analyses, normality of the data was tested by Shapiro–Wilk's W test and homoscedasticity by Levene's test. To check the homogeneity of covariance matrices Box M test and the Bartlet's test was used to check for sphericity. Where necessary, data were ln-transformed to correct for deviations from these assumptions. Differences in RGR were submitted to a one-way ANOVA, using Tukey's *post hoc* test ($P < 0.05$). To minimize the influence of outliers and reduce the within-harvest-variation, prior to growth analysis data were trimmed by the removing the smallest and the largest plant from each treatment (Barnett and Lewis 1978).

Regression analysis was used to determine whether phenotypic plasticity in response to light (dependent variable) and phenotypic integration of traits in shady and sunny conditions (independent variable) are inversely related in *C. equisetifolia*. Values of RDPI were log-transformed before analysis [log($x + 1$)]. To test the statistical significance between phenotypic integration indices across light conditions, 95 % confidence intervals for the overall R obtained in each environment were calculated by bootstrapping 1000 times (García-Verdugo *et al.* 2009).

Survival analysis was done using the 'survival' package (Therneau 2015) and phenotypic integration index and percentage of maximum possible integration were calculated using the 'PHENIX' package (Torices and Muñoz-Pajares 2015) in R version 3.0.3 (R Development Core Team 2014). The other analyses were done in Statistica (version 7.0, Statsoft Inc., Tulsa, OK). Graphical display was performed with R and Origin (version 8.0, OriginLab, MA, Cary, NC).

Results

Seed longevity in the soil

Casuarina equisetifolia seeds remained viable in the soil for at least 24 months, germinated under light and under dark (Fig. 2A) and had a predicted seed viability of 51.1 months ($y = 71.53 - 1.40x$). The interaction between storage condition and storage time was significant for germination rate (ANCOVA, $F = 90.19$, $P < 0.001$) but not for germination percentage (ANCOVA, $F = 1.18$, $P = 0.28$). There were no significant interactions between light conditions and storage time for germination percentage

(ANCOVA, $F = 6.72$, $P = 0.12$) and rate (ANCOVA, $F = 2.89$, $P = 0.09$) **[see Supporting Information—Table S2]**.

Germination percentage decreased over time ($R^2 = 0.55$, $P < 0.001$), but germination rate was not affected by the storage time ($R^2 < 0.001$, $P = 0.95$). In relation to the two storage conditions, there was no significant difference in germination percentage (ANCOVA, $F = 1.98$, $P = 0.16$; Fig. 2C). Nevertheless, germination rate was significantly higher in seeds stored in the soil than at -18 °C (ANCOVA, $F = 104.34$, $P < 0.001$; Fig. 2D). For seeds buried in the soil, germination percentage and rate were significantly higher under light than under dark conditions (ANCOVA, $F = 25.62$, $P < 0.001$; $F = 55.08$, $P < 0.001$, respectively; Fig. 2A and B) **[see Supporting Information—Table S2]**.

Seed traits and germination tests

Casuarina equisetifolia samaras had a dry weight of 0.75 ± 0.12 mg, moisture content of 10.8 ± 1.7 %, length of 5.9 ± 0.5 mm and width of 3.1 ± 0.3 mm. Under light, there were no significant differences in relation to constant and alternating temperature regimes, except for the constant temperature of 40 °C, which completely inhibited germination. The conditions that promoted the highest values of germination rates were 30 and 35 °C (Fig. 3). Thus, 30 °C was chosen as optimal germination temperature for *C. equisetifolia* and was used in the other germination experiments. Germination percentage at 25 and 30 °C was significantly reduced under dark compared to the light conditions (Table 1). Nevertheless, an alternating temperature of 30/20 °C did not have significant differences between the two regimes of light. The absence of luminosity reduced germination rate at all temperatures.

Casuarina equisetifolia seeds responded significantly to the treatments involving exposure to the various R:FR ratios (Fig. 4). Seeds were considered neutral photoblastic and showed higher germination percentages in light than in dark conditions. Seed germination increased slightly up to the higher R:FR, as indicated by the good fit to the data ($R^2 = 0.981$; $P < 0.01$) provided by the regression analysis. Germination was also sensitive to water and salt stresses, but the decrease in germination percentage and rate was higher in PEG 8000 than in NaCl solution (Table 2). Significant decreases in germination percentages were observed from the water and salt potential of -0.5 and -0.75 MPa, respectively. In both osmotic solutions germination was null from -1.0 MPa. Germination rate dropped as water and salt potentials decreased. After the seeds were transferred to distilled water (recovery treatment), total germination percentage in all treatments showed no significant differences from the control (Table 2).

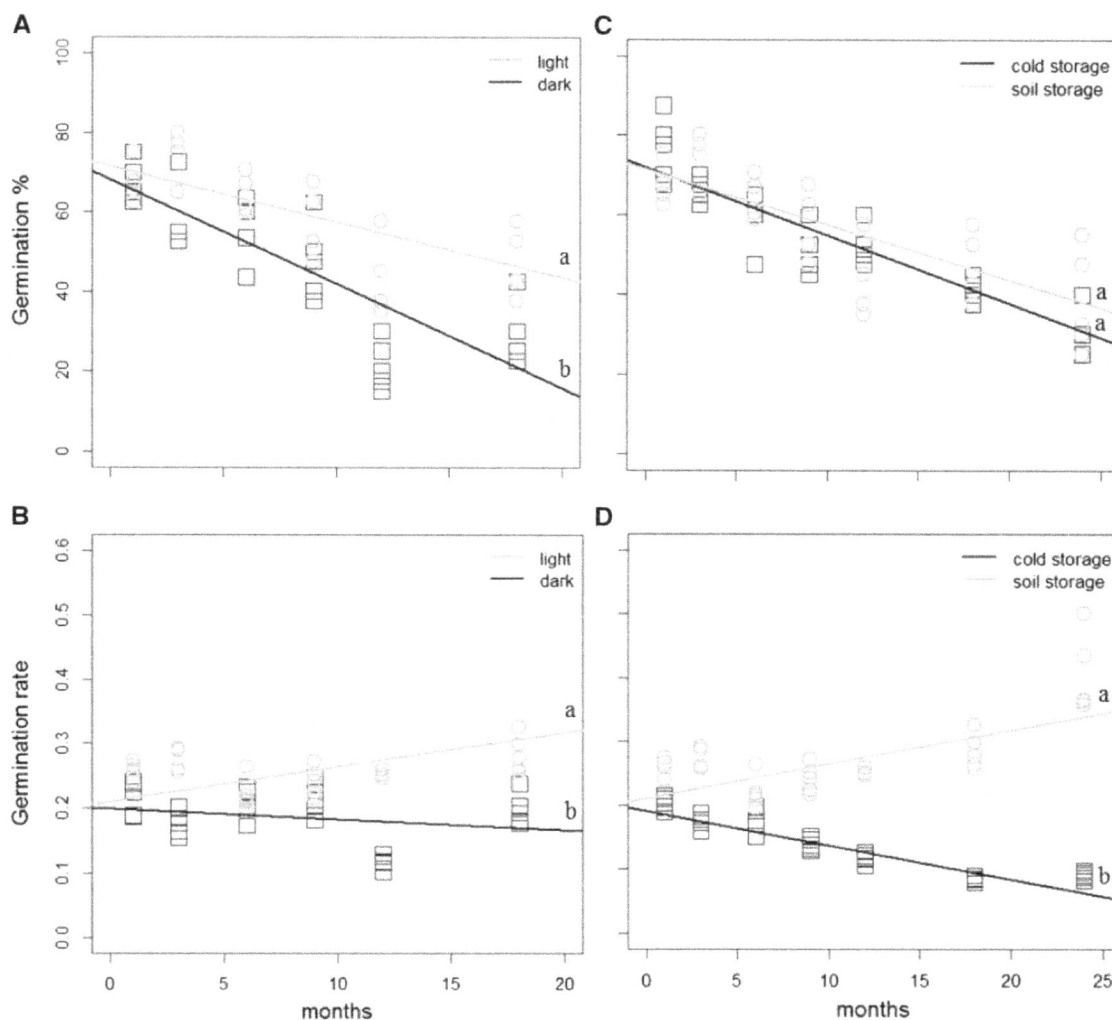

Figure 2. Relationships between storage period (months) and germination percentage (A and C) or germination rate (B and D) for *Casuarina equisetifolia* L seeds. (A and B) Germination percentage and rate of the seeds buried in the soil under light (photoperiod of 8/16 h; grey circles) and under dark (black squares) conditions; (C and D) germination percentage and rate of the seeds stored in cold conditions (black squares) and buried in the soil (grey circles) under light conditions. Data points were fitted with a linear regression function. Germination percentage: light = soil storage ($y = -1.50x + 72.11$; $R^2 = 0.52$; $P < 0.001$); dark ($y = -2.63x + 68.17$; $R^2 = 0.67$; $P < 0.001$); cold storage ($y = -1.71x + 71.87$; $R^2 = 0.77$; $P < 0.001$). Germination rate: light = soil storage ($y = 0.0009x + 0.25$; $R^2 = 0.005$; $P = 0.29$); dark ($y = -0.002x + 0.20$; $R^2 = 0.03$; $P = 0.20$) and cold storage ($y = -0.005x + 0.20$; $R^2 = 0.84$; $P < 0.001$). Different letters denote significant differences between the curves with ANCOVA ($P < 0.05$) [**see Supporting Information—Table S2**].

Survival and growth

Survival rates of the young plants had a different response to the combined effect of light and water stress (Fig. 5). Survival was improved under high water conditions [**see Supporting Information—Fig. S3A**] and the probability of death was ~46 times higher under low water than under high water conditions (Hazard Ratio = 45. 97, Wald's P value < 0.001). Similarly, 2 % light conditions had a negative effect on survival rates. Deep shade increased the risk of mortality almost 4 times (Hazard Ratio = 3.70, Wald's P value = 0.03). There were no significant differences between survival rates at 15, 70 and

100 % of light [**see Supporting Information—Fig. S3B**]. Under high water regime, survival was significantly lower at 2 % light, while there were no significant differences in survival between the light regimes under dry conditions (Fig. 5). The interaction between light and water was significant (Wald's P value = 0.008) because the effect of drought was higher under high light (70 and 100 % of light) than under low light (2 and 15 % of light) [**see Supporting Information—Fig. S3C and D**].

Light intensity had a significant effect in all morphological and biomass allocation traits that were measured (Table 3). Shade conditions (15 % of light) led to

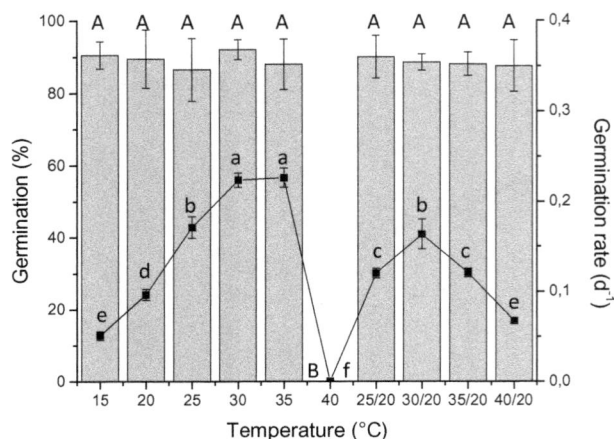

Figure 3. Seed germination (mean ± SD) of *Casuarina equisetifolia* at constant and alternating temperatures. Different upper case letters indicate significant differences in germination percentage (bars, left *y*-axis) and different lower case letters indicate significant differences between germination rate (line, right *y*-axis). ANOVA, *post hoc* Tukey's test ($P < 0.05$).

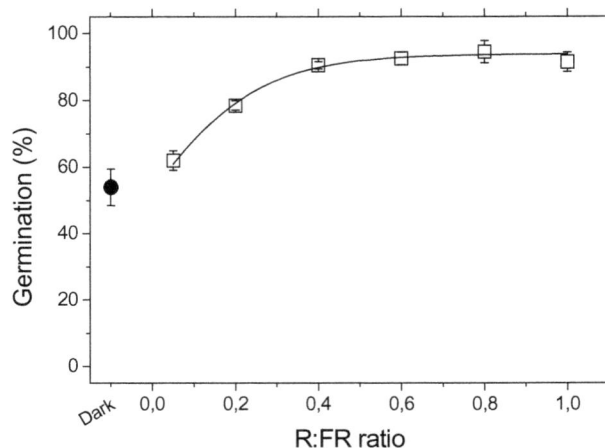

Figure 4. Effect of red/far-red light (R:FR) ratios on mean final germination percentage (±SD) of *Casuarina equisetifolia* L. Data points were fitted with a sigmoidal regression function (solid line; $R^2 = 0.983$; $P < 0.05$).

Table 1 Light and temperature effects on seed germination (mean ± SD) of *Casuarina equisetifolia*.

Temperature (°C)	Germination (%)		Germination rate (d^{-1})	
	Light	Dark	Light	Dark
25	86.5±8.6 a	16.5±6.5 b	17.1±1.2 a	12.7±0.9 b
30	92.0±2.7 a	54.0±5.5 b	22.4±0.8 a	11.6±0.3 b
30/20	88.5±2.2 a	92.5±4.0 a	16.4±1.7 a	14.1±0.6 b

Letters denote significant differences between the treatments (Student's *t*-test, $P < 0.05$).

significantly lower values of relative growth rates in total biomass and total leaf area and root mass fraction. At high light (70 and 100 % of light), leaf and shoot mass fraction, shoot: root fraction and slenderness index, were significantly lower than under shade. Young plants growing under shady conditions had higher values of specific leaf area and leaf area ratio than plants under sunny conditions. The only trait that differed significantly between 70 and 100 % light was specific stem length (Table 3).

Phenotypic plasticity and phenotypic integration

The overall value of RDPI was 0.32. Phenotypic plasticity in response to light changed in relation to the trait. The value of RDPI ranged between 0.08 (LMF) until 0.59 (SRL) (Fig. 6). Trait plasticity could be ranked as: SRL > SSL > TDM > TLM > SI > LAR > SR > SLA > TLA > RMF > SMF > LMF.

Table 2 Mean (± SD) germination percentage and germination rate of *Casuarina equisetifolia* seeds in response to osmotic (sodium chloride—NaCl) and water (polyethylene glycol 8000—PEG 8000) potential and recovery treatments.

Treatment	Potentials (MPa)	Germination (%)	Germination rate (10^{-2})	Recovery germination (%)
NaCl	0.00	91.5±4.2 a	20.3±1.8 a	91.5±4.2 ns
	−0.25	88.5±4.5 a	13.7±2.6 b	88.5±4.5
	−0.50	86.5±1.4 a	8.0±0.3 c	92.0±4.1
	−0.75	43.5±5.2 b	6.3±0.3 c	86.0±2.8
	−1.00	0 c	0 d	84.0±5.8
	−1.25	0 c	0 d	94.0±2.9
	−1.50	0 c	0 d	85.0±7.3
PEG 8000	0.00	91.5±4.2 a	20.3±1.8 a	91.5±4.2 ns
	−0.25	84.5±7.8 a	9.9±1.2 b	84.5±7.8
	−0.50	57.0±6.9 b	5.3±0.2 c	93.5±5.8
	−0.75	3.0±1.1 c	4.3±0.3 c	90.0±5.0
	−1.00	0 c	0 d	92.5±4.7
	−1.25	0 c	0 d	93.0±2.1
	−1.50	0 c	0 d	90.5±4.1

The letter codes indicate homogeneous groups among treatments, ns, not significant (Tukey's test, $P < 0.05$).

The phenotypic integration index and 95 % confidence intervals overlap between light conditions (15 % = 2.78 ± 1.97; 70 % = 2.53 ± 1.83; 100 % = 2.20 ± 1.40). The magnitude of individual correlations between the

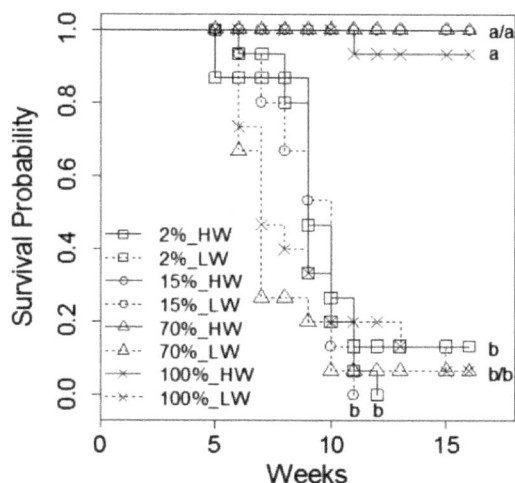

Figure 5. Survival curves of young plants of *Casuarina equisetifolia* (*n* = 15) under combined effects of light (2, 15, 70 and 100 %) and water regimes (HW—high water, LW—low water) over 16 weeks. Survival analysis was performed with the Kaplan–Meier product limit method. The letter codes indicate homogeneous groups (log-rank test, $P < 0.05$).

Table 3 Mean ± SD, *F* and *P* values (one-way ANOVA) on data for 12 morphological and *biomass allocation* traits of young *plants of Casuarina equisetifolia* in response to three light levels (15, 70 and 100 % of photosynthetic photon-flux density) after 16 weeks.

Traits	15 %	70 %	100 %	F	P
RGRb	0.02±0.00 b	0.04±0.00 a	0.04±0.00 a	168.88	<0.01
RGRa	0.02±0.02 b	0.03±0.00 a	0.03±0.00 a	24.85	<0.001
LMF	0.58±0.02 a	0.48±0.02 b	0.48±0.04 b	62.15	<0.001
SMF	0.20±0.01 a	0.14±0.01 b	0.16±0.02 b	35.27	<0.001
RMF	0.22±0.02 b	0.38±0.03 a	0.36±0.05 a	76.68	<0.001
SR	0.88±0.11 a	0.39±0.05 b	0.45±0.12 b	95.40	<0.001
SI	170.85±19.07 a	69.01±6.77 b	63.02±5.44 b	385.02	<0.01
SSL	129.09±21.33 a	33.97±3.96 b	28.95±3.12 c	480.98	<0.01
SRL	99.78±20.46 a	13.21±3.46 b	13.67±3.39 b	28.79	<0.001
SLA	227.65±20.93 a	116.62±20.96 b	113.07±15.08 b	153.26	<0.01
LAR	132.77±14.43 a	56.56±11.57 b	54.23±7.88 b	195.87	<0.01

The traits shown are relative growth rate in total biomass (RGRb) and total leaf area (RGRa), leaf mass fraction (LMF), stem mass fraction (SMF), root mass fraction (RMF), shoot: root ratio (SR), slenderness index (SI), specific stem length (SSL), specific root length (SRL), specific leaf area (SLA) and leaf area ratio (LAR). the letter codes indicate homogeneous groups among treatments for light intensities (Tukey's test, $P < 0.05$).

traits changed from one environment to another [see **Supporting Information—Table S4**]. Phenotypic plasticity was positively associated with phenotypic integration under shade ($R^2 = 0.51$, $P = 0.006$) (Fig. 7). Under sunny conditions, plasticity and integration of the traits showed no significant relationship ($R^2 = 0.011$, $P = 0.74$).

Discussion

Although *Casuarina equisetifolia* invades mainly coastal regions (Wheeler *et al.* 2011), abiotic conditions in the restingas can limit the naturalization of introduced populations of this species. High temperatures prevent seed germination and low light affects the survival and growth of young plants. As the impact of drought negatively affects the performance of both seeds and seedlings, water stress is the main environmental factor that limits its naturalization in open scrub vegetation, which covers large areas of sandy coastal plains in many parts of the world (Araújo and Pereira 2002). The different formations of the restingas along the Brazilian coast have different percentage of cover and variation in the water availability (Lacerda *et al.* 1993). As high light conditions and high water availability increase its seed germination and young plants survival, *C. equisetifolia* naturalization may be favoured in the formations of the restingas that has mainly open areas near water bodies.

Persistent soil seed bank may favour invasion

Seed longevity under both storage (buried in the soil and cold/dry laboratory) conditions over 24 months was similar. These results, together with the small seed mass and the low moisture content at maturity, suggest that its seeds exhibit long-lived (orthodox) storage behaviour. The capacity to form a persistent soil seed bank for potentially up to 50 months are likely due to the dry climate and low rainfall in the restinga of State Park of Costa do Sol (Barbiére 1984); these conditions inhibit seed deterioration, soil microbial activity and decomposition processes (Cuneo *et al.* 2010). As *C. equisetifolia* seeds can remain viable in the soil for almost 4 years, they may germinate whenever environmental conditions are favourable for germination (Baskin and Baskin 2014). All these features increase the overall probability of recruitment and further naturalization of this species on sandy coastal plains of Brazil.

Seeds can germinate across a broad range of temperatures and light conditions

Casuarina equisetifolia seeds had a fast physiological response when in contact with water, and germination started in 3–4 days after water uptake in optimal germination temperatures (30 and 35 °C). For small-seeded species, high germination rate is crucial for the recruitment of new individuals, mainly in environments with water restrictions, as is the case in the restingas (Martins *et al.* 2012). The capacity to germinate under a wide range of temperature conditions, including low (15 °C) and high alternating temperatures (40/20 °C), although with decrease in the germination rate, is also an important factor for a population to become naturalized in the restinga, where the temperatures can range from 21 to 31 °C (mean of 25 °C) inside the patches, and from 19 to 44 °C (mean of 30 °C) in open areas (Matos 2014).

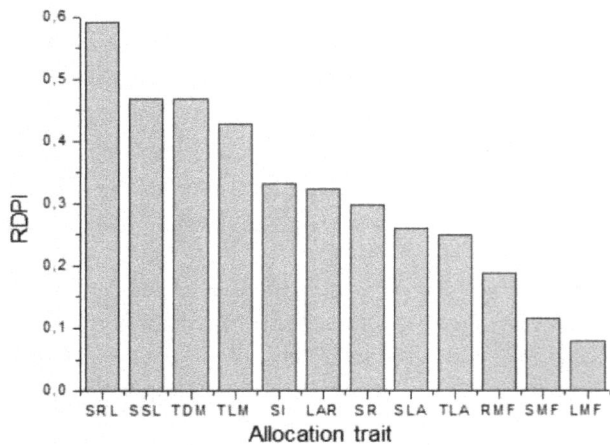

Figure 6. Relative Distance Plasticity Index (RDPI) for 12 allocation traits of young plants of *Casuarina equisetifolia* in response to three levels of light (15, 70 and 100 % of photosynthetic photon-flux density) after 16 weeks. The traits shown are specific root length (SRL), specific stem length (SSL), total dry mass (TDM), total leaf mass (TLM), slenderness index (SI), leaf area ratio (LAR), shoot: root ratio (SR), specific leaf area (SLA), total leaf area (TLA), root mass fraction (RMF), stem mass fraction (SMF) and leaf mass fraction (LMF). The RDPI values range from 0 (no plasticity) to 1 (maximal plasticity).

Nonetheless, the bare sand of the restinga may reach temperatures as high as 70 °C at the peak of radiation during mid-summer, in which the recruitment via seeds is restricted to a few species (Scarano 2002). In relation to the light conditions, small seeds of some species often require light for germination (Milberg *et al.* 2000), however, *C. equisetifolia* seeds are negatively photoblastic, and darkness only partially prevents its germination, although it depends on the interaction of the light with the temperature (Baskin and Baskin 2014).

The high germination percentage of *C. equisetifolia* seeds across a wide range of temperature and light conditions was evidence of its robustness (i.e. the constant expression of a particular phenotype despite genotypic and environmental variation; Waddington 1942). This increases its capacity to become naturalized in a high heterogeneous environment of temperature and light conditions, such as the restinga (Scarano 2002; Matos 2014). In addition, germination rate increased in response to favourable conditions of temperature and water availability, indicating that this species displays germination plasticity. A potential advantage of germination plasticity is the opportunistic germination response to favourable environmental conditions (Richards *et al.* 2006). Germination plasticity may have adaptive value if it enables a species to establish in variable environments where resource levels fluctuate (Wainwright and Cleland 2013), as occurs in the restingas (Matos 2014). Both robustness of germination to a range of conditions and plastic fitness response to the environment may enhance the ability of alien species to invade new ecosystems (Richards *et al.* 2006; Wainwright and Cleland 2013).

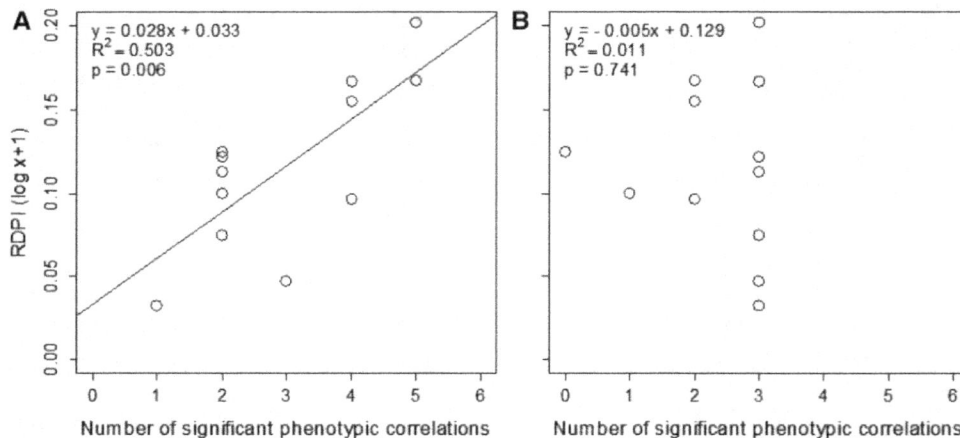

Figure 7. Regression analysis between Mean Relative distance plasticity index (RDPI) in response to light and Phenotypic Integration (PI) in response to shady (15 % of light; A) and sunny conditions (100 % of light; B) among 12 morphological traits of young plants of *Casuarina equisetifolia*. Each point in the regression analysis corresponds to a single trait.

Salinity and drought reduce seed germination

Salinity and drought tolerance are also two important environmental determinants for plant recruitment on sandy coastal plains (Martins *et al.* 2012; Lai *et al.* 2015). Although *C. equisetifolia* colonizes extensive sandy areas (Morton 1980), its germinability (percentage and rate) was very sensitive to both salt and water stresses. Germination sensitivity to salt stress has been reported previously for this species (Tani and Sasakawa 2003) and for other 10 *Casuarina* species (Clemens *et al.* 1983). The germination pattern of this species is typical of halophyte species (*sensu* Woodell 1985), where seeds retain viability under saline soils and germinate in favourable conditions (e.g. after a rainy period, when the salt is leached from the substrate). In addition to halophyte seed behaviour, *C. equisetifolia* seedlings show salt stress tolerance related to physiological and biochemical mechanisms (Clemens *et al.* 1983; Tani and Sasakawa 2003). Therefore, the halophyte behaviour allows *C. equisetifolia* seeds to become quiescent in response to salt–water stresses and ensure a fast and high germination when these limiting factors are overcome. This may be another important adaptive strategy for *C. equisetifolia* to become naturalized in the restingas.

Drought and shade reduce survival and growth of young plants

Young plants showed lower tolerance to shade and water stress than seed germination. Although its seeds have the capacity to germinate in environments with low levels of light, young plants are shade-intolerant and will not survive. Thus, even if *C. equisetifolia* seeds germinate inside vegetation patches, seedlings will not establish (T.G. Zimmermann *et al.* unpubl. data). In areas with high availability of water, young plants of *C. equisetifolia* can survive in a broad range of light conditions, except under deep shade (< 2 % of light), a condition that is often found inside vegetation patches (Matos 2014). Mainly in the restinga, tolerance of high light intensities may enhance plant survival. As for germination, water availability is crucial for the survival of young plants of *C. equisetifolia*. This species can tolerate dry climates only if the roots can grow down to the water table (Whistler and Elevitch 2006). Therefore, this tree has the capacity to become naturalized mainly in areas adjacent to watercourses. As in *C. equisetifolia*, distance to water bodies was also one of the main determinants of naturalization of *C. cunninghamiana* in South Africa (Potgieter *et al.* 2014*b*).

In contrast to *C. equisetifolia*, shaded microsites beneath the canopy in vegetation patches is the most favourable niche for regeneration for many restinga species (Matos 2014). As fluctuation in resource availability is a key factor controlling invasibility (Davis *et al.* 2000), alien species will be more successful at invading communities if they do not encounter intense competition from resident species for available resources such as light. Therefore, following a disturbance, a light increment followed by a rainy event will increase the susceptibility of the restinga to the invasion of *C. equisetifolia*.

Casuarina equisetifolia showed differences in growth rate and biomass allocation in response to changes in light intensity. Although plant survival was high at 15 % light levels under high water conditions, shading decreased growth and the young plants exhibited shade avoidance responses, such as high shoot: root ratio, slenderness index, stem mass fraction and specific stem length (Ryser and Eek 2000). Under high water, *C. equisetifolia* exhibits similar growth between conditions of 100 % of light and in the *Casuarina* stand (70 % of light), which improves its potential to become naturalized in open areas. In attempt to minimize evaporative demand (Bloor and Grubb 2004), *C. equisetifolia* showed changes in leaf morphological traits under high light conditions, which results in lower specific leaf area and leaf area ratio. This adaptation is important for an alien species to become naturalized in habitats with low water availability, such as the restinga. In addition, specific leaf area is a plant trait that has shown to be associated with invasive success across a broad range of species (van Kleunen *et al.* 2010; Leishman *et al.* 2014).

Low phenotypic plasticity and high phenotypic integration of traits

Although *C. equisetifolia* showed germination plasticity, young plants exhibited low morphological plasticity in response to light. Low phenotypic plasticity has also been reported in other invasive species in habitats with multiple stress factors, such as in *Acacia longifolia* in Mediterranean dunes (Peperkorn *et al.* 2005), indicating that morphological plasticity may be advantageous in favourable environments, whereas stability is more beneficial under adverse conditions (e.g. Valladares *et al.* 2000, 2007).

Several studies have shown that phenotypic integration tends to increase with environmental stress, and the higher levels of integration observed in these habitats should constrain the plastic responses of plants (Gianoli 2004; García-Verdugo *et al.* 2009; Gianoli and Palacio-López 2009). Nevertheless, in the stressful environment (shade) occurred a positive effect of phenotypic integration on the plastic expression of *C. equisetifolia* morphological traits. As long as environmental conditions ameliorate it is likely that this alien species does not need to coordinate the phenotype to exhibit plasticity. Therefore, phenotypic integration may not constrain phenotypic plasticity of plants in adverse conditions. The values of phenotypic integration index for *C. equisetifolia* was similar between shady (2.20) and sunny (2.78) conditions, even though the magnitude of individual correlations often changed from one environment to another. These values may be considered high, since studies showed that ranges from 0.77 to 1.63 (Waitt and Levin 1993; Boucher *et al.* 2013). A high degree of phenotypic integration may thus be a facilitator of adaptation, because it can reduce maladaptive variation (Armbruster *et al.* 2014), which is an important factor in the evolutionary ecology of this species. This appears to be an important strategy for an alien species to become naturalized in environments with multiple stress conditions. Nonetheless, the role of phenotypic integration in invasiveness remains poorly understood (Godoy *et al.* 2012), and more work is needed to elucidate the function of the trait correlations along the naturalization–invasion continuum.

The large production of small seeds (Apfelbaum *et al.* 1983), associated with anemochory and hydrochory dispersal syndromes (Morton 1980, Rentería 2007, Wheeler *et al.* 2011), the long-term persistence of seeds in the soil, high germination, survival and growth under high light, higher efficiency in allocating biomass on structures for water absorption (low shoot: root ratio) and light-capturing (high leaf mass fraction), together with the low phenotypic plasticity and high phenotypic integration, are crucial factors that allow *C. equisetifolia* to overcome barriers to reproduction and survival and to become naturalized on sandy coastal plains. These traits, coupled with the salt tolerance and symbiotic associations (Zhong *et al.* 1995, Diagne *et al.* 2013) enable this species to invade mainly open, sandy habitat, adjacent to watercourses, especially along coastlines, where disturbances have occurred.

Management strategies

To limit further naturalization of *C. equisetifolia* and to prevent it from becoming invasive in the restingas planting of the species should be avoided, especially in open areas near water bodies. Removal of *C. equisetifolia* is difficult, because of its capacity for vigorous regrowth (Morton 1980), and seeds can remain viable in the soil for almost 4 years. Thus, we recommend the periodic removal of cones and seeds especially at the edge of the *Casuarina* stands, to prevent recruitment and further invasion in the restinga. As *C. equisetifolia* does not tolerate shade and drought and invades mainly degraded areas, one of the best ways of hampering its

naturalization in the restinga is to conserve the remaining fragments. Therefore, habitat disturbance should be minimized to reduce opportunities for the colonization of this species. Where habitats are disturbed, immediate replanting with native vegetation is required. Nevertheless, restingas have been severely threatened mainly by anthropogenic disturbances which altering the key processes that naturally make restingas resistant to *C. equisetifolia* invasion. Further degradation is sure to lead to the status of this species changing from naturalized to invasive in large areas in Brazil.

Conclusions

The long-term persistence of seeds in the soil, the capacity to germinate across a wide range of temperature and light conditions and the high survival rate of the young plants in conditions with moderate and high irradiance with high soil moisture are key factors that favour the naturalization of *C. equisetifolia*. Thus, areas in the restingas and on sandy coastal plains that present high-light conditions and are near water bodies are prone to naturalization of the introduced population of this species. As young plants showed lower tolerance to shade and water stress than seed germination, even if the seeds can germinate, young plants will not survive under low light (e.g. vegetation patches). Although this species exhibited high germination plasticity, young plants showed low phenotypic plasticity, which is important in habitats with multiple stress factors (Valladares *et al.* 2000, 2007). The high phenotypic integration is an important factor in the evolutionary ecology of this species because can facilitate adaptation, thereby improving the chances of this species becoming naturalized in environments with harsh conditions. As *C. equisetifolia* does not tolerate shade and drought and invades mainly degraded areas, conservation of the restingas is crucial to limit invasion of this species.

Sources of Funding

This study was supported by the Rio de Janeiro Botanic Garden Research Institute (JBRJ), Coordenação de Aperfeiçoamento de Pessoal de Nível Superior (CAPES) and Fundação de Amparo a Pesquisa do Estado do Rio de Janeiro (FAPERJ).

Contributions by the Authors

T.G.Z. and A.C.S.A. conceived the idea. T.G.Z. conducted the experiments and ran the statistics. T.G.Z and A.C.S.A. led the writing with assistance of D.M.R.

Acknowledgements

We thank F. Silva for his field assistance, A.P.M. Cruz and M. Fernandes for help with the laboratory experiments, L.L. Leal for her assistance in the nursery, I.S. Matos for constructive advice and R.D. Zenni and O. Godoy for discussion and comments on parts of the article. We thank the guest editor Heidi Hirsch and two anonymous reviewers for comments that improved the quality of this article. D.M.R. acknowledges funding from the DST-NRF Centre of Excellence for Invasion Biology and the National Research Foundation, South Africa (grant 85417). This article is part of the first author's PhD thesis.

Supporting Information

The following additional information is available in the online version of this article —

Figure S1. Patchy structure of the restinga (patchy shrub vegetation).

Table S2. *F*-test of significance for main effects and interactions in an analysis of covariance (ANCOVA) for the effects of storage and light conditions in germination of *Casuarina equisetifolia* seeds.

Figure S3. Survival curves of young plants of *Casuarina equisetifolia* in response to light (A), water stress (B), drought under high light (C) and drought under low light (D).

Table S4. Spearman's rank correlation matrices among 12 morphological traits of young plants of *Casuarina equisetifolia*.

Literature Cited

Apfelbaum SI, Ludwig JP, Ludwig CE. 1983. Ecological problems associated with disruption of dune vegetation dynamics by *Casuarina equisetifolia* L. at Sand Island, Midway Atoll. *Atoll Research Bulletin* **261**:1–19.

Araújo DSD. 1992. Vegetation types of sandy coastal plains of tropical Brazil: a first approximation. In: Seeliger U, ed. *Coastal plant communities of Latin America*. San Diego: Academic Press, 337–347.

Araújo DSD, Pereira MCA. 2002. *Sandy coastal vegetation. International commission on tropical biology and natural resources*. Oxford: Eolss Publishers.

Araújo DSD, Sá CFC, Fontella-Pereira J, Garcia DS, Ferreira MV, Paixão RJ, Schneider SM, Fonseca-Kruel VS. 2009. Área de proteção ambiental de Massambaba, Rio de Janeiro: caracterização fitofisionômica e florística. *Rodriguésia* **60**:67–96.

Armbruster WS, Pélabon C, Bolstad GH, Hansen TF. 2014. Integrated phenotypes: understanding trait covariation in plants and animals. *Philosophical Transactions of the Royal Society B* **369**:1–16.

Barbiére EB. 1984. Cabo Frio e Iguaba Grande, dois microclimas distintos a um curto intervalo espacial. In: Lacerda LD, Araújo DDD,

Cerqueira R, Turcq B, eds. *Restingas: origem, estrutura, processos*. Niterói: CEUFF, 3–13.

Barnett V, Lewis T. 1978. *Outliers in statistical data*. New York: John Wiley & Sons.

Baskin CC, Baskin JM. 2014. *Seeds: ecology, biogeography, and evolution of dormancy and germination*, 2nd edn. San Diego: Elsevier/Academic Press.

Blackburn TM, Pyšek P, Bacher S, Carlton JT, Duncan RP, Jarošík V, Wilson JRU, Richardson DM. 2011. A proposed unified framework for biological invasions. *Trends in Ecology & Evolution* **26**:333–339.

Bloor JMG, Grubb PJ. 2004. Morphological plasticity of shade-tolerant tropical rainforest tree seedlings exposed to light changes. *Functional Ecology* **18**:337–348.

Bonner FT. 2008. Storage of seeds. In: Bonner FT, Karrfalt RP, ed. *The woody plant seed manual. Agriculture handbook 727*. Washington: Department of Agriculture, 85–95.

Boucher FC, Thuiller W, Arnoldi C, Albert CH, Lavergne S. 2013. Unravelling the architecture of functional variability in wild populations of *Polygonum viviparum* L. *Functional Ecology* **27**: 382–391.

Callaway RM, Walker LR. 1997. Competition and facilitation: a synthetic approach to interactions in plant communities. *Ecology* **78**:1958–1965.

Clemens J, Campbell LC, Nurisjah S. 1983. Germination, growth and mineral ion concentrations of *Casuarina* species under saline conditions. *Australian Journal of Botany* **31**:1–9.

Cosendey BN, Rocha CFD, Menezes VA. 2016. Population density and conservation status of the teiid lizard *Cnemidophorus littoralis*, an endangered species endemic to the sandy coastal plains (restinga habitats) of Rio de Janeiro state, Brazil. *Journal of Coastal Conservation*. doi:10.1007/s11852-016-0421-4.

Cuneo P, Offord CA, Leishman MR. 2010. Seed ecology of the invasive woody plant African Olive (*Olea europaea* subsp. *cuspidata*): implications for management and restoration. *Australian Journal of Botany* **58**:342–348.

Davis MA, Grime JP, Thompson K. 2000. Fluctuating resources in plant communities: a general theory of invasibility. *Journal of Ecology* **88**:528–534.

Diagne N, Diouf D, Svistoonoff S, Kane A, Noba K, Franche C, Bogusz D, Duponnois R. 2013. *Casuarina* in Africa: distribution, role and importance of carbuncular mycorrhizal, ectomycorrhizal fungi and *Frankia* on plant development. *Journal of Environmental Management* **128**:204–209.

Evans GC. 1972. *The quantitative analysis of plant growth*. Berkley: University of California Press.

Ferriter A, Doren B, Winston R, Thayer D, Miller B, Thomas B, Barrett M, Pernas T, Hardin S, Lane J, Kobza M, Schmitz D, Bodle M, Toth L, Rodgers L, Pratt P, Snow S, Goodyear C. 2007. The status of nonindigenous species in the south Florida environment. *South Florida Environmental Report*, **Vol. I**, 1–9.

Flores-Moreno H, García-Trevino ES, Letten AD, Moles AT. 2015. The beginning: phenotypic change in three invasive species through their first two centuries since introduction. *Biological Invasions* **17**:1215–1225.

Funk JL. 2008. Differences in plasticity between invasive and native plants from a low resource environment. *Journal of Ecology* **96**: 1162–1173.

García-Verdugo C, Granado-Yela C, Manrique E, Casas RR, Balaguer L. 2009. Phenotypic plasticity and integration across the canopy of *Olea europaea* subsp. *guanchica* (Oleaceae) in populations with different wind exposures. *American Journal of Botany* **96**: 1454–1461.

Gianoli E. 2004. Plasticity of traits and correlations in two populations of *Convolvulus arvensis* (Convolvulaceae) differing in environmental heterogeneity. *International Journal of Plant Science* **165**:825–832.

Gianoli E, Palacio-López K. 2009. Phenotypic integration may constrain phenotypic plasticity in plants. *Oikos* **118**:1924–1928.

Godoy O, Valladares F, Castro-Díez P. 2011. Multispecies comparison reveals that invasive and native plants differ in their traits but not in their plasticity. *Functional Ecology* **25**:1248–1259.

Godoy O, Valladares F, Castro-Díez P. 2012. The relative importance for plant invasiveness of trait means, and their plasticity and integration in a multivariate framework. *New Phytologist* **195**: 912–922.

Gómez-Aparicio L, Valladares F, Zamora R, Quero JL. 2005. Response of tree seedlings to the abiotic heterogeneity generated by nurse shrubs: an experimental approach at different scales. *Ecography* **28**:757–768.

Gómez-Aparicio L, Valladares F, Zamora R. 2006. Differential light responses of Mediterranean tree saplings: linking ecophysiology with regeneration niche in four co-occurring species. *Tree Physiology* **26**:947–958.

Hamilton MA, Murray BR, Cadotte MW, Hose GC, Baker AC, Harris CJ, Licari D. 2005. Life-history correlates of plant invasiveness at regional and continental scales. *Ecology Letters* **8**:1066–1074.

Hastwell GT, Facelli JM. 2003. Differing effects of shade-induced facilitation on growth and survival during the establishment of a chenopod shrub. *Journal of Ecology* **91**:941–950.

Hata K, Kato H, Kachi N. 2010. Litter of an alien tree, *Casuarina equisetifolia*, inhibits seed germination and initial growth of a native tree on the Ogasawara Islands (subtropical oceanic islands). *Journal of Forest Research* **15**:384–390.

Hesp PA, Martínez ML. 2007. Disturbance processes and dynamics in coastal dunes. In: Johnson EA, Miyanishi K, eds. *Plant disturbance ecology. The process and the response.* Cambridge: Elsevier, 215–247.

I3N Brazil 2015. *Base de dados nacional de espécies exóticas invasoras, I3N Brasil, Instituto Hórus de desenvolvimento e Conservação Ambiental.* http://i3n.institutohorus.org.br (20 May 2015).

ISTA. 1999. *International rules for seed testing.* Zürich: Seed Science Technology, ISTA.

Kueffer C, Pyšek P, Richardson DM. 2013. Integrative invasion science: model systems, multi-site studies, focused meta-analysis and invasion syndromes. *New Phytologist* **200**:615–633.

Labouriau LG, Pacheco A. 1978. On the frequency of isothermal germination in seeds of *Dolichos biflorus* L. *Plant and Cell Physiology* **19**:507–512.

Lacerda LD, Araújo DSD, Maciel NC. 1993. Dry coastal ecosystems of the tropical Brazilian coast. In: Van der Maarel E, ed. *Dry coastal ecosystems: Africa, America, Asia, Oceania.* Amsterdam: Elsevier, 477–493.

Lai L, Tian Y, Wang Y, Zhao X, Jiang L, Baskin JM, Baskin CC, Zheng Y. 2015. Distribution of three congeneric shrub species along an aridity gradient is related to seed germination and seedling emergence. *AoB PLANTS* **7**:plv071; doi:10.1093/aobpla/plv071.

Leishman LR, Cooke J, Richardson DM. 2014. Evidence for shifts to faster growth strategies in novel ranges of invasive alien plants. *Journal of Ecology* **102**:1451–1461.

Martins LST, Pereira TS, Carvalho ASR, Barros CF, Andrade ACS. 2012. Seed germination of *Pilosocereus arrabidae* (Cactaceae) from a semiarid region of south-east Brazil. *Plant Species Biology* **27**: 191–200.

Matos IS. 2014. *Crescimento, sobrevivência e plasticidade fenotípica de plântulas de espécies de restinga sob gradientes experimentais de intensidade de luz e de disponibilidade hídrica. Masters dissertation,* Rio de Janeiro Botanical Garden Research Institute, Brazil.

Maun MA. 1994. Adaptations enhancing survival and establishment of seedlings on coastal dune systems. *Vegetatio* **1**:59–70.

Milberg P, Anderson L, Thompson K. 2000. Large seeded species are less dependent on light for germination than small seeded ones. *Seed Science Research* **10**:99–104.

Molina-Montenegro MA, Peñuelas J, Munné-Bosch S, Sardans J. 2012. Higher plasticity in ecophysiological traits enhances the performance and invasion success of *Taraxacum officinale* (dandelion) in alpine environments. *Biological Invasions* **14**:21–33.

Morton JF. 1980. The Australian pine or beefwood (*Casuarina equisetifolia* L.) an invasive "weed" tree in Florida. *Proceedings of the Florida State Horticultural Society* **93**:87–95.

Palacio-López K, Gianoli E. 2011. Invasive plants do not display greater phenotypic plasticity than their native or non-invasive counterparts: a meta-analysis. *Oikos* **120**:1393–1401.

Parrotta JA. 1993. *Casuarina equisetifolia* L. ex J.R. & G. Forst. SO-ITF-SM-46. In: *U.S. Department of agriculture, forest service.* International Institute of Tropical Forestry. Puerto Rico: Río Piedras, 1–11.

Pearson TRH, Burslem DFRP, Mullins CE, Dalling JW. 2003. Functional significance of photoblastic germination in neotropical pioneer trees: a seed's eye view. *Functional Ecology* **17**: 394–402.

Peperkorn R, Werner C, Beyschlag W. 2005. Phenotypic plasticity of an invasive acacia versus two native Mediterranean species. *Functional Plant Biology* **32**:933–944.

Pigliucci M. 2003. Phenotypic integration: studying the ecology and evolution of complex phenotypes. *Ecology Letters* **6**: 265–272.

Potgieter LJ, Richardson DM, Wilson JRU. 2014a. *Casuarina*: biogeography and ecology of an important tree genus in a changing world. *Biological Invasions* **16**:609–633.

Potgieter LJ, Richardson DM, Wilson JRU. 2014b. *Casuarina cunninghamiana* in the Western Cape, South Africa: determinants of naturalisation and invasion, and options for management. *South African Journal of Botany* **92**:134–146.

Potgieter LJ, Wilson JRU, Strasberg D, Richardson DM. 2014c. *Casuarina* invasion alters primary succession on lava flows on La Réunion Island. *Biotropica* **4**:268–275.

Pyšek P, Richardson DM. 2007. Traits associated with invasiveness in alien plants: where do we stand? In: Nentwig W, ed. *Biological invasions, ecological studies.* Berlin: Springer-Verlag, 97–125.

Pyšek P, Richardson DM, Pergl J, Jarošík V, Sixtová Z, Weber E. 2008. Geographical and taxonomic biases in invasion ecology. *Trends in Ecology & Evolution* **23**:237–244.

R Core Team. 2014. *R: a language and environment for statistical computing.* R Foundation for Statistical Computing Vienna. http://www.R-project.org/. (19 Jul 2015).

Reinert F, Roberts A, Wilson JM, Ribas L, Cardinot G, Griffiths H. 1997. Gradation in nutrient composition and photosynthetic pathways across the restinga vegetation of Brazil. *Botanica Acta* **110**: 135–142.

Rejmánek M, Richardson DM. 2013. Trees and shrubs as invasive alien species – 2013 update of the global database. *Diversity and Distributions* **19**:1093–1094.

Rentería JL. 2007. *Plan de manejo para la erradicación de Casuarina equisetifolia L. (Casuarinaceae), especie invasora de limitada distribución en la isla Santa Cruz, Galápagos. Estación Científica Charles Darwin, Galápagos, Ecuador:* Estación Científica Charles Darwin.

Ribeiro MC, Metzger JP, Martensen AC, Ponzoni F, Hirota MM. 2009. Brazilian Atlantic forest: how much is left and how is the remaining forest distributed? Implications for conservation. *Biological Conservation* **142**:1141–1153.

Richards CL, Bossdorf O, Muth NZ, Gurevitch J, Pigliucci M. 2006. Jack of all trades, master of some? On the role of phenotypic plasticity in plant invasions. *Ecology Letters* **9**:981–993.

Richardson DM, Pyšek P. 2012. Naturalization of introduced plants: ecological drivers of biogeographical patterns. *New Phytologist* **196**:383–396.

Rocha CFD, Van Sluys M, Alves MS, Jamel CE. 2007. The remnants of restinga habitats in the Brazilian Atlantic Forest of Rio de Janeiro state, Brazil: habitat loss and risk of disappearance. *Brazilian Journal of Biology* **67**:263–273.

Rouget M, Robertson MP, Wilson JRU, Hui C, Essl F, Renteria J, Richardson DM. 2016. Invasion debt – quantifying future biological invasions. *Diversity and Distributions* **22**:445–456.

Ryser P, Eek L. 2000. Consequences of phenotypic plasticity vs. interspecific differences in leaf and root traits for acquisition of above-ground and below-ground resources. *American Journal of Botany* **87**:402–411.

Salisbury FB, Ross CW. 1992. *Plant physiology*. Belmont: Wadsworth Publishing Company Inc.

Scarano FR. 2002. Structure, function and floristic relationships of plant communities in stressful habitats marginal to the Brazilian Atlantic Rainforest. *Annals of Botany* **90**:517–524.

Scarano FR. 2009. Plant communities at the periphery of the Atlantic rain forest: rare-species bias and its risks for conservation. *Biological Conservation* **142**:1201–1208.

Schlichting CD. 1989. Phenotypic plasticity in Phlox II. *Plasticity of character correlations. Oecologia* **78**:496–501.

Smith H. 2000. Phytochromes and light signal perception by plants – an emerging synthesis. *Nature* **407**:585–591.

Tani C, Sasakawa H. 2003. Salt tolerance of *Casuarina equisetifolia* and *Frankia* Ceq1 strain isolated from the root nodules of *C. equisetifolia. Soil Science & Plant Nutrition* **49**:215–222.

Therneau T. 2015. A package for survival analysis in S. *R package*. http://CRAN. R-project. org/package=survival (25 Jul 2015).

Torices R, Muñoz-Pajares AJ. 2015. PHENIX: an R package to estimate a size-controlled phenotypic integration index. *Applications in Plant Sciences* **3**:1–4.

U.S. Research Council. 1984. *Casuarinas: nitrogen-fixing trees for adverse sites*. Washington: National Academy Press.

Valladares F, Gianoli E, Gómez JM. 2007. Ecological limits to plant phenotypic plasticity. *New Phytologist* **176**:749–763.

Valladares F, Sánchez-Gómez D, Zavala MA. 2006. Quantitative estimation of phenotypic plasticity: bridging the gap between the evolutionary concept and its ecological applications. *Journal of Ecology* **94**:1103–1116.

Valladares F, Wright SJ, Lasso E, Kitajima K, Robert WP. 2000. Plastic phenotypic response to light of 16 congeneric shrubs from a Panamanian rainforest. *Ecology* **81**:1925–1936.

van Kleunen M, Weber E, Fischer M. 2010. A meta-analysis of trait differences between invasive and non-invasive plant species. *Ecology Letters* **13**:235–245.

Villela FA, Beckert OP. 2001. Potencial osmótico de soluções aquosas de polietileno glicol 8000. *Revista Brasileira De Sementes* **23**: 267–275.

Waddington CH. 1942. Canalization of development and the inheritance of acquired characters. *Nature* **150**:563–565.

Wagner GP. 1984. On the eigenvalue distribution of genetic and phenotypic dispersion matrices: evidence for a nonrandom organization of quantitative character variation. *Journal of Mathematical Biology* **21**:77–95.

Wainwright CE, Cleland EE. 2013. Exotic species display greater germination plasticity and higher germination rates than native species across multiple cues. *Biological Invasions* **15**: 2253–2264.

Waitt DE, Levin DA. 1993. Phenotypic integration and plastic correlations in *Phlox drumondii* (Polemoniaceae). *American Journal of Botany* **80**:1224–1233.

Wheeler GS, Taylor GS, Gaskin JF, Purcell MF. 2011. Ecology and management of sheoak (*Casuarina* spp.), an invader of coastal Florida, U.S.A. *Journal of Coastal Research* **27**: 485–492.

Whistler WA, Elevitch CR. 2006. *Casuarina equisetifolia* (reach sheoak) and *C. cunninghamiana* (river she-oak). In: Elevitch CR, ed. *Species profiles for pacific island agroforestry*. Holualoa, Hawaii: Permanent Agriculture Resources (PAR).

Woodell SRJ. 1985. Salinity and seed germination patterns in coastal plants. *Vegetatio* **61**:223–229.

Zar JH. 1999. *Biostatistical analysis*, 4th edn. Upper Saddle River: Prentice Hall.

Zenni RD, Ziller SR. 2011. An overview of invasive plants in Brazil. *Revista Brasileira De Botânica* **34**:431–446.

Zhong C, Gong M, Chen Y, Wang F. 1995. Inoculation of *Casuarina* with mycorrhizal fungi and *Frankia*. In: Brundrett M, Dell B, Malajczuk N, Gong M, eds. *Mycorrhizas for plantation forests in Asia*. Canberra: CSIRO, 122–126.

16

Rapid increase in growth and productivity can aid invasions by a non-native tree

Rafael Dudeque Zenni*, Wanderson Lacerda da Cunha, and Guilherme Sena
Department of Ecology, University of Brasília, Campus Universitário Darcy Ribeiro, Brasília CEP 70910-900, Brazil

Guest Editor: Johannes Le Roux

Abstract. Research on biological invasions has produced detailed theories describing range expansions of introduced populations. However, current knowledge of evolutionary factors associated with invasive range expansions, especially those related to rapid evolution of long-lived organisms, is still rudimentary. Here, we used a system of six 40-year-old invasive pine populations that originated from replicated introduction events to study evolution in productivity, growth, and chemical defence traits. We tested the hypotheses that invasive populations were undergoing rapid phenotypic change as populations spread, that populations exhibit trade-offs between evolution in growth and chemical defences, and that rates of rapid evolution in plant growth and productivity effect rates of invasion. Although all invasions started from replicated pools of genetic material and equal propagule pressure, we found divergence in mean values for the six invasive populations in the six traits measured. Not only were there between-population variations but also invasive populations were also rapidly changing along each invasive population expansion. Two populations displayed greater leaf areas (LAs) and smaller specific LAs (SLAs) during range expansion. Four populations had faster growth rates at the leading edge of the invasion front in comparison with plants at the rear edge. In terms of total plant defences, non-volatile resin increased in plants along one invasion gradient and decreased in a second, total needle phenolics increased in plants along one invasion gradient and total wood phenolics increased in plants along the one invasion gradient and decreased in a second. We found no trade-offs between investments in growth and chemical defence. Also, faster rates of change in growth rate and LA were positively associated with greater dispersal distances of invasive populations, suggesting rapid evolution may increase invasiveness. Understanding the roles of both natural and human-mediated ecological and evolutionary processes in population-level dynamics is key to understanding the ability of non-native species to invade.

Keywords: Biological invasions; contemporary evolution; exotic species; growth-defence trade-offs; invasion biology; invasiveness; *Pinus taeda*; range expansion; tree invasions.

* Corresponding author's e-mail address: rafaeldz@gmail.com

Introduction

Biological invasions are a leading cause of environmental degradation, are a main focus of concern for conservation practitioners and provide important insights on species responses to climate change (Pyšek et al. 2012; Caplat et al. 2013; Moran and Alexander 2014; Kuebbing and Simberloff 2015). Invasive organisms alter ecosystem properties and community dynamics, impacting native species and ecosystem functioning (Wardle et al. 2011; Yelenik and D'Antonio 2013). Driving the spread and impact of non-native populations are dynamic ecological and evolutionary processes acting at levels ranging from genes to global scale (Zenni 2014; Zenni et al. 2014; Zenni and Hoban 2015b). Several decades of ecological research have produced detailed frameworks and theories describing range expansions of introduced populations (e.g. Blackburn et al. 2011; Gurevitch et al. 2011). More recently, researchers started to disentangle the evolutionary mechanisms driving spread of non-native populations and to incorporate them into invasion theory (Parker et al. 2003; Prentis et al. 2008; Sargent and Lodge 2014). However, the role of contemporary evolution in invasive range expansions is still poorly understood (Colautti and Barrett 2013; Zenni et al. 2014), especially for long-lived organisms such as trees.

Climate has been hypothesized to be a major selection force on adaptation of invasive plant populations, changing plant phenology and size (Colautti and Barrett 2013; Colomer-Ventura et al. 2015). However, evolution can also occur in growth rates and defence traits (Buswell et al. 2011). Investments in defence against herbivory is costly for plants, and resource allocation for chemical defences can limit plant growth or reproduction (Coley et al. 1985; Mithöfer and Boland 2012). The well-established relationship of constant trade-offs among investment in growth, reproduction and defence traits has led to the formulation of the 'evolution of increased competitive ability' hypothesis, which states that introduced plants are liberated from natural enemies and, thus, can allocate towards growth and reproduction resources previously required for defence (Blossey and Nötzold 1995). By increasing their resource investment in growth and reproduction, non-native populations may become abundant and widespread (Keane and Crawley 2002). Indeed, many invasives, including pines, escape from natural enemies when introduced to a new range (Liu and Stiling 2006). In their native ranges, several species of Pinus exhibit trade-offs between growth rates and chemical defences; slow-growing species and populations invest more in constitutive defences (Moreira et al. 2014). For non-native populations, the absence of significant herbivory pressure may reduce plant resource allocation towards defence, possibly increasing fitness of non-native populations. Consequently, invasive populations may evolve increased growth rates and reproduction during the invasion process.

Besides local adaptation, genetic drift and phenotypic plasticity, the human-mediated introduction of pre-selected and adapted genetic lineages may also produce fit organisms in non-native ranges and benefit range expansions (Zenni et al. 2014). The introduction of highly variable groups of individuals may produce the same effect (Forsman et al. 2012; Zenni and Simberloff 2013; Forsman and Wennersten 2015). Further, the co-introduction of previously allopatric populations can lead to genetic recombination and novelty, possibly increasing levels of heterozygosity and polymorphism, which could trigger invasions that may not occur based on the original genotypes alone. However, the importance of admixture during invasions has not received much support from the literature (Rius and Darling 2014). While local adaptation, drift and phenotypic plasticity act during the naturalization stage and are part of all natural systems, the latter two mechanisms (introduction of pre-selected and adapted genetic lineages, and introduction of highly variable groups of individuals) act prior and during the introduction stage and may be exclusive to human mediated biological invasions (as opposed to natural range expansions, even those owned to human-mediated climate change). The processes acting at each stage of the invasion process can be different, and most studies have lumped together species at different stages of the introduction–naturalization–invasion continuum, or have looked only at processes occurring at the invasion stage (Blackburn et al. 2011; Richardson and Pyšek 2012; Moodley et al. 2013). Thus, understanding the roles of both natural and human-mediated agents prior to and during naturalization is key to understanding the invasion process. For this, researchers need to know the history of the introduction, including number and diversity of propagule sources, and follow the fate of the non-native populations during naturalization and invasion (Burton et al. 2010; Donaldson et al. 2014).

For animals, several studies have shown association between evolutionary dynamics and invasive range expansions. Invasive cane toads (Rhinella marina), for instance, evolved longer legs in the 70 years since introduction in Australia, increasing 5-fold the annual rate of expansion of the toad invasion front (Phillips et al. 2006). Also, for invasive European starlings (Sturnus vulgaris) in South Africa, unfavourable environmental conditions enhanced dispersal, which preserves genetic diversity during range expansion and reduces potentially detrimental founder effects (Berthouly-Salazar et al. 2013). For lizards (Podarcis muralis) invading in Germany,

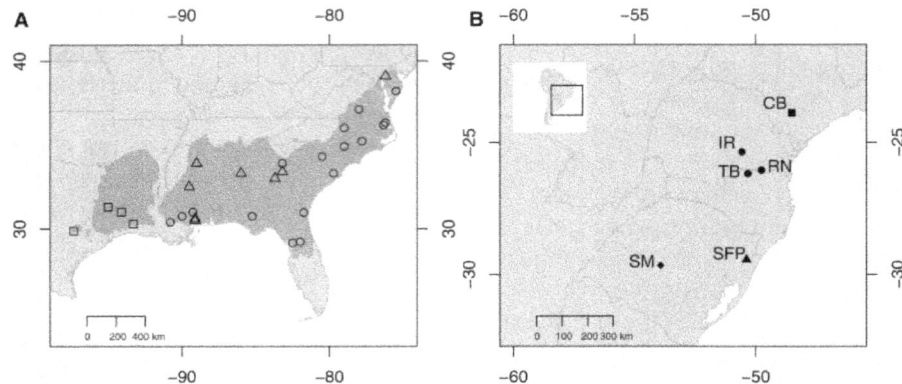

Figure 1. Origin of the *Pinus taeda* seed sources in the USA and location of common gardens and invasive populations in Brazil. (A) Light grey represents the continental USA, dark grey represents *P. taeda* native range, and dots are the location of the 32 seed sources planted in the six common gardens. Open symbols represent the three genetic provenances. (B) Location of the six common gardens in Brazil. Solid symbols (circles, triangle, diamond and square) represent different climates. CB, Capão Bonito; IR, Irati; RN, Rio Negro; SFP, São Francisco de Paula; SM, Santa Maria; TB, Três Barras.

a study found a trend for an increase in genetic differentiation and a decrease in genetic diversity from the invasion centre to the expanding range edge, suggesting genetic drift as a major factor in the structuring of populations (Schulte *et al.* 2013). For plants, virtually all examples of rapid evolution during invasive range expansions come from short-lived species. For example, in North America, local adaptation of purple loosestrife (*Lythrum salicaria*) in response to climate allowed populations to spread northward (Colautti and Barrett 2013). Further, the great majority of studies on evolution of invasive plants have compared invasive and native populations, or have compared distinct invasive populations. Very few studies have studied populations along invasion gradients.

In order to evaluate the role of human-mediated and natural evolutionary processes (natural selection and genetic drift) in the invasion success of non-native tree species, we designed a study using six fully replicated common garden experiments in southern Brazil where *Pinus taeda* L. (loblolly pine) was introduced at the same time (1973 and 1975), in the same numbers, from the same seed sources, and has formed invasive populations expanding outward from the plantations (i.e. invasive range expansions). *Pinus taeda* is a long-lived forest tree that has multi-generational populations, reproduces early (5 years) and yearly, and is wind-dispersed with viable seed dispersal distances of less than 20 m (Zenni *et al.* 2014). The common garden experiments were originally planted to serve as forestry provenance trials for silvicultural purposes (Shimizu and Higa 1981). The experiments encompass a north-south transect covering about 850 km or 6° of latitude (Fig. 1). Previous work on the system showed provenance-by-environment interactions where genetic lineages of *P. taeda* exhibited differential naturalization success depending on the climate

of the location into which they were introduced (Zenni *et al.* 2014). Furthermore, 25 genes were undergoing significant shifts in allele frequencies along the invasion gradients (Zenni and Hoban 2015b). Although the genes evolving were mostly population specific, many were associated to important plant functions, such as phloem sugar transport, nitrate uptake, and pollen tube growth (Zenni and Hoban 2015b). Taken together, both studies make it evident that *P. taeda* is undergoing rapid evolution in all six invasive populations (Zenni *et al.* 2014).

Here, we aimed to identify phenotypic changes undergone by the six populations of *P. taeda* in order to see how rapid evolution and genotype-by-environment interactions altered these invasive populations in growth, productivity, and chemical defence traits. We tested the following hypotheses: (i) the observed genotypic changes resulted in phenotypic changes for growth, productivity, and chemical defence traits; (ii) patterns of phenotypic changes match patterns of genetic change across populations; (iii) populations exhibit trade-offs between evolution in growth and chemical defences; and (iv) rates of rapid evolution in plant growth and productivity effect rates of invasion.

Methods

Study system

In 1973 and 1975, six common garden experiments for *P. taeda* were established in Brazil at the Santa Maria Experimental Farm (53.92°W 29.66°S), São Francisco de Paula National Forest (50.38°W 29.43°S), Três Barras National Forest (50.32°W 26.19°S), Rio Negro Experimental Station (49.76°W 26.05°S), Irati National Forest (50.57°W 25.36°S), and Capão Bonito National

Forest (48.51°W 23.88°S). The six locations represent four climates and two of these four climates (Santa Maria and Capão Bonito) are different from the climate in *P. taeda* native range. Annual precipitations vary from 1212 mm to 2068 mm, whereas mean annual temperatures vary from 15.2 °C to 19.1 °C. Both precipitation and temperature patterns exhibit clinal variations from north to south. Climate types were determined using multivariate clustering analyses with data from Worldclim (see Zenni *et al.* 2014 for details). Furthermore, Irati, Três Barras, Rio Negro and São Francisco de Paula original ecosystem are forest (Araucaria ombrophilous forest), Capão Bonito is a neotropical savanna (cerrado), and Santa Maria is a grassland ecosystem. Previous work showed no effect of original ecosystem type on the genetic changes observed for *P. taeda* (Zenni *et al.* 2014; Zenni and Hoban 2015b).

The common gardens were planted with 29 or 32 seed sources of which 20 were present in all gardens (Zenni *et al.* 2014). Each seed source corresponded to a seed lot collected from between 5 and 10 trees in natural stands in the species' native range (Fig. 1) [**see Supporting Information**]. In each Brazilian common garden, seed sources were planted in randomized blocks with four repetitions—a total of 144 trees from each seed source were planted in each common garden in four randomly placed squares of 6×6 trees (Shimizu and Higa 1981). Over the years, each common garden and its surroundings received circumstantial and haphazard management (e.g. some of the common gardens were thinned and plants growing on fire breaks and along roads were cut). The invasive populations themselves received only minor and dispersed management interventions. There were also high mortality rates for some seed sources in the common gardens probably owed to the high density of the plantations. In November and December 2014, all seed sources were still represented by at least 10 trees at any given garden, but the mean number of trees per seed source per site was usually higher. Previous work showed the seed sources form three native range genetic provenances (a provenance is an environmentally explicit genetic delineation of seed sources, and may be defined by a genetic cluster) and that the spatial distribution of these provenances across *P. taeda* native and invasive ranges correlates with temperature and precipitation patterns (Zenni *et al.* 2014). The common gardens are considered parallel replicated introduction pools resulting in identical propagule pressures and residence times for these six locations (Zenni *et al.* 2014; Zenni and Hoban 2015b).

Since introduction, the common gardens have produced invasive populations expanding between ca. 100 and 450 m from the common garden. By sampling plants at different distances from each of the common gardens, from the border of the plantations to the leading edge of the invasion fronts, we were able to track changes in genotype and phenotype frequencies in these six naturalized populations over multiple generations encompassing 40 years of population growth. This approach already showed that the six naturalized populations are undergoing contemporary evolution and significant genetic changes are occurring along the invasive range expansions. At the provenance level, genetic changes were associated with climate (Zenni *et al.* 2014), whereas at the gene level, genetic changes were mostly population specific (Zenni and Hoban 2015b).

Data collection and trait measurements

The experiment design, description of sample and data collection, and specifications of the laboratory and genotyping methods used for the genetic work have been detailed previously (Zenni *et al.* 2014; Zenni and Hoban 2015b). Briefly, 50 plants were haphazardly sampled from each of six invasive *P. taeda* populations (a total of 300 plants) and genotyped for 94 single nucleotide polymorphisms (SNPs) using Fluidigm® SNPtype Assays. Plants sampled were at least 1.3 m tall and had between 3 and 34 years of age. For this study, the original plantings were not included because traits from trees growing at high-density monocultures may not be comparable with trees naturally growing in heterogeneous habitats. All the SNPs chosen were a subset of the ones used by Eckert *et al.* (2010) and were located in functional genes. Complete genotype data is available at Dryad data repository (Zenni and Hoban 2015a, b). The distances between the plants and the edge of the common gardens were measured using the function "Hub distance" in the package MMQGIS for QGIS (Quantum GIS Development Team 2015), and distances were normalized for each population (divided by the maximum distance of spread) to account for variations in spread rates across sites (D_{NORM}). For the phenotypic part of the study, fieldwork was carried out in November 2014 and January 2015. To collect phenotypic data, we visited the same trees previously sampled for genotyping in all six invasive populations. However, between 2012 (original sampling) and 2014 (sampling for this study), some trees died owing to idiosyncratic factors (i.e. tree falls and establishment of firebreaks). The number of surviving individuals was 47 for Capão Bonito, 48 for Irati, 49 for Rio Negro, 49 for Três Barras, 40 for São Francisco de Paula and 50 for Santa Maria, for a total of 283 plants. We used the same individuals from previous studies and did not include new plants because the sampled plants had already been genotyped.

For each plant, we collected an increment core at the base of the tree (~ 20 cm from the ground) using a 30 cm increment borer (Haglöf Sweden), a handful of fully developed and healthy needles from the tip of the lowest branch, and a $\sim$15 cm segment of wood from the tip of the lowest branch. We also measured circumference at breast height (CBH; 1.30 m), tree total height, and bark thickness. All fresh wood and needle samples collected were stored in a freezer (-20 °C) immediately upon arrival at the laboratory. Increment cores were air dried, mounted in wooden supports and sandpapered to improve measurement accuracy. Lengths of growth rings were measured using a digital calliper to determine annual growth, and growth rings were counted to estimate the number of years contained in the extracted core (plant age). For all the analyses, we used the plants' mean annual growth (MAG). We measured the area of 20 fully developed and healthy needles using a table scanner (Epson Perfection™ V700 Photo) and Digimizer v. 4.3.0 (MedCalc Software), and divided the total area by the number of needles to calculate mean leaf area (LA) of each plant. To determine specific LA (SLA), we dried the previously measured needles at 70 °C for 72 h and weighted them (leaf mass; LM) immediately after removing from the drying oven (SLA = LA×LM^{-1}). To estimate inductive and constitutive plant defences in our studied plants, we determined non-volatile wood resin content, needle total phenolic content, and wood total phenolic content. Resin and phenolic extraction and determination followed the procedure described by Moreira et al. (2012). Briefly, resin was extracted from 15 cm long × $\sim$ 0.5 cm wide pieces of wood by mixing them with a solvent (hexane) in an ultrasonic bath at 45 °C for 20 min followed by a 24 h rest (performed twice to maximize extraction outputs). The solution was filtered in paper filters and left to dry. The residual content was weighted to the nearest 0.0001 g, and divided by the weight of the wood dry mass (oven dried at 80 °C for 24 h) to be expressed as mg g^{-1}. Phenolic content was determined colorimetrically by the Folin–Ciocalteu method in a Bel Photonics 2000UV spectrophotometer (Bel Photonics do Brasil Ltda., Osasco, SP, Brazil) at 760 nm, using tannic acid as a standard (Moreira et al. 2014). Needles were oven-dried for 72 h at 60 °C, ground in a Wiley mill, and sieved in a 5 mm mesh. For each sample, we used 100 mg of ground needles. The ground samples were transferred to centrifuge tubes, mixed with 5 ml of 70% acetone and left in the refrigerator at 4 °C for 1 h. The solutions were centrifuged for 10 min at 7800 RPM and 0.3 ml of supernatant was removed and mixed with deionized water to make 1 ml of solution. To this solution, we added 0.1 N NaOH in Na$_2$CO$_3$ and the Folin–Ciocalteu reagent. After 2 h in the refrigerator, we took the spectrophotometer readings.

Statistical analyses

First, we performed six analyses of variance with Bonferroni correction of P values to test for differences in mean trait values of LA, SLA, MAG, non-volatile resin, and total needle and wood phenolic contents among the six locations. Second, we built generalized linear models with Gamma probability distribution and the inverse link function to test the effects of climate (mean annual temperature and annual precipitation) on each of the traits measured. Climate data were obtained from the Worldclim database (Hijmans et al. 2005).

Next, we used linear mixed-effect models to test how LA, SLA, MAG, non-volatile resin, and total needle and wood phenolic contents varied as a factor of distance from the introduction point (D$_{NORM}$) and/or were genetically controlled at each location. The genetic clustering results (provenances) from Zenni et al. (2014) were used as the genetic makeup of individual plants. The dispersal model assumed for this study is that spread occurs as new generations establish farther from the point of introduction with some degree of back dispersal from the invasion front to the rear edge. Thus, the studied populations represent six unique invasions set out by six fully replicated introduction events. Distance of the invasive plant to the common garden and genetic clustering data were considered fixed effects, whereas plant age was added as a random effect (this was done to remove the effect of plant age from the model). For these analyses, we tested traits separately. For each trait, we built four models: a full model including distance, provenance and age; a distance model including distance as fixed and age as random factor; a provenance model including provenance as fixed and age as random factor; and a null model including only an intercept and age as random factor. We compared the four models for each trait using a likelihood ratio test and models were considered significant if statistically different from the null model at $\alpha = 0.05$. Coefficients of determination were calculated using likelihood-ratio based pseudo-r^2 and represent the variance explained by fixed effects (Nakagawa and Schielzeth 2013). These analyses were done in R 3.2 using the packages "lme4" v. 1.1-7 for mixed-effect models, and "MuMIn" v. 1.14.0 for pseudo-r^2.

To test the association between mean annual growth rates and chemical defences, we built three linear models for each naturalized population using non-volatile wood resin content, needle total phenolic content or wood total phenolic content as a dependent variable and MAG as an independent variable.

We also used the slopes of the linear mixed-effect models to test if rate of change in the measured traits resulted in increased invasive potential. For this, we built

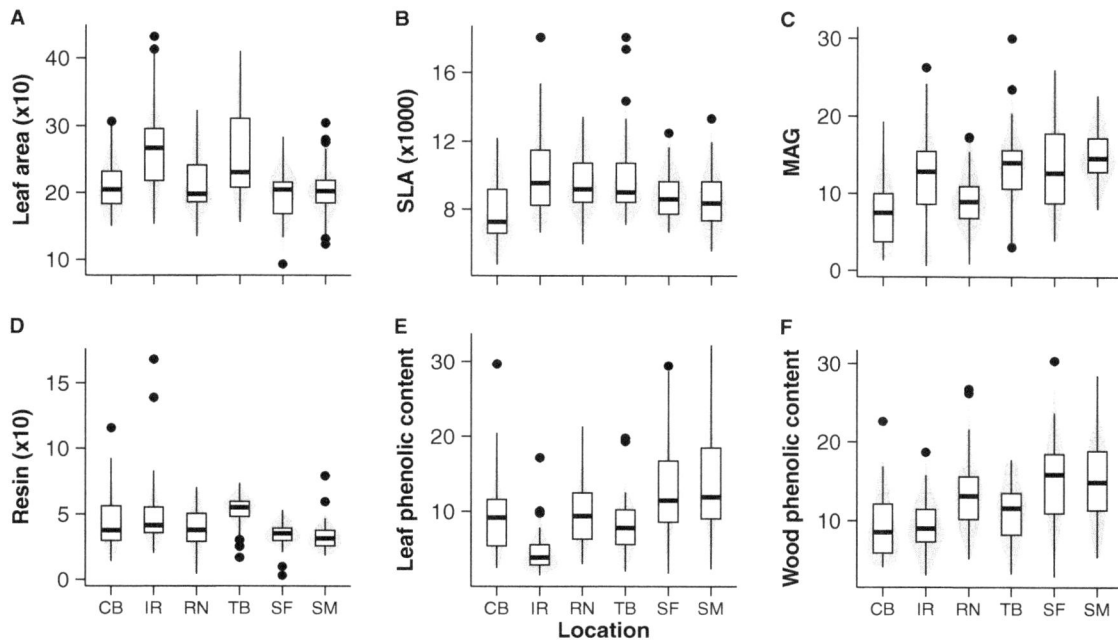

Figure 2. Trait variations among the studied *Pinus taeda* invasive populations. Violin plots and boxplots of (A) leaf area (mm²), (B) specific leaf area (SLA, mm²/g), (C) mean annual growth (MAG, mm), (D) non-volatile wood resin content (mg/g), (E) total needle phenolic content (mg/g) and (F) total wood phenolic content (mg/g) for each invasive population (CB, Capão Bonito; IR, Irati; RN, Rio Negro; SF, São Francisco de Paula; SM, Santa Maria; TB, Três Barras). Violins (grey areas) are kernel probability densities of the data at different values. Boxplots are median (bold black line), quartiles (white rectangles), standard deviation (black lines) and possible outliers (dots).

linear models using the mixed-model slopes for each trait at each location as the independent variable and maximum distance of spread at each location as dependent variable. The slope coefficients are the mean slope for each trait of all plants in the model.

Finally, we performed a genome-wide association analysis to identify genes associated with the measured phenotypes. We tested each trait separately and assumed a codominant genetic model. The 94 unlinked SNPs were added as factors to the models and a Bonferroni correction was done to counteract the multiple comparisons problem. The analysis was done in R 3.2 using the package "SNPassoc" v. 1.9-2. SNPs identified as having associations with traits were further investigated for specific functions in the Dendrome database (http://dendrome.ucdavis.edu/DiversiTree/) and in Genbank.

Results

Although all invasions started from replicated pools of genetic material and equal propagule pressure (Zenni et al. 2014), we found divergence in mean trait values in the six invasive populations for the six traits measured (Fig. 2 and Table 1): LA ($F_{5,273} = 16.69$, $P < 0.001$), SLA ($F_{5,273} = 7.976$, $P < 0.001$), mean annual growth (MAG;

$F_{5,276} = 15.56$, $P < 0.001$), non-volatile resin content ($F_{5,273} = 9.945$, $P < 0.001$), needle total phenolic content ($F_{5,273} = 19.36$, $P < 0.001$) and wood total phenolic content ($F_{5,264} = 16.63$, $P < 0.001$). Climate (mean annual temperature and annual precipitation) explained divergence in LA ($P < 0.02$), SLA ($F_{3,264} = 12.1$, $P < 0.001$ for MAT), MAG ($F_{3,264} = 22.5$, $P < 0.05$), resin content ($P < 0.02$), needles phenolic content ($P < 0.001$) and wood phenolic content ($P < 0.001$ for AP) (Table 2 and Fig. 3). While LA and SLA were smaller in hotter and wetter locations, MAG was higher in locations with higher annual precipitations.

Not only were there between-population variations but also the invasive populations were rapidly changing along each invasive range expansion (Fig. 4 and Table 3). We tested rapid evolution along the six invasive range expansions using linear mixed-effect models with distance from plantation and provenance as fixed effects and plant age as random effect. Two populations (Rio Negro and Três Barras) showed increases in LAs (Fig. 4 and Table 3, pseudo-$r^2 = 0.08$ and 0.26, $P = 0.045$ and $P < 0.001$, respectively) and decreases in SLA (Fig. 4 and Table 3, pseudo-$r^2 = 0.1$ and 0.2, $P = 0.02$ and 0.005, respectively) during range expansion. Four populations (Capão Bonito, Rio Negro, Três Barras and Santa Maria) showed faster growth rates at the leading edge of the

Table 1. Mean trait values and standard deviation for each invasive population. LA, leaf area; SLA, specific leaf area; MAG, mean annual growth.

	Capão Bonito	Irati	Rio Negro	Três Barras	São Francisco	Santa Maria
LA	210.96±38.64	264.82±61.45	211.45±44.06	257.65±63.79	191.42±39	206.39±40.73
SLA	7904.98±1812.52	10085.93±2657.68	9572.39±1728.83	9849.64±2351.04	9156.25±1682.43	8535.52±1726.33
MAG	7.43±4.18	12.05±6.92	9.2±3.45	13.23±5.39	13.48±6.08	15.27±4.07
Wood resin	43.4±21.79	49.09±26.59	37.93±16.31	52.57±11.12	33.78±10.14	33.16±10.82
Needle phenolics	9.47±5.38	4.75±2.8	10.06±4.56	8.05±3.55	13.07±6.48	13.78±7.07
Wood phenolics	9.2±4.06	9.25±3.44	13.55±4.68	11.33±3.87	15.49±5.52	15±5.33

Table 2. Summary results of generalized linear models (Gamma family distribution) for the relationship between plant traits and climate. Null deviance shows how well the response variable is predicted by a model that includes only the intercept, whereas residual deviance shows how well the response variable is predicted by the alternative model. SLA, specific leaf area; MAG, mean annual growth; MAT, mean annual temperature; AP, annual precipitation.

Model	Variable	t	P	df_{Null}	$df_{Residual}$	$Deviance_{Null}$	$Deviance_{Residual}$
Leaf area	MAT	2.40	0.017	267	265	15.282	14.614
	AP	3.18	0.002				
SLA	MAT	3.72	0.000	267	265	13.692	12.936
	AP	1.86	0.065				
MAG	MAT	−2.03	0.043	267	265	97.28	89.144
	AP	−6.02	0.000				
Wood resin	MAT	2.48	0.014	267	265	51.766	48.595
	AP	3.96	0.000				
Needle phenolics	MAT	−3.73	0.000	267	265	98.674	88.972
	AP	−5.38	0.000				
Wood phenolics	MAT	−1.05	0.295	267	265	49.618	44.821
	AP	−5.26	0.000				

invasion front in comparison with plants at the rear edge (Fig. 4 and Table 3, pseudo-r^2 = 0.1, 0.1, 0.3, 0.1, and P = 0.03, 0.03, <0.001, 0.02, respectively). In two of these cases (Irati and Três Barras), MAG increased 2-fold in 40 years (Fig. 4). None of the populations showed decreases in MAG. In terms of constitutive and inductive plant defences, non-volatile resin content increased in plants along one invasion gradient (Rio Negro, pseudo-r^2 = 0.1, P = 0.03) and decreased in a second (São Francisco, pseudo-r^2 = 0.1, P = 0.04), total needle phenolic content increased in plants along one invasion gradient (Capão Bonito, pseudo-r^2 = 0.1, P = 0.02) and total wood phenolic content increased in plants along the Capão Bonito invasion gradient (pseudo-r^2 = 0.1, P = 0.03) and decreased in a second (Três Barras, pseudo-r^2 = 0.1, P = 0.04).

Some instances revealed that evolutionary forces did not produce phenotypic changes along the invasive range expansions for the traits measured. Some trait variations were explained by the genetic provenance of the plants (i.e. genetically conserved) (Table 3). In Irati, LA (pseudo-r^2 = 0.02, P < 0.001), SLA (pseudo-r^2 = 0.09, P < 0.001), MAG (pseudo-r^2 = 0.09, P < 0.001) and total phenolic (pseudo-r^2 = 0.02, P < 0.001) values were explained by the plants genetic provenance. Ancestry also explained SLA in Santa Maria (pseudo-r^2 = 0.004, P < 0.001), wood resin in Capão Bonito, Três Barras and Santa Maria (pseudo-r^2 = 0.009, pseudo-r^2 = 0.08, pseudo-r^2 = 0.004, all P < 0.001), needle total phenolic content in São Francisco and Santa Maria (pseudo-r^2 = 0.02 and 0.03, and P < 0.001 for both) and wood

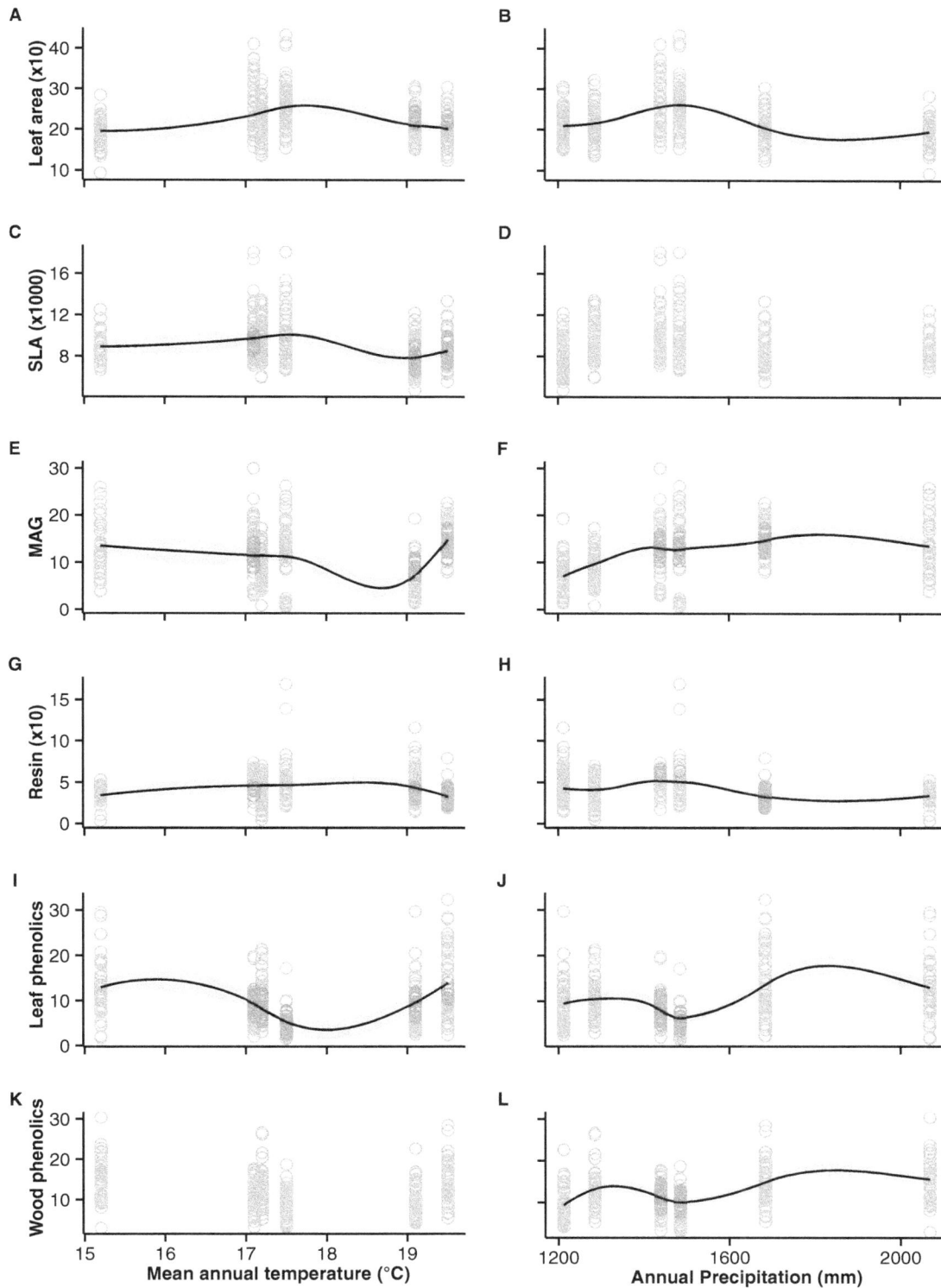

Figure 3. Relationship between leaf area (mm^2; A and B), specific leaf area (SLA, mm^2/g; C and D), mean annual growth (MAG, mm/year; E and F), resin (mg/g; G and H), needle phenolic content (mg/g; I and J) and wood phenolic content (mg/g; K and L) and climate (mean annual temperature and annual precipitation) of the six studied locations. Solid lines represent local polynomial regression fitting and grey shades are 95% confidence intervals. Panels D and K show no relationship between variables.

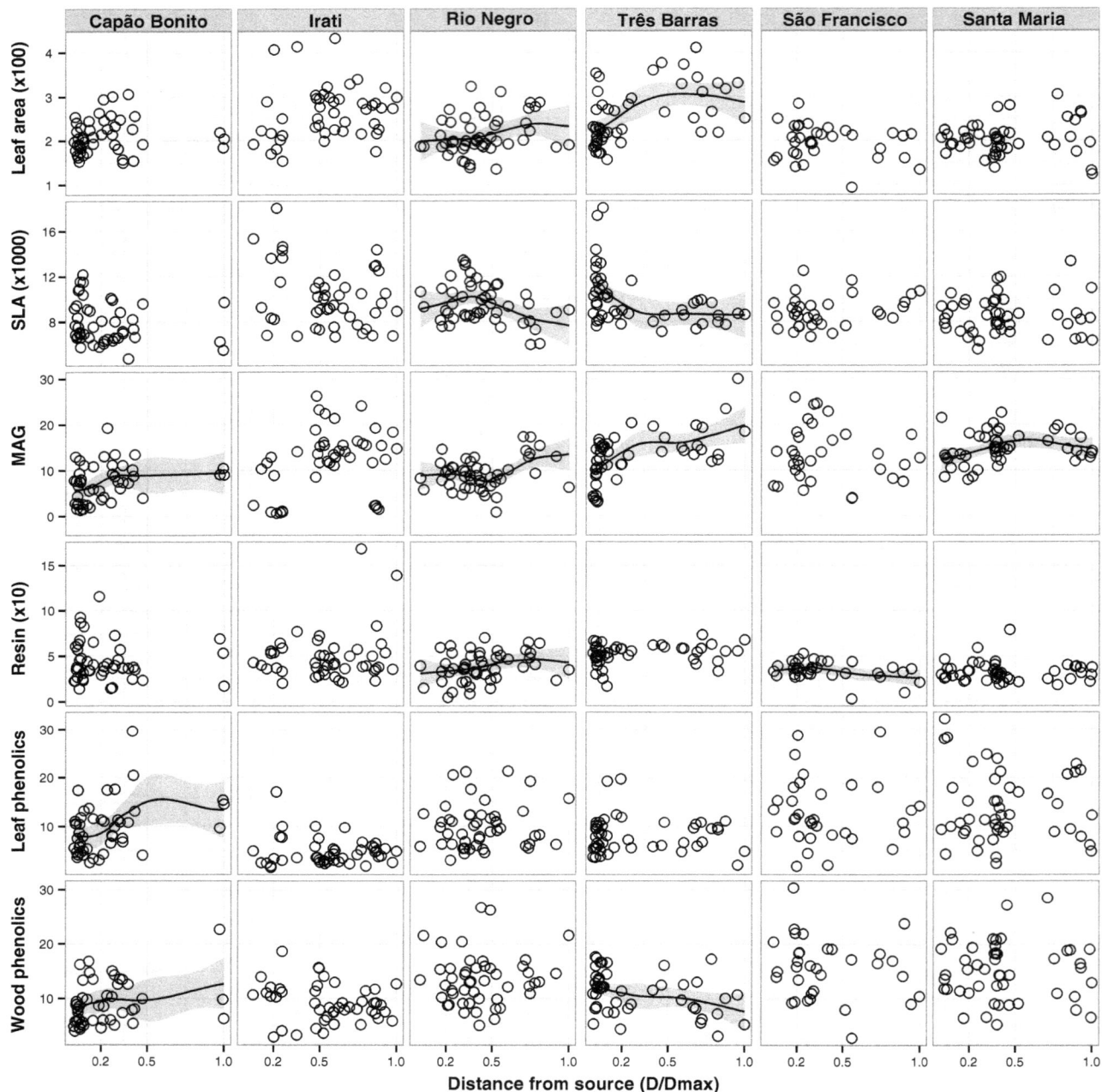

Figure 4. Relationship between productivity, growth, and defence traits and dispersal distance (D/D_{max}, see Methods section for details). Scatterplots of leaf area (mm^2), specific leaf area (SLA, mm^2/g), mean annual growth (MAG, mm/year), non-volatile wood resin content (mg/g) and total needle phenolic content (mg/g) across the invasion gradient (normalized spread distance) for each invasive population. Lines represent local polynomial regression fit and grey shades are the 95% confidence intervals.

total phenolic content in Rio Negro (pseudo-$r^2 = 0.006$, $P < 0.001$).

Our analyses of potential trade-offs between growth and chemical defences yielded no significant relationships ($P > 0.05$) among MAG and wood resin, wood total phenolics, or needle total phenolics in all but two models (Fig. 5). In Irati, concentrations of total phenolics in needles were lower in plants with higher MAG ($r^2 = 0.12$, $P = 0.02$), but a single outlier (IRT6) with very high phenolic content drove this result. Removing the outlier from the model yielded a non-significant relationship ($P = 0.09$). In Capão Bonito, contrary to all expectations, concentrations of total phenolics in needles and MAG were positively associated ($r^2 = 0.22, P < 0.001$).

Table 3. Summary results of likelihood ratio test among linear mixed-effect models. Pseudo-r^2 represents the variance explained by fixed effects. Degrees of freedom (df) are reported for the likelihood ratio tests. SLA, specific leaf area; MAG, mean annual growth.

		Capão Bonito			Irati			Rio Negro			Três Barras			São Francisco de Paula			Santa Maria		
		df	Pseudo-r^2	P	df	Pseudo-r^2	P	df	Pseudo-r^2	P	df	Pseudo-$r2$	P	df	Pseudo-r^2	P	df	Pseudo-r^2	P
Leaf area	Full	5	0.012	0.551	5	0.017	0.811	5	0.179	0.016	5	0.283	<0.001	5	0.050	0.198	5	0.024	0.447
	Invasion	4	0.006	0.615	4	0.003	0.702	4	0.080	0.045	4	0.262	<0.001	4	0.049	0.196	4	0.016	0.368
	Provenance	4	0.003	1.000	4	0.016	<0.001	4	0.074	1.000	4	0.005	1.000	4	0.002	1.000	4	0.012	1.000
SLA	Full	5	0.084	0.108	5	0.110	0.268	5	0.103	0.024	5	0.303	0.003	5	0.056	0.232	5	0.004	0.965
	Invasion	4	0.040	0.169	4	0.030	0.184	4	0.103	0.023	4	0.165	0.005	4	0.039	0.235	4	0.000	0.969
	Provenance	4	0.032	1.000	4	0.088	<0.001	4	0.002	1.000	4	0.136	1.000	4	0.017	1.000	4	0.004	<0.001
MAG	Full	5	0.191	0.010	5	0.158	0.113	5	0.098	0.032	5	0.338	<0.001	5	0.049	0.205	5	0.092	0.026
	Invasion	4	0.096	0.032	4	0.082	0.070	4	0.098	0.030	4	0.335	<0.001	4	0.049	0.199	4	0.092	0.017
	Provenance	4	0.068	1.000	4	0.093	<0.001	4	0.002	1.000	4	0.002	1.000	4	0.002	1.000	4	0.016	1.000
Wood resin	Full	5	0.020	0.473	5	0.034	0.211	5	0.135	0.014	5	0.117	0.179	5	0.146	0.037	5	0.004	0.971
	Invasion	4	0.007	0.557	4	0.034	0.203	4	0.100	0.025	4	0.041	0.166	4	0.114	0.040	4	0.000	0.896
	Provenance	4	0.009	<0.001	4	0.001	1.000	4	0.020	1.000	4	0.081	<0.001	4	0.033	1.000	4	0.004	<0.001
Leaf phenolics	Full	5	0.110	0.021	5	0.021	0.527	5	0.008	0.549	5	0.001	0.850	5	0.034	0.760	5	0.036	0.694
	Invasion	4	0.109	0.021	4	0.007	0.581	4	0.007	0.562	4	0.001	0.844	4	0.002	0.770	4	0.008	0.521
	Provenance	4	0.000	1.000	4	0.015	<0.001	4	0.000	1.000	4	0.000	1.000	4	0.032	<0.001	4	0.033	<0.001
Wood phenolics	Full	5	0.116	0.048	5	0.041	0.243	5	0.010	0.663	5	0.147	0.012	5	0.023	0.468	5	0.005	0.616
	Invasion	4	0.094	0.034	4	0.040	0.257	4	0.005	0.634	4	0.131	0.012	4	0.019	0.453	4	0.005	0.615
	Provenance	4	0.036	1.000	4	0.005	1.000	4	0.006	<0.001	4	0.019	1.000	4	0.008	1.000	4	0.000	1.000

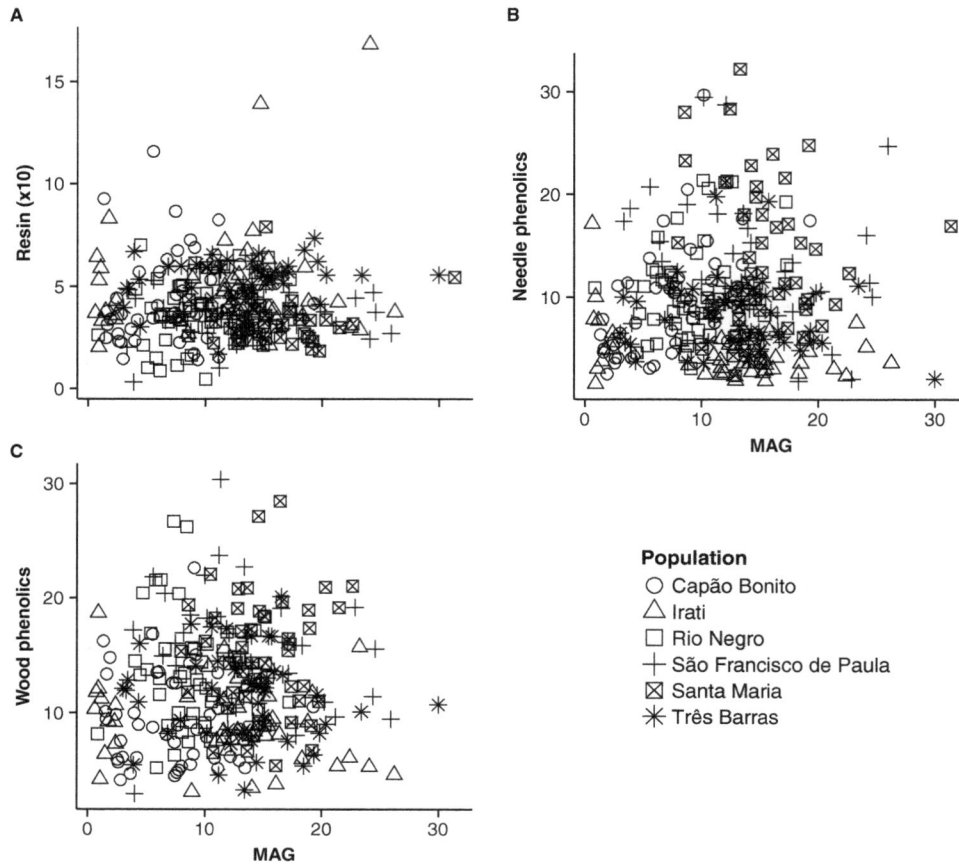

Figure 5. Relationship between chemical defences and mean annual growth in plants across the six studied naturalized populations. (A) Relationships between resin contents and mean annual growth, (B) relationships between needle total phenolic contents and mean annual growth and (C) relationships between wood total phenolic contents and mean annual growth. Dots of different shapes represent different naturalized populations. None but one of the relationships were statistically significant (needle phenolics and MAG in Capão Bonito; see text for details).

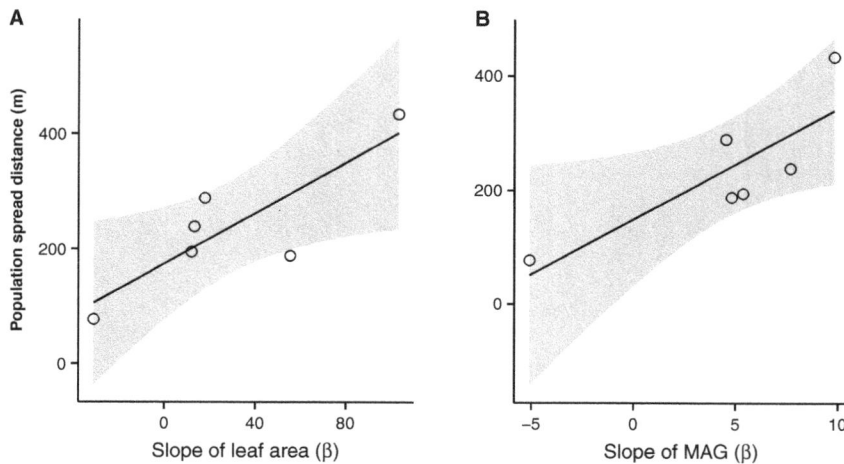

Figure 6. Relationship between evolution in leaf area and growth and invasive range expansion. Regression line and 95% confidence intervals (grey shade) of the relationship between the rate of evolution in (A) leaf area and (B) mean annual growth (MAG) for the six invasive populations. Evolution is represented by the mean slope of the mixed models between normalized distance of spread and trait value (Fig. 4).

When testing how rate of evolutionary responses affected invasive ranges expansions, we found that rate of change of increased MAG and rate of change of increased LA were positively associated to the total spread distances of the invasive populations ($r^2 = 0.7$ and $P = 0.04$ for both models, Fig. 6). However, although plant MAG and LA were not correlated ($P = 0.07$) and both were positively related to total spread distance, the full model was not significant ($P = 0.2$). This was probably due to the small sample size ($n = 6$ populations).

In a genome-wide association analysis, we detected one gene associated with LA (SNP_210094-Pita) and two genes associated with MAG (SNP_215552-Pita and SNP_225711-Pita). SNP_210094-Pita is related to phosphorus assimilation by the roots (GenBank Acc: BE496394) and phosphorous has been experimentally shown to affect LA in forest trees (Herbert and Fownes 1995). SNP_215552-Pita is related to root development (GenBank Acc: CO198215), drought resistance (GenBank Acc: CF392949) and fungal resistance (GenBank Acc: DR093744). SNP_225711-Pita is associated with wood formation (GenBank Acc: BE496394).

Discussion

Our results show strong evidences for rapid phenotypic change in all six invasive populations, providing support for our first hypothesis. However, contrary to our second hypothesis, patterns of phenotypic changes did not match patterns of genetic change across populations, and each population showed a unique pattern of phenotypic change along the invasion gradient. Surprisingly, also contrary to our initial expectations, populations did not exhibit trade-offs between evolution in growth and chemical defences. In one instance (Capão Bonito), there was a positive association between MAG and leaf phenolic content. Finally, supporting our last hypothesis, we found positive associations between rates of rapid change in plant growth (MAG) and productivity (LA) and the rates of population spread. The mismatches observed between phenotypic changes, genetic makeup of the plants and climate patterns suggest part of the variation found may be caused by phenotypic plasticity or genetic drift. Although populations are changing genetically along the invasion gradients (Zenni et al. 2014), the traits measured might be responding to the environment without a underlying genetic change. Also, the genetic and phenotypic changes might not be the result of selection, but the result of genetic drift and founder effects (van Kleunen and Fischer 2008; Monty et al. 2013).

A previous study in this system showed that climate was one of the factors determining the patterns of genetic change across the six populations (Zenni et al. 2014). Accordingly, climate (mean annual temperature and annual precipitation) also partially explained patterns of variation in LA, SLA, MAG, wood resin and leaf and wood phenolic contents among populations (Fig. 3). The strongest association was between MAG and annual precipitation, where growth rates were higher in wetter locations. Although statistically significant, mean annual temperature did not strongly affect the growth rates of plants. Relationships between P. taeda phenotypic patterns and climatic factors had also been reported in the species native range (e.g. Eckert et al. 2010). For other systems, climate has also been shown to be an important factor effecting population growth, spread, or retraction (Chen et al. 2011; Pfeifer-Meister et al. 2013; Kerr et al. 2015). In our study, the fact that all populations started from replicated introduction pools, and genotypic and phenotypic responses to each climate were observed during range expansions in less than 40 years, suggests climate may be an important driver of evolution for invasive populations. Under new climates, either owing to human-meditated introductions to a new range or to climate change, invasive populations may be the ones able to adapt quickly to the new environment.

Climate not only can drive ecological patterns of plant chemical defences but also the evolution of constitutive chemical defences in pines may be driven by it (Moreira et al. 2014). Our results seem to support this notion (Fig. 3), although we measured total phenolic content, which includes both constitutive and inducible defences. In the current study, we found differences in chemical defence traits both among and within populations (Table 2 and Figs 2 and 4). The among population differences may be due to ecological responses of populations to each environment (climate and habitat), whereas the changes along each invasion gradient may be showing evolutionary responses of each population to each environment. Considering that the among population differences are stronger than the within population changes, it is not possible to determine if adaptive evolution is occurring in chemical defences in the six studied populations or if differences are due to genetic drift caused by founder effects (Monty et al. 2013). However, it is possible evolution in the studied defence traits is occurring, but the effects were not detected. Another study, disentangling constitutive and inducible defences would be necessary to address the genotypic (constitutive), phenotypic (constitutive and inducible) or epigenetic components of defence responses.

The production of chemical defences is costly, and increased resource allocation to defences tends to result in disproportionally high decreases in growth rates for plants (Mithöfer and Boland 2012). Consequently, if non-native plants escape from natural enemies during intro-duction and are able to afford neglecting their chemical defences to favour growth, reproduction, and dispersal traits, they may become invasive (Callaway and Ridenour 2004; Felker-Quinn et al. 2013). However, in our study, we generally found no relationship between investments in plant growth, productivity and chemical defences. Plants showed increased or kept existing resource use strategies, high growth rates and levels of chemical de-fences. The quantities of resin and total phenolics we found were similar to quantities reported for the species by other studies in other regions (Moreira et al. 2014), in-dicating that P. taeda plants in the six studied popula-tions had levels of chemical defences similar to the species' average. Although investments in chemical defences varied across populations, we did not find a consistent trend towards increased or decreased invest-ments along the invasion gradients as we did for MAG. On one hand, resin and phenolics total contents might be a plastic response of populations and, given the lack of important natural enemies in the introduced ranges, there are no selection pressures for increased chemical defences. Or populations show a slower pace of change for chemical defences than for growth rates given the complexity of the metabolic processes involved in pro-ducing chemical defences (Mithöfer and Boland 2012). On the other hand, there may be no limiting resources requiring a trade-off between chemical defences and growth and increase growth rates are simply a result of favourable climatic conditions for P. taeda (e.g. Nuñez and Medley 2011).

Although the invasion literature advocates that high SLAs and fast growth have positive associations with woody plant invasiveness (Grotkopp and Rejmánek 2007; Zenni and Simberloff 2013), it has also been suggested that natural selection can reduce a population invasive potential (Lankau et al. 2009). Our data suggest that rapid evolution, either via natural selection or genetic drift, can result in populations with increased invasive po-tential in comparison to the original introduction pool. In most of our cases, plants at the leading edges of the in-vasion fronts grew faster and were more productive, on an average, than plants at the rear edges (Fig. 4). These effects were mediated by climate, suggesting that in con-ditions adverse for the species, evolutionary changes may not be so pronounced. However, in none of the cases did populations showed slower growth rates or lower productivities at the leading edges. We found two populations (Rio Negro and Três Barras) that produced lower SLA along the invasion gradients, besides being ac-tively spreading and invading. Four populations increased growth rates along the invasion gradients. Although our results seem to contradict the established literature

(Tecco et al. 2010; Porté et al. 2011), we believe that the decrease in SLA combined with increases in LA may be explained as a response of P. taeda plants to the forest ecosystems of Rio Negro and Três Barras. MAG results were in agreement with the established literature that fast growth is a key invasion trait (Lamarque et al. 2011).

Also supporting the idea that rapid evolutionary change in growth rates and productivity can increase the invasive potential of populations (Lamarque et al. 2011; Monty et al. 2013; Sargent and Lodge 2014), we found strong relationships between MAG and LA and total spread of populations. Populations with greater rate of change towards increased growth rates and LA spread farther in the 40 years period encompassed by this study (Fig. 6). However, it is also possible that greater spread could lead to higher growth if plants are experiencing less competition at the leading edge of the invasion front, or if growth and spread were both correlated to a third factor. Nevertheless, the strong effect rapid evolu-tion can have on populations' invasiveness suggests that invasions must be dealt with sooner rather than latter.

Several of the statistical models including only genetic provenance as a factor were significantly different than the null models, but overall they had low explanatory power (Table 2). It is possible that genetic provenance is a coarse scale to test fine evolutionary changes in popu-lations. The fine-scaled genome-wide association analy-ses, where trait values were tested for each one of the 94 genes included in this study, we found three genes associated with the traits measured. Given that we used putatively functional markers, we expected that more of the genes would be associated with traits. A possible ex-planation for the few significant associations found is that most traits result from the expression of many genes, instead of only a few genes of large effect. Also, because we measured plants growing in natural environ-ments and not in a controlled homogenous setting (e.g. greenhouse), phenotypic plasticity may be masking the role of specific genes in trait expression. Taken together with previous findings on the study system that several genes were undergoing rapid changes in genotype fre-quencies (Zenni and Hoban 2015b), the results support our claim that evolution is occurring in this system and changes were not caused only by phenotypic plasticity.

Because our study took place in a natural setting, and not a controlled environment, it is subject to several ca-veats. It is possible some of the variations in traits values could be explained by the age and size of the trees, as well as by micro-environmental factors that differ for each individual tree, and that may vary along each of the invasion gradients. We tried to address the effect of tree size and age in the statistical models by having tree age as a random effect in the mixed models and by using only the variance explained by the fixed effects in subsequent analyses and interpretations of the model results. We were unable to disentangle the relative importance of adaptive evolution, genetic drift and phenotypic plasticity for the traits measured. However, given the structure of our statistical models, it is likely that plastic responses only decreased the chances of us finding significant relationships between trait evolution and invasion spread. We still found a number of significant associations. Further, although we acknowledge the existence and the potential importance of micro-environmental factors in affecting trait values on each plant (adaptive and plastic), we think they make our findings even more remarkable. We showed that despite all potential micro-environmental variations in each of six naturalized population spanning an 840 km transect, four climate zones, and four decades of natural processes, rapid evolution in several functional traits could still be detected along invasion gradients.

Conclusions

In summary, our study provides strong support for the role of rapid evolution on the success of non-native range expansions of an invasive species. We also show that evolutionary changes are in part associated with climate, but are mostly fine-scaled and context-specific, even though invasive populations started from fully replicated introduction events. The results of this study highlight the unique nature of the ecological and evolutionary dynamics of each population, suggesting that predicting invasion trajectories may be daunting. Further, we demonstrate that the potential for rapid evolution can affect the capacity of populations to cope with new environments. Taken together, the capacity for adaptation to different conditions, evolution of increased growth rates, and the lack of trade-offs between growth and defences make *P. taeda* an invasive species that requires management and control before its spread reach large areas.

Sources of Funding

Our work was funded by CNPq-Brazil (313926/2014-0).

Contributions by the Authors

R.D.Z. conceived the study, collected data, measured resin contents, did statistical analyses and wrote the manuscript. W.L.C. collected data, measured SLAs and mean annual growths. G.S. measured phenolic contents. All authors discussed the results and commented on the manuscript.

Acknowledgements

The authors thank Leandro Falleiros and Wesley Rocha for lab assistance, and Dr. Sarah Oliveira, Dr. José Gonçalves Jr., and Dr. John Hay for proving lab spaces and infrastructure. The authors also thank Dr. Sean Hoban for comments on the manuscript and the administrators of the study sites for logistical support.

Literature Cited

Berthouly-Salazar C, Hui C, Blackburn TM, Gaboriaud C, Rensburg BJ, Vuuren BJ, Roux JJ. 2013. Long-distance dispersal maximizes evolutionary potential during rapid geographic range expansion. *Molecular Ecology* **22**:5793–5804.

Blackburn TM, Pyšek P, Bacher S, Carlton JT, Duncan RP, Jarošík V, Wilson JRU, Richardson DM. 2011. A proposed unified framework for biological invasions. *Trends in Ecology & Evolution* **26**: 333–339.

Blossey B, Nötzold R. 1995. Evolution of increased competitive ability in invasive nonindigenous plants: a hypothesis. *Journal of Ecology* **83**:887–889.

Burton OJ, Phillips BL, Travis JMJ. 2010. Trade-offs and the evolution of life-histories during range expansion. *Ecology Letters* **13**: 1210–1220.

Buswell JM, Moles AT, Hartley S. 2011. Is rapid evolution common in introduced plant species? *Journal of Ecology* **99**:214–224.

Callaway RM, Ridenour WM. 2004. Novel weapons: invasive success and the evolution of increased competitive ability. *Frontiers in Ecology and the Environment* **2**:436–443.

Caplat P, Cheptou PO, Diez J, Guisan A, Larson BMH, Macdougall AS, Peltzer DA, Richardson DM, Shea K, van Kleunen M, Zhang R, Buckley YM. 2013. Movement, impacts and management of plant distributions in response to climate change: insights from invasions. *Oikos* **122**:1265–1274.

Chen I-C, Hill JK, Ohlemüller R, Roy DB, Thomas CD. 2011. Rapid range shifts of species associated with high levels of climate warming. *Science* **333**:1024–1026.

Colautti RI, Barrett SCH. 2013. Rapid adaptation to climate facilitates range expansion of an invasive plant. *Science* **342**: 364–366.

Coley PD, Bryant JP, Chapin FS. 1985. Resource availability and plant antiherbivore defense. *Science* **230**:895–899.

Colomer-Ventura F, Martínez-Vilalta J, Zuccarini P, Escolà A, Armengot L, Castells E. 2015. Contemporary evolution of an invasive plant is associated with climate but not with herbivory. *Functional Ecology.* **29**:1475–1485.

Donaldson JE, Hui C, Richardson DM, Robertson MP, Webber BL, Wilson JR. 2014. Invasion trajectory of alien trees: the role of introduction pathway and planting history. *Global Change Biology* **20**:1527–1537.

Eckert AJ, van Heerwaarden J, Wegrzyn JL, Nelson CD, Ross-Ibarra J, González-Martínez SC, Neale DB. 2010. Patterns of population structure and environmental associations to aridity across the range of loblolly pine (*Pinus taeda* L., Pinaceae). *Genetics* **185**:969–982.

Felker-Quinn E, Schweitzer JA, Bailey JK. 2013. Meta-analysis reveals evolution in invasive plant species but little support for Evolution of Increased Competitive Ability (EICA). *Ecology and Evolution* **3**:739–751.

Forsman A, Wennersten L. 2015. Inter-individual variation promotes ecological success of populations and species: evidence from experimental and comparative studies. *Ecography.* doi:10.1111/ecog.01357:n/a-n/a.

Forsman A, Wennersten L, Karlsson M, Caesar S. 2012. Variation in founder groups promotes establishment success in the wild. *Proceedings of the Royal Society B: Biological Sciences.* doi:10.1098/rspb.2012.0174.

Grotkopp E, Rejmánek M. 2007. High seedling relative growth rate and specific leaf area are traits of invasive species: phylogenetically independent contrasts of woody angiosperms. *American Journal of Botany* **94**:526–532.

Gurevitch J, Fox GA, Wardle GM, Inderjit TD. 2011. Emergent insights from the synthesis of conceptual frameworks for biological invasions. *Ecology Letters* **14**:407–418.

Herbert DA, Fownes JH. 1995. Phosphorus limitation of forest leaf area and net primary production on a highly weathered soil. *Biogeochemistry* **29**:223–235.

Hijmans RJ, Cameron SE, Parra JL, Jones PG, Jarvis A. 2005. Very high resolution interpolated climate surfaces for global land areas. *International Journal of Climatology* **25**:1965–1978.

Keane RM, Crawley MJ. 2002. Exotic plant invasions and the enemy release hypothesis. *Trends in Ecology & Evolution* **17**:164–170.

Kerr JT, Pindar A, Galpern P, Packer L, Potts SG, Roberts SM, Rasmont P, Schweiger O, Colla SR, Richardson LL, Wagner DL, Gall LF, Sikes DS, Pantoja A. 2015. Climate change impacts on bumblebees converge across continents. *Science* **349**:177–180.

Kuebbing S, Simberloff D. 2015. Missing the bandwagon: nonnative species impacts still concern managers. *NeoBiota* **25**:73–86.

Lamarque L, Delzon S, Lortie C. 2011. Tree invasions: a comparative test of the dominant hypotheses and functional traits. *Biological Invasions* **13**:1969–1989.

Lankau RA, Nuzzo V, Spyreas G, Davis AS. 2009. Evolutionary limits ameliorate the negative impact of an invasive plant. *Proceedings of the National Academy of Sciences* **106**: 15362–15367.

Liu H, Stiling P. 2006. Testing the enemy release hypothesis: a review and meta-analysis. *Biological Invasions* **8**:1535–1545.

Mithöfer A, Boland W. 2012. Plant defense against herbivores: chemical aspects. *Annual Review of Plant Biology* **63**:431–450.

Monty A, Bizoux J-P, Escarré J, Mahy G. 2013. Rapid plant invasion in distinct climates involves different sources of phenotypic variation. *PLoS One* **8**:e55627.

Moodley D, Geerts S, Richardson DM, Wilson JRU. 2013. Different traits determine introduction, naturalization and invasion success in woody plants: Proteaceae as a test case. *PLoS One* **8**: e75078.

Moran EV, Alexander JM. 2014. Evolutionary responses to global change: lessons from invasive species. *Ecology Letters* **17**: 637–649.

Moreira X, Mooney KA, Rasmann S, Petry WK, Carrillo-Gavilán A, Zas R, Sampedro L. 2014. Trade-offs between constitutive and induced defences drive geographical and climatic clines in pine chemical defences. *Ecology Letters* **17**:537–546.

Moreira X, Zas R, Sampedro L. 2012. Differential allocation of constitutive and induced chemical defenses in pine tree juveniles: a test of the optimal defense theory. *PLoS One* **7**:e34006.

Nakagawa S, Schielzeth H. 2013. A general and simple method for obtaining R2 from generalized linear mixed-effects models. *Methods in Ecology and Evolution* **4**:133–142.

Nuñez MA, Medley KA. 2011. Pine invasions: climate predicts invasion success; something else predicts failure. *Diversity and Distributions* **17**:703–713.

Parker IM, Rodriguez J, Loik ME. 2003. An evolutionary approach to understanding the biology of invasions: local adaptation and general-purpose genotypes in the weed Verbascum thapsus. *Conservation Biology* **17**:59–72.

Pfeifer-Meister L, Bridgham SD, Little CJ, Reynolds LL, Goklany ME, Johnson BR. 2013. Pushing the limit: experimental evidence of climate effects on plant range distributions. *Ecology* **94**: 2131–2137.

Phillips BL, Brown GP, Webb JK, Shine R. 2006. Invasion and the evolution of speed in toads. *Nature* **439**:803–803.

Porté AJ, Lamarque LJ, Lortie CJ, Michalet R, Delzon S. 2011. Invasive Acer negundo outperforms native species in non-limiting resource environments due to its higher phenotypic plasticity. *BMC Ecology* **11**:28.

Prentis PJ, Wilson JRU, Dormontt EE, Richardson DM, Lowe AJ. 2008. Adaptive evolution in invasive species. *Trends in Plant Science* **13**:288–294.

Pyšek P, Jarošík V, Hulme PE, Pergl J, Hejda M, Schaffner U, Vilà M. 2012. A global assessment of invasive plant impacts on resident species, communities and ecosystems: the interaction of impact measures, invading species' traits and environment. *Global Change Biology* **18**:1725–1737.

Richardson DM, Pyšek P. 2012. Naturalization of introduced plants: ecological drivers of biogeographical patterns. *New Phytologist* **196**:383–396.

Rius M, Darling JA. 2014. How important is intraspecific genetic admixture to the success of colonising populations? *Trends in Ecology & Evolution* **29**:233–242.

Sargent LW, Lodge DM. 2014. Evolution of invasive traits in nonindigenous species: increased survival and faster growth in invasive populations of rusty crayfish (*Orconectes rusticus*). *Evolutionary Applications* **7**:949–961.

Schulte U, Veith M, Mingo V, Modica C, Hochkirch A. 2013. Strong genetic differentiation due to multiple founder events during a recent range expansion of an introduced wall lizard population. *Biological Invasions* **15**:2639–2649.

Shimizu JY, Higa AR. 1981. Variação racial do Pinus taeda L. no sul do Brasil até o sexto ano de idade. *Boletim De Pesquisa Florestal* **2**:1–25.

Tecco PA, Díaz S, Cabido M, Urcelay C. 2010. Functional traits of alien plants across contrasting climatic and land-use regimes: do aliens join the locals or try harder than them? *Journal of Ecology* **98**:17–27.

van Kleunen M, Fischer M. 2008. Adaptive rather than non-adaptive evolution of Mimulus guttatus in its invasive range. *Basic and Applied Ecology* **9**:213–223.

Wardle DA, Bardgett RD, Callaway RM, Van der Putten WH. 2011. Terrestrial ecosystem responses to species gains and losses. *Science* **332**:1273–1277.

Yelenik SG, D'Antonio CM. 2013. Self-reinforcing impacts of plant invasions change over time. *Nature* **503**:517–520.

Zenni RD. 2014. Analysis of introduction history of invasive plants in Brazil reveals patterns of association between biogeographical origin and reason for introduction. *Austral Ecology* **39**: 401–407.

Zenni RD, Bailey JK, Simberloff D. 2014. Rapid evolution and range expansion of an invasive plant are driven by

17

Differential plant invasiveness is not always driven by host promiscuity with bacterial symbionts

Metha M. Klock*[1], Luke G. Barrett[2], Peter H. Thrall[2] and Kyle E. Harms[1]

[1] Department of Biological Sciences, Louisiana State University, Baton Rouge, LA 70803, USA
[2] CSIRO Agriculture Flagship, Canberra, ACT 2601, Australia

Guest Editor: David Richardson

Abstract. Identification of mechanisms that allow some species to outcompete others is a fundamental goal in ecology and invasive species management. One useful approach is to examine congeners varying in invasiveness in a comparative framework across native and invaded ranges. *Acacia* species have been widely introduced outside their native range of Australia, and a subset of these species have become invasive in multiple parts of the world. Within specific regions, the invasive status of these species varies. Our study examined whether a key mechanism in the life history of *Acacia* species, the legume-rhizobia symbiosis, influences acacia invasiveness on a regional scale. To assess the extent to which species varying in invasiveness correspondingly differ with regard to the diversity of rhizobia they associate with, we grew seven *Acacia* species ranging in invasiveness in California in multiple soils from both their native (Australia) and introduced (California) ranges. In particular, the aim was to determine whether more invasive species formed symbioses with a wider diversity of rhizobial strains (i.e. are more promiscuous hosts). We measured and compared plant performance, including aboveground biomass, survival, and nodulation response, as well as rhizobial community composition and richness. Host promiscuity did not differ among invasiveness categories. *Acacia* species that varied in invasiveness differed in aboveground biomass for only one soil and did not differ in survival or nodulation within individual soils. In addition, acacias did not differ in rhizobial richness among invasiveness categories. However, nodulation differed between regions and was generally higher in the native than introduced range. Our results suggest that all *Acacia* species introduced to California are promiscuous hosts and that host promiscuity *per se* does not explain the observed differences in invasiveness within this region. Our study also highlights the utility of assessing potential mechanisms of invasion in species' native and introduced ranges.

Keywords: Acacia; biological invasions; interactions; invasive; legume; mutualisms; rhizobia.

Introduction

Non-native species are a threat to native ecosystems, particularly when they colonize new areas and rapidly expand in abundance. Collectively, invasive species have negative impacts at both local and global scales, threatening biodiversity, accelerating global change and

* Corresponding author's e-mail address: klockmm@gmail.com

causing economic losses (D'Antonio and Vitousek 1992; Vitousek *et al.* 1996; Mack *et al.* 2000; Pimentel *et al.* 2000). Although not all introduced species become invasive, those that do variously alter food sources for native wildlife, change fire regimes, outcompete native species, and impact soil communities, for example, by altering microbial structure and soil nitrogen levels (Mack and D'Antonio 1998; Mack *et al.* 2000; Brooks *et al.* 2004). To better understand how species become invasive in new environments, in-depth investigations of mechanisms driving species invasions are needed.

Diverse mechanisms and hypotheses have been proposed for why introduced species become invasive. Many of the better-investigated drivers of invasiveness are based on antagonistic or competitive interactions (Blossey and Notzold 1995; Callaway and Aschehoug 2000; Keane and Crawley 2002; Levine *et al.* 2003). Much work to date has investigated the role of enemy-release in facilitating species invasions (i.e. invaders that prosper in new environments because they leave their parasites, pests, and predators behind [Keane and Crawley 2002]). The Evolution of Increased Competitive Ability Hypothesis predicts that adaptive evolution of invaders provides a competitive advantage in novel ranges (Blossey and Notzold 1995). Although overcoming adversity imposed by antagonists and competitors may be the driver of invasiveness for some species, mutualistic interactions may also play a key alternate or synergistic role in some invasions (Richardson *et al.* 2000).

A growing body of work has examined the role of mutualisms in the invasion of non-native species (Richardson *et al.* 2000; Birnbaum *et al.* 2012; Wandrag 2012). The Enhanced Mutualism Hypothesis proposes that species encounter novel beneficial symbionts in their native range, which enhance their ability to survive and spread abroad (Richardson *et al.* 2000). The Accompanying Mutualist Hypothesis suggests that invasive species are introduced concurrently with their native mutualistic partners, thereby enhancing their ability to survive in novel habitats (Rodríguez-Echeverría 2010). Mutualisms such as those between legumes and their symbiotic nitrogen-fixing soil bacteria (i.e. rhizobia) may be particularly important in explaining the ability of this group of species to establish and expand abroad. Elucidating the potential role that mutualistic interactions play in species establishment and colonization may point towards mechanisms driving differential levels of species invasion.

Australian *Acacia* species (Family: Fabaceae) are a diverse group of legumes that form symbiotic relationships with rhizobia. They have been introduced throughout the world for a variety of purposes, including ornamental use, fuel wood, erosion control, and forestry (Kull and Rangan 2008; Carruthers *et al.* 2011). Many *Acacia*

Table 1. Definition of the terms "invasive," "naturalized" and "casual" as they relate to the invasiveness categories of *Acacia* species introduced to novel ranges.

Term	Definition	Reference
Invasive	Non-native species that (1) have self-sustaining populations which, for a minimum of 10 years have reproduced by seed or ramets without (or despite) human intervention, and (2) have spread and established reproductive populations at large distances from parent plants	Richardson *et al.* (2011)
Naturalized	Non-native species that have escaped cultivation and established self-sustaining populations but have not spread to the extent of invasive species	Richardson *et al.* (2011)
Casual	Non-native species that do not establish populations without the aid of humans (also 'waifs')	Richardson *et al.* (2000); Jepson Flora Project (2015)

species that have been introduced outside their native range have become invasive abroad (Richardson *et al.* 2011). Of the more than 1000 *Acacia* species occurring in Australia (Miller *et al.* 2011), ~400 species have been introduced outside their native range, with ~6 % becoming invasive, ~12 % becoming naturalized and ~82 % remaining as casuals (Richardson *et al.* 2011; Rejmánek and Richardson 2013) (see Table 1 for definition of invasiveness categories).

Globally, acacias vary in the number of regions they have invaded [regions defined by Richardson and Rejmánek (2011) and Rejmánek and Richardson (2013) include North America, Europe, Middle East, Asia, Indonesia, Pacific Islands, New Zealand, Australia, Indian Ocean Islands, Africa (southern), Africa (rest), Atlantic Islands, South America, Caribbean Islands, and Central America]. Differences in acacia invasiveness among regions may be due to variation in invasive capacity of these species, lower propagule pressure in particular regions or differences in incidence reports among regions (Richardson and Rejmánek 2011).

Within geographic regions, there is also evidence that acacias vary in invasiveness. For example, sixteen Australian *Acacia* species have been introduced to California and differ in their invasive status in this region (Jepson Flora Project 2015) (Table 2). Whereas all these

Table 2. *Acacia* species occurring in California. California invasiveness status compiled from CalFlora (CalFlora 2015), Cal-IPC (Cal-IPC 2006) and Jepson herbarium (Jepson Flora Project 2015). Regions invaded globally compiled from Richardson *et al.* (2011) and Rejmanek *et al.* (2013) [Regions include: North America, Europe, Middle East, Asia, Indonesia, Pacific Islands, New Zealand, Indian Ocean Islands (including Madagascar), Africa (southern), Africa (rest), Atlantic islands, South America, Caribbean islands, and Central America]. *Acacia* species included in this study are noted with an *.

Species	California status	Regions invaded globally
A. baileyana*	Naturalized	2
A. cultriformis*	Casual	0
A. cyclops	Naturalized	4
A. dealbata*	Invasive	6
A. decurrens	Casual	3
A. elata	Casual	1
A. longifolia*	Naturalized	7
A. mearnsii	Casual	12
A. melanoxylon*	Invasive	10
A. paradoxa	Casual	4
A. podalyriifolia	Casual	2
A. pycnantha*	Casual	2
A. redolens	Naturalized	0
A. retinodes	Casual	2
A. saligna	Naturalized	4
A. verticillata*	Casual	2

species except for two (*A. cultriformis* and *A. redolens*) are invasive in at least one part of the world, they vary markedly in their ability to invade and expand population sizes in California. *Acacia* species were first introduced to California for ornamental purposes and sold through the nursery trade beginning in the mid-1800s (Butterfield 1938). Two species, *A. dealbata* and *A. melanoxylon*, are currently designated as invasive in California (Cal-IPC 2006), five species as naturalized and nine species as casuals (Jepson Flora Project 2015) (Table 2). Definitions of invasiveness categories used for the purpose of this study can be found in Table 1. Understanding the mechanisms that enable multiple closely related species to differentially establish and colonize natural areas in one particular region is important for understanding what controls and promotes species establishment in general (Klock *et al.* 2015).

One mechanism that may be an important determinant of invasion success for acacias is their symbiotic

relationship with rhizobia. The legume–rhizobia interaction has been long recognized as critical for the growth and establishment of many legumes (Sprent 2001), including acacias (Thrall *et al.* 2005). Rhizobia are Gram-negative bacteria that convert atmospheric nitrogen to a form usable by the plant (Bauer 1981; Sprent and Sprent 1990). Within nodules, the plant provides rhizobia access to carbon substrates and micronutrients, and also protects them from desiccation (Sprent 2001). When legumes form an association with compatible symbiotic bacteria they obtain a direct source of nitrogen unavailable to other plants. Soil nitrogen availability for plants is often low (Masclaux-Daubresse *et al.* 2010), so species that are more readily able to form such associations may have a competitive advantage over other plant species, particularly in low-resource environments (Funk and Vitousek 2007).

The selectivity of different plant hosts for particular rhizobial symbionts (hereafter referred to as "host promiscuity") may contribute to the differential ability of *Acacia* species to establish and expand abroad. Hosts that are more promiscuous (i.e. are able to effectively associate with a wider range of rhizobial strains) may have a competitive advantage when introduced to novel areas, where they are likely to encounter unfamiliar nitrogen-fixing bacteria (Richardson *et al.* 2000; Rodríguez-Echeverría 2010; Birnbaum *et al.* 2012). Previous research suggests that widely distributed acacias in their native range are more promiscuous rhizobial hosts, whereas those with more limited distribution are more specific hosts (Thrall *et al.* 2000). In addition, acacias that have become invasive in multiple regions of the globe appear to be more promiscuous hosts than naturalized or casual acacias (Klock *et al.* 2015). Variation in host promiscuity among *Acacia* species introduced to California may help explain why certain species have differentially invaded this region.

The goal of this study was to characterize the nodulation ability of a suite of *Acacia* species that have become differentially invasive within California. To examine this, we used multiple *Acacia* species representing different invasiveness categories and performed whole soil inoculation experiments with a range of soils from different environments and two different continents. Examining species in their native and introduced ranges can provide essential information for understanding the context-dependent mechanisms influencing the invasion of non-native species (Shea *et al.* 2005). By better understanding the biological attributes of species in their home range, we can predict and compare their responses abroad, thereby gaining insight into which mechanisms are influencing species survival and expansion in different ranges (Hierro *et al.* 2005). Using species of *Acacia* and their rhizobial mutualists, we aimed to assess whether the mechanisms promoting establishment and survival at home are the same that facilitate invasion abroad. The purpose of conducting this experiment in the native range was to challenge acacias with unfamiliar rhizobial communities in areas where they naturally occur. This mimics the conditions legume hosts face when introduced abroad (although potential rhizobial mutualists are likely to be more closely related to those they typically associate with). Our approach also allowed us to determine if observed patterns are maintained in the invasive range, where rhizobial mutualists may be more distantly related.

In particular, we evaluated aboveground plant growth (biomass), survival, and nodulation responses. We examined whether treatment of acacias with different soil inoculants influenced plant performance. We also used terminal restriction length polymorphism (T-RFLP), to examine the composition and richness of rhizobial strains associating with acacias in different invasiveness categories. We hypothesized that invasiveness of non-native acacias in California would be influenced by host promiscuity with rhizobial strains, with the following predictions: (1) invasive acacias would have higher biomass, survival and nodulation responses (i.e. plant performance) in both native and introduced ranges across a greater number of soils than naturalized or casual acacias; and (2) invasive acacias would associate with a greater number of rhizobial strains (as measured by number of ribotypes, or unique terminal restriction fragment lengths) in both native and introduced ranges than naturalized or casual species.

Methods

Study species

The genus *Acacia* (Fabaceae: Mimosoideae) is native to Australia, with over 1000 species occurring variously across the continent (Miller *et al.* 2011) (Fig. 1). We focused on seven species that have been introduced to California and have become invasive (*A. dealbata* and *A. melanoxylon*), naturalized (*A. baileyana* and *A. longifolia*) or remained casual aliens (*A. cultriformis*, *A. pycnantha* and *A. verticillata*) in this region (Cal-IPC 2006; Jepson Flora Project 2015) (see Fig. 1 for *Acacia* range distributions in Australia and California). Five of these species have been previously characterized for levels of host promiscuity (*A. dealbata*, *A. cultriformis*, *A. longifolia*, *A. melanoxylon* and *A. pycnantha*) using pure rhizobial cultures (Thrall *et al.* 2000; Bever *et al.* 2013; Klock *et al.* 2015), whereas two species have not (*A. baileyana* and *A. verticillata*). All species examined here are native to southeastern Australia and range from broadly distributed to narrowly restricted within their native region (AVH 2015) (Fig. 1). Previous research has provided at least some evidence that more widely distributed acacias in their native range are more promiscuous rhizobial hosts than those that are narrowly distributed (Thrall *et al.* 2000), and that globally invasive acacias are more promiscuous hosts than those that are naturalized or casual aliens (Klock et al. 2015). Given the analogous variation in the occurrence of our selected species within their novel range, we used these species to examine whether invasiveness in California might also be linked to variation in host promiscuity.

Soil inoculant collection and preparation

Soil samples were collected from multiple sites in *Acacia* species' native (Australia) and introduced (California) ranges to obtain a diverse suite of rhizobial communities for use in glasshouse inoculation studies (Fig. 2 **[see Supporting Information—Table S1]**). Whole soil inoculations were used rather than individual rhizobial cultures to challenge acacias with rhizobial communities

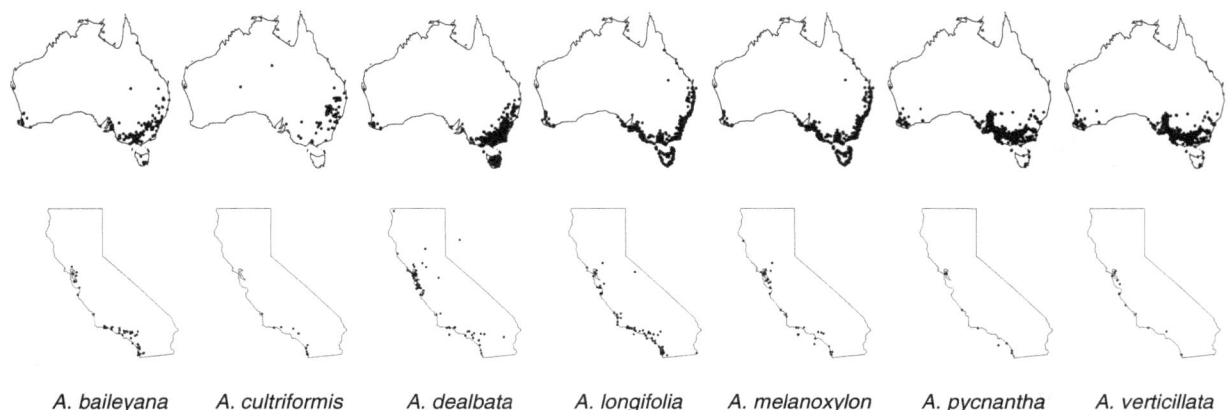

| *A. baileyana* | *A. cultriformis* | *A. dealbata* | *A. longifolia* | *A. melanoxylon* | *A. pycnantha* | *A. verticillata* |

Figure 1. Distribution maps for *Acacia* species used in this experiment in their native continent of Australia (top row) (based on herbarium records from the Australian National Herbarium, Canberra, Australia [AVH 2015]) and introduced range of California (bottom row) (Data provided by the participants of the Consortium of California Herbaria [ucjeps.berkeley.edu/consortium/, last accessed 04 August 2016.]).

they have not previously been exposed to, thereby reflecting more accurately the conditions acacias may face when introduced abroad. Soils likely contained organisms other than just rhizobia; however, all soils were bulked and mixed within soil collection site, and all *Acacia* species inoculated with soils from each site to achieve homogenous treatment conditions.

In Australia, we collected soils from ten sites within a 150 km radius of Canberra, ACT, during July 2011. Sites varied in disturbance regimes, from a highly disturbed agricultural field, to an abandoned paddock, to an undisturbed diverse native legume site. In California, we collected soils from ten sites within a 50 km radius of San Francisco, CA, during December 2011 (Fig. 2; **[see Supporting Information—Table S1]**). Weather conditions

in Australia and California were very similar during the sampling periods (high temp 11.2 °C vs. 14 °C; low temp −1.4 °C vs. 1.7 °C; precipitation 0.04 cm vs. nil) (www.ncdc.noaa.gov, last accessed 03 August 2016).

In both ranges, we chose sites that did not contain any of the *Acacia* species used in this study to challenge all of the study species with unfamiliar rhizobial communities. This was done to mimic conditions that hosts might encounter when introduced to a new area. Soils were collected over the course of one week. Soil samples were excavated using a clean shovel and stored in paper bags until processing. We collected multiple samples from within each site and then bulked them within replicates, with site as the level of replication, to make a single composite for each of the 10 sites. Following

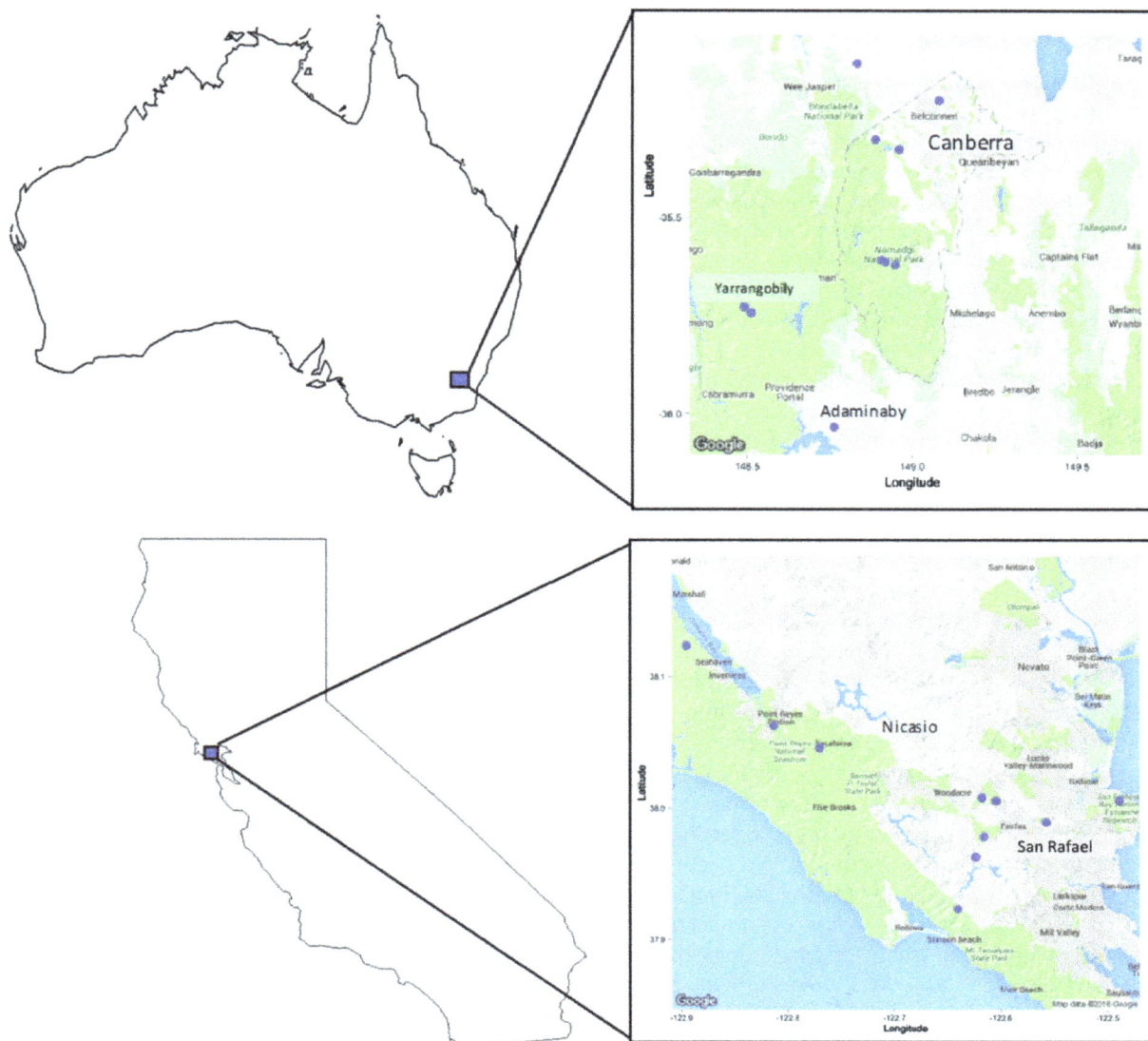

Figure 2. Soil inoculation collection sites in Australia (top) and California (bottom). Maps created using the R statistical package "ggmap" version 2.5.2 (Kahle and Wickham 2013).

collection, soils were dried for up to 6 days. Once dry, they were sieved through 3-mm mesh to remove rocks and other debris and stored in paper bags until use. Soils collected in California were shipped to Louisiana State University (LSU) for use in the introduced range glasshouse experiment. Temperatures at which soils were stored fluctuated due to transport and handling requirements but otherwise were held constant at 4°C. Previous research has shown that the abundance of rhizobial strains can decline over time in dry soil storage; however, rhizobial strains are still abundant in soils after 6 months (Martyniuk and Oroń 2008; Thrall and Barrett pers. obs.). In addition, as each *Acacia* species was subject to each soil treatment, exposure to available rhizobial strains was the same among species.

Glasshouse experiments

We conducted two glasshouse experiments to examine the promiscuity of *Acacia* species in different invasiveness categories. For the first experiment (hereafter called the "native experiment"), glasshouse facilities were located at CSIRO's Black Mountain site in Canberra, ACT, Australia. For the second experiment (hereafter called the "introduced experiment"), glasshouse facilities were located at LSU in Baton Rouge, Louisiana. The native experiment was conducted from July to November 2011. Seeds of all *Acacia* species used in this experiment were collected within Australia and obtained from the Australian Seed Company. The introduced experiment was conducted from March to July 2012. For this component, seeds of all *Acacia* species were collected directly from plants in California in September 2011, shipped to LSU, and stored in paper bags until use.

For both experiments, seeds were subjected to a boiling water treatment to induce germination (boiling water was poured over seeds and they were left to imbibe water for 24 h). No further seed sterilization methods were undertaken; however, seedlings were observed for nodule presence at time of planting and none were nodulated. In addition, while the native experiment control did experience a moderate level of contamination at final harvest, samples in the introduced range control treatment showed no contamination, suggesting that the source of contamination for the native experiment was not the vertical transmission of rhizobia. Seeds were transferred to trays of steam-sterilized vermiculite and watered daily with sterile water for 14–20 days, or until germination occurred. Seedlings were grown in the glasshouse under local natural light conditions.

Once germinated, seedlings were transferred to individual pots inoculated with soils collected from each of the 10 sites. In the native experiment, for each of the

bulked soils, 10 replicates of each *Acacia* species were planted in 8 × 15 cm pots filled ¾ with sterilized sand and vermiculite (1:1 volume), 50 g of an individual soil treatment as a live inoculant and topped with additional sterilized sand and vermiculite (1:1 volume) to avoid cross contamination. For the introduced experiment, seedlings were similarly planted and inoculated, however replication varied due to availability of seed for individual species (10 replicates of *A. baileyana, A. longifolia, A. melanoxylon* and *A. verticillata*; 5 replicates of *A. dealbata* and *A. pycnantha*; 4 replicates of *A. cultriformis*). A rhizobia-free (N⁻) control was also included in both experiments in which plants were not inoculated. For both experiments, *Acacia* species × soil combinations were spatially randomized by glasshouse bench such that each bench contained one replicate of each species × soil combination. Pot placement on the bench was randomized. All plants were watered twice weekly with sterile N-free McKnight's solution (McKnight 1949) and sterile water as needed. Plants were spaced well apart on glasshouse benches to minimize cross-contamination during watering.

Plants were grown for 16 weeks in a temperature-controlled glasshouse (~20 °C) and harvested in November 2011 (native experiment) and July 2012 (introduced experiment), respectively. At harvest, seedlings were clipped at the soil surface and aboveground material was stored in paper bags. For the native experiment, aboveground material was oven dried at 70 °C for 48 h and weighed. A malfunction with the drying oven destroyed aboveground material for the introduced experiment, therefore biomass data were lost. Belowground material for both experiments (roots and attached nodules) of each plant was stored individually in plastic bags and frozen at –20 °C until processing for molecular analysis. Roots were scored at harvest for nodulation quantity (0, <10, 10–50, >50) and quality (none, ineffective [black or very small white nodules], intermediate [mixture of small to medium white/pink nodules] and good [pink nodules]) (Thrall *et al.* 2011).

Isolation of DNA and T-RFLP

We used terminal restriction length polymorphism (T-RFLP) to identify community composition and genotypic richness of rhizobia nodulating with *Acacia* species in the glasshouse experiments. This technique is frequently used for examining taxon richness of bacterial communities (Liu *et al.* 1997). To extract DNA from root nodules collected during harvest, 2–10 intact nodules per plant (depending on availability) were first snipped from roots stored at –20 °C. Nodules were surface sterilized by immersion in 90 % ethanol for five to ten seconds, transferred to 3 % sodium hypochlorite and soaked for 2–4 minutes, and rinsed in five changes of sterile water. Nodules were crushed using liquid nitrogen, and DNA was extracted using Mo Bio PowerPlant® DNA Isolation kits following the protocol of the manufacturer (Mo Bio Laboratories, Inc., Carlsbad, CA, USA). Nodule processing and DNA extractions for the native experiment were conducted at CSIRO laboratories in Canberra, Australia, and for the introduced experiment at LSU in Baton Rouge, LA. DNA extractions from the introduced experiment were shipped to CSIRO laboratories where all additional molecular analyses were conducted. For all samples, we amplified the 16S rRNA gene using the primers GM3 (5'-AGA GTT TGA TCM TGG C-3') and GM4 (5'-TAC CTT GTT ACG ACT T-3') and the following PCR program: initial denaturation at 95 °C for 2 min, followed by 35 cycles of 95 °C for 30 s, 50 °C for 30 s and 72 °C for 90 s, followed by a final extension step at 72 °C for 10 min and a final holding temperature of 4 °C. We digested the PCR product using the restriction enzyme MspI (New England BioLabs) in 30 μl reaction mixtures, and analysed the fragment sizes using a 3130×l genetic analyzer (Applied Biosystems, Warrington, United Kingdom). We used GeneMapper version 5 (Life Technologies, Grand Island, NY, USA) to examine T-RFLP profiles and included peaks over 50 bp for further analysis. We quantified resulting peaks using the local southern method (Southern 1979). Peaks were binned using Ramette's interactive binner script (Ramette 2009) in the R statistical programming language version 3.2.0 (R Core Team 2015).

DNA extracted from nodules contained both acacia plastid and rhizobial DNA. While the GM3/GM4 primers can also amplify mitochondrial and chloroplast DNA, in-silico analyses of restriction-fragment polymorphisms for all *Acacia* species plastid sequences obtained from Genbank indicated that polymorphisms in plastid DNA were unlikely to contribute any variation to our T-RFLP dataset. Specifically, to identify which restriction-fragments corresponded to acacia plastid DNA, we conducted an in-silico T-RFLP analysis by searching for the primer sequences and restriction enzyme cut sites in acacia plastid DNA sequences downloaded from GenBank. We found that the restriction enzyme MspI cut sites for acacia plastid sequences generated DNA fragments greater in size than the cut-off for fragments used in our analysis (i.e. the largest restriction-fragment in our analysis was 545.3 bp, whereas the smallest restriction fragment for acacia plastid DNA was 553 bp). Because our cut-off was lower than the largest acacia plastid restriction-fragment, any peaks corresponding to acacia plastid DNA were excluded from our analysis. In addition, review of polymorphisms attributable to individual host species showed there were no polymorphisms unique to all replicates of a host species (or group of host

species), further indicating that acacia plastid DNA did not explain variation in the dataset.

Plant growth, survival and nodulation response

We examined the responses of acacias representing three invasiveness categories to inoculation with 20 different soils (10 soils each in the native and introduced ranges) collected from habitats in which the acacias used in this experiment do not occur. We measured differences among the invasiveness categories by assessing aboveground biomass (native range only), survival, nodulation presence/absence and nodulation index of effectiveness. The nodulation index of effectiveness categorizes the number of nodules found on the roots of plant specimens, and is divided into levels of none, low, medium, and high, delineated as follows: 0 nodules = score of 0; 1–10 nodules = score of 1; 11–50 nodules = score of 2; >50 nodules = score of 3.

We examined these four variables for the entire data set using generalized linear mixed models (GLMM) and used AIC to select the best models (Burnham and Anderson 2002). *Acacia* species was included in the models as a random effect to include individual variation of species in each invasiveness category. Aboveground biomass and nodulation index were modelled using a Gaussian distribution, and nodule presence/absence and survival were modelled using a binomial distribution with a logit link function. Negative control samples were not included in models for the native experiment, as almost all control specimens did not survive; however, they were included in models for the introduced experiment.

We used the R statistical package "lme4" version 1.1-9 (Bates *et al.* 2012) to determine whether main effects (soil, invasiveness category and *Acacia* species) contributed significantly to the models of interest, and whether there were interactions among main effects. *Acacia* species was maintained in all models as a random effect. Models with the lowest AIC score were selected for further analysis; models with a difference in AIC values of <2 were considered equally likely (Burnham and Anderson 2002; Bolker *et al.* 2009). Further analysis consisted of conducting multiple comparisons of means (MCMs) with Tukey contrasts using the R statistical package "multcomp" version 1.4-1 (Hothorn *et al.* 2008), which allowed us to determine whether there were significant differences among invasiveness categories for the response variables of interest (i.e. biomass, nodulation presence and nodulation index) for individual soils, while maintaining *Acacia* species in the model as a random variable.

We also examined biomass (native experiment only), nodulation presence/absence, and survival for individual

Acacia species to assess species-specific responses to individual soil inoculants. We used ANOVA to compare biomass among species x soil combinations and logistic regression to analyse survival and nodulation presence. We used a post-hoc Tukey's HSD test to compare biomass of different species to each soil inoculant using the R statistical package "agricolae" version 1.1-2 (De Mendiburu 2009). Analyses were conducted using the R statistical programming language version 3.2.0 (R Core Team 2015).

Rhizobial community composition and richness

We analysed binary data obtained from T-RFLP analysis using non-metric multidimensional scaling (NMDS) based on a Jaccard similarity matrix. We used the R statistical package "vegan" version 2.3-0 (Oksanen *et al.* 2015) to conduct ordination and Permutational ANOVA (PerManova; function "ADONIS") to test for differences in rhizobial community composition among invasiveness categories and soil types. If differences were detected we ran pairwise comparisons between groups using "ADONIS" with a Holm correction. We used ANOVA to examine whether there were differences in ribotype richness among invasiveness categories. Analyses were conducted using the R statistical programming language version 3.2.0 (R Core Team 2015).

Results

Native experiment

We detected a significant interaction between soil and invasiveness category for aboveground biomass (ΔAIC =

19.1, $w_i = 1.00$) (i.e. the best fitting model had an AIC value >2 than all other models), indicating that the growth response of species in different invasiveness categories was influenced by the soil in which they were grown [**see Supporting Information—Table S2**]. We, therefore, examined each soil individually using MCMs with Tukey contrasts and found that plants in different invasiveness categories differed significantly in average biomass response for only one soil (Fig. 3 [**see Supporting Information—Table S3**])

ANOVA results indicated that biomass varied for individual *Acacia* species across soil treatments ($F_{9,605} = 470.21$, $P < 0.001$); we also found a significant difference in biomass across *Acacia* species ($F_{6,608} = 346.80$, $P < 0.001$), and an interaction between species and soil treatment ($F_{54,545} = 135.01$, $P < 0.001$) (Table 3). From here on, individual *Acacia* species are indicated in the text as I (invasive), N (naturalized) and C (casual). Using as a comparison the soil where biomass was lowest for each species, post-hoc Tukey's HSD test showed that *A. longifolia* (N) and *A. melanoxylon* (I) had significantly greater biomass for three soils, *A. baileyana* (N), *A. cultriformis* (C), *A. dealbata* (I) and *A. verticillata* (C) for two soils, and *A. pycnantha* (C) for one soil [**see Supporting Information—Fig. S1 and Table S4**].

The model with the best support for plant survival included soil inoculation as a main effect with species as a random variable (ΔAIC = 3.87, $w_i = 0.87$) [**see Supporting Information—Table S5A**], indicating that variation in survival was driven by individual soils rather than invasiveness category. Survival across soils was generally high for all invasiveness categories

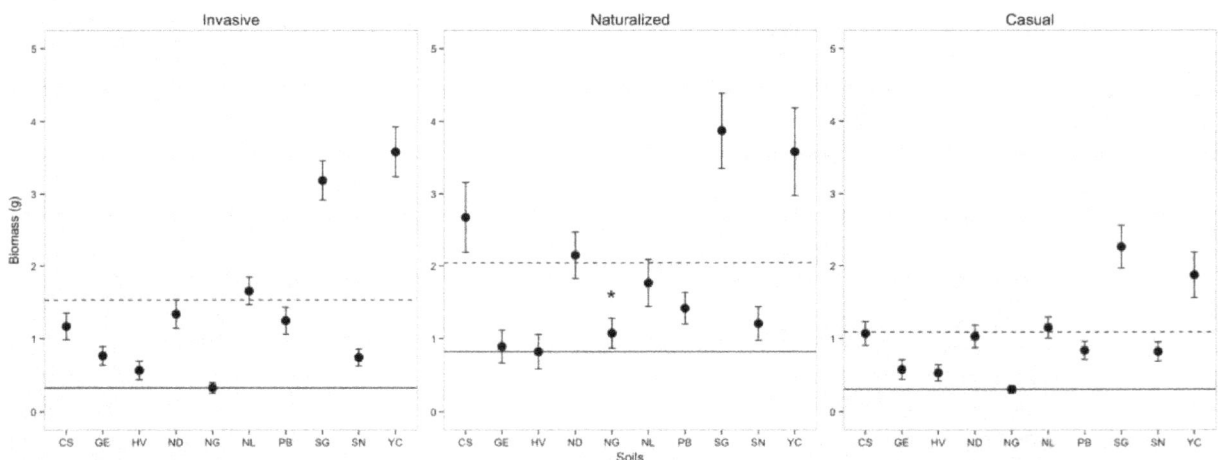

Figure 3. Average aboveground biomass (g) response of all *Acacia* species/replicates in each invasiveness categories to different soil inoculants in the native experiment (Australia). The horizontal solid line indicates the point at which host species within a given invasiveness category have the same biomass response as their least effective soil. The dashed line is the average biomass response for all host species within a given invasiveness category combined across all soils. The "*" indicates the soil in which there was a significant difference ($P < 0.05$) in biomass response of the invasiveness categories. Error bars represent standard errors (SE) of the means.

(>50 % for all soils for the naturalized and casual categories and nine out of ten soils for the invasive category) (Fig. 4A **[see Supporting Information—Table S6A]**).

Table 3. Summary of analysis of variance results testing the effects of host species and soil treatment on the aboveground biomass response.

Source	df	SS	F	P
Host species	69	346.80	61.93	<0.001
Soil	9	470.21	55.98	<0.001
Host x Soil	54	135.01	2.68	<0.001
Residual	545	508.63		

Survival for individual species was also generally high across soils. We observed over 50 % survival for each species in a minimum of seven soils (*A. pycnantha* [C]) and a maximum of all ten soils (*A. longifolia* [N] and *A. verticillata* [I]) **[see Supporting Information—Fig. S2 and Table S7A]**.

There was a moderate level of contamination in the negative controls (nodules were found on ~33 % of samples), and very few samples that were not contaminated survived, therefore, they were excluded from all native experiment analyses.

The model with best support for nodulation presence included soil inoculation as a main effect with species as a random variable (Native experiment: ΔAIC = 3.92, $w_i = 0.88$) **[see Supporting Information—Table S8A]**,

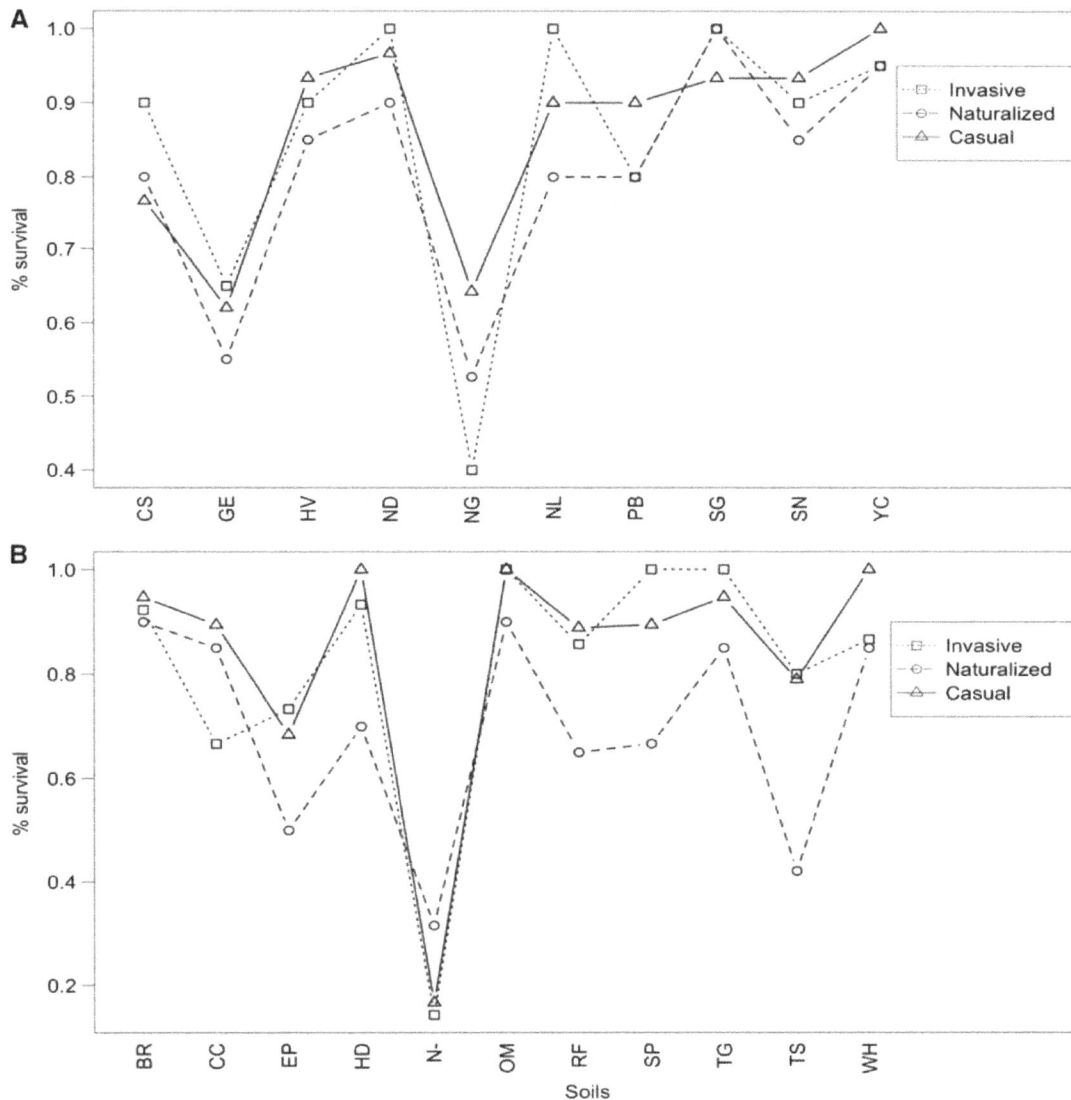

Figure 4. Average percent survival of all *Acacia* species/replicates in each invasiveness category in the (A) native and (B) introduced experiments among soil treatments.

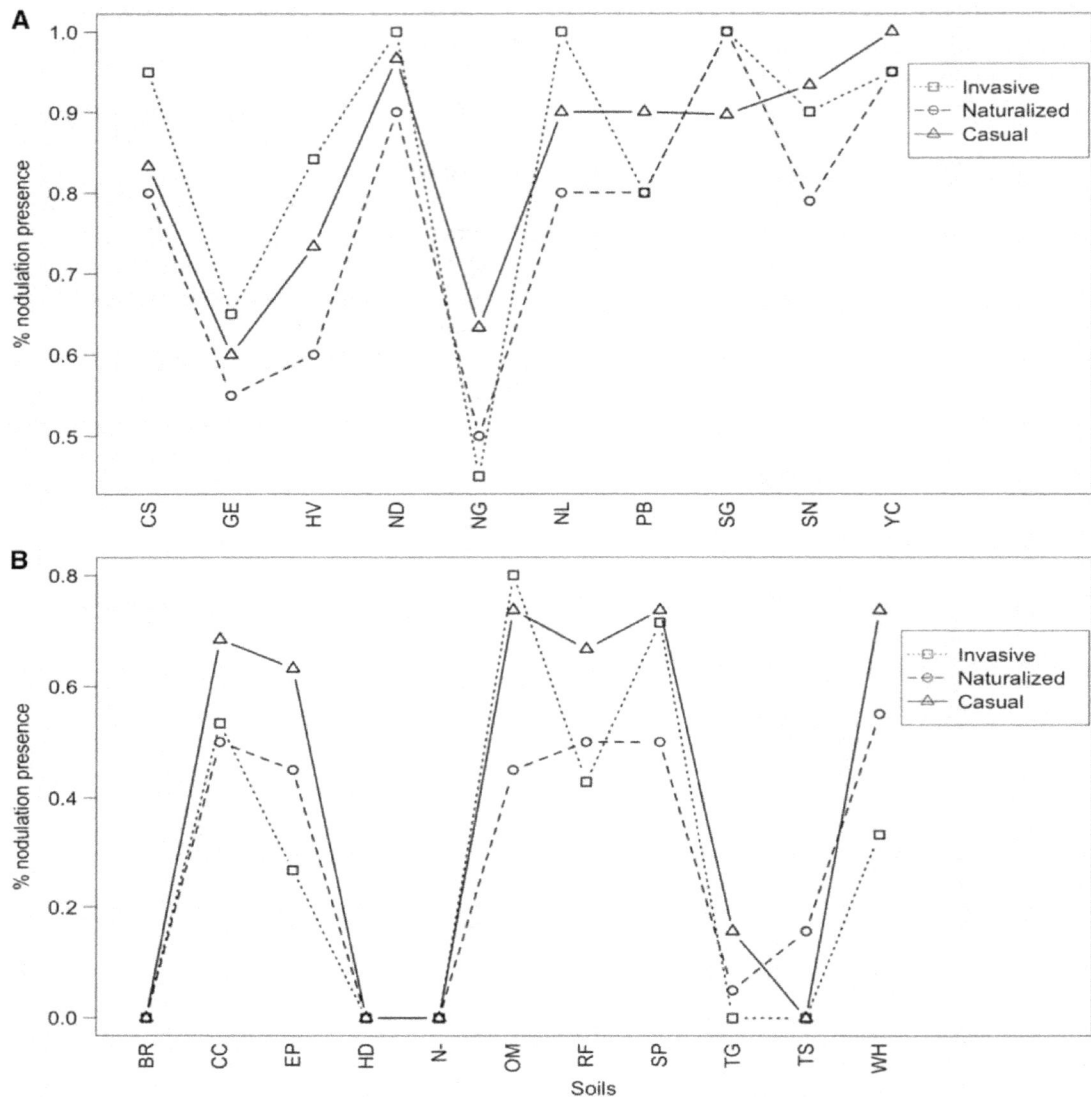

Figure 5. Average percent nodulation of all *Acacia* species/replicates in each invasiveness category in the (A) native and (B) introduced experiments among soil treatments.

indicating that differences in nodulation presence were driven by individual soils rather than invasiveness category. The presence of nodules across soils was generally high for all invasive categories (>50 % in ten soils for the casual category and nine soils for the naturalized and invasive categories) (Fig. 5A **[see Supporting Information—Table S9A]**).

Nodulation presence for individual species was also generally high across soils, with over 50 % nodulation presence for each species in a minimum of seven soils (*A. baileyana* [N]) and a maximum of all ten soils (*A. cultriformis* [C], *A. longifolia* [N], *A. melanoxylon* [I] and *A. verticillata* [C]) (**see Supporting Information—Fig. S4 and Table S10A]**).

We found a significant interaction between soil and invasiveness category for nodulation index of effectiveness (ΔAIC = 9.7, w_i = 0.97) **[see Supporting Information—Table S11A]**. This indicates that there was an effect of individual soils on nodulation index, such that the number of nodules on plants belonging to different invasiveness categories depended on the soil in which they were grown. We, therefore, could not generalize nodulation index response for invasiveness categories across all soils, and examined nodulation index for each soil individually using MCMs with Tukey contrasts. When soils were examined individually, we found no significant difference in nodulation index among invasiveness categories (Fig. 6A).

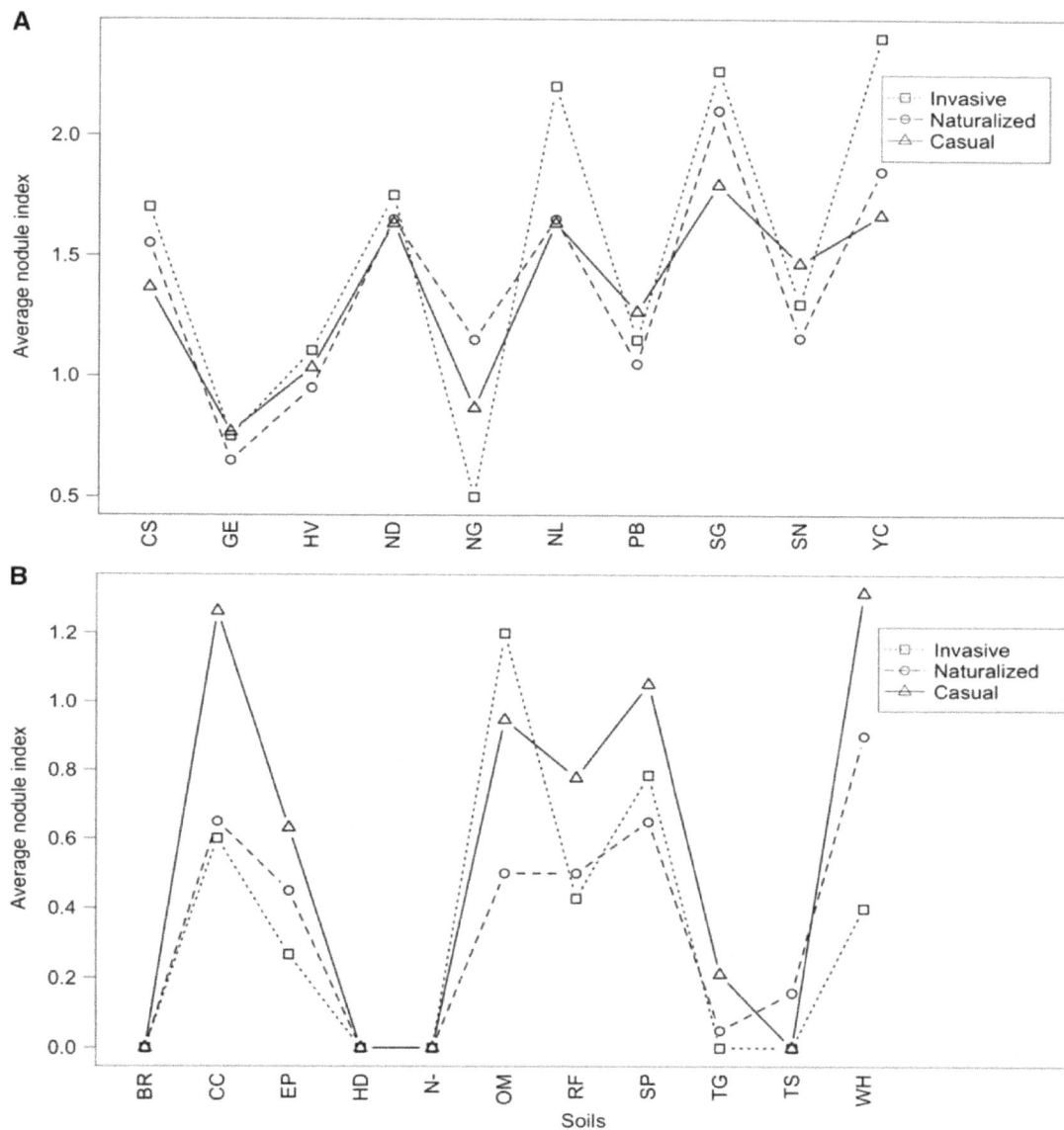

Figure 6. Average nodulation index of all *Acacia* species/replicates in each invasiveness category in the (A) native and (B) introduced experiments among soil treatments. The different shapes depict different invasiveness categories (square = invasive, circle = naturalized, triangle = non-invasive).

Visual assessment of ordination diagrams did not indicate a clear difference in rhizobial community composition among *Acacia* species invasiveness categories (Fig. 7A). However, PerManova results from T-RFLP analyses indicated a small but significant difference in rhizobial community composition between the invasive and casual categories (ADONIS, $R = 0.08$, adjusted $P = 0.024$). Despite this slight difference in community composition, there was no significant difference in rhizobial richness among invasiveness categories ($F = 1.287$, $P = 0.284$).

Introduced experiment

No contamination occurred in the introduced range experiment so all control samples were retained for all analyses. The best-supported model for survival in the introduced range experiment included soil inoculation as a main effect with species as a random variable ($\Delta AIC = 2.73$, $w_i = 0.79$) **[see Supporting Information—Table S5B]**. Similar to the native experiment, survival across soils was generally high for all invasiveness categories (>50% for all soils for the invasive and casual categories, and 8 out of 10 soils for

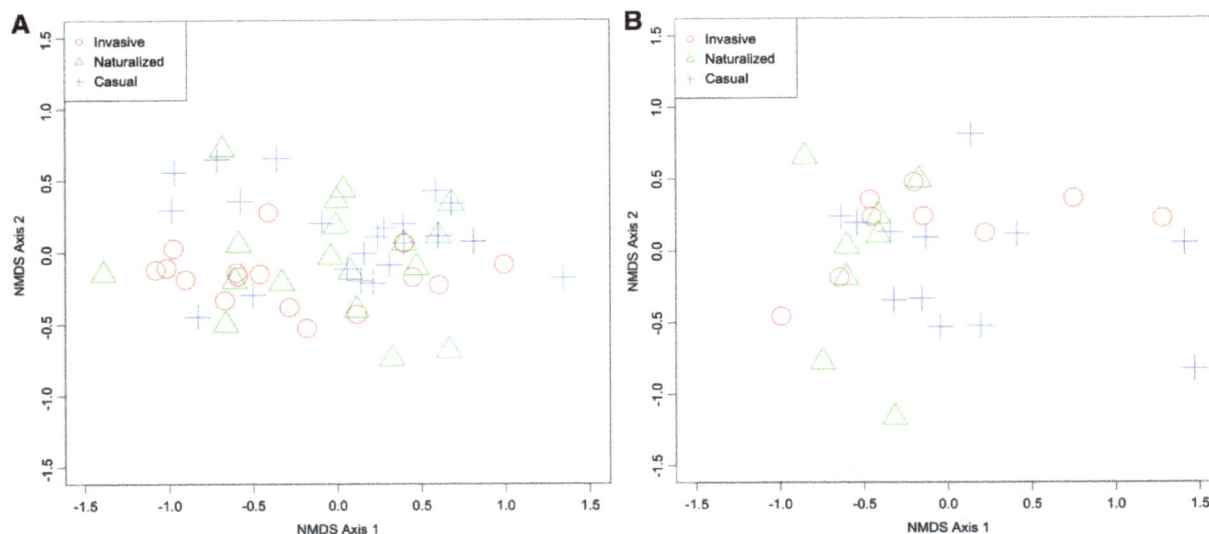

Figure 7. Ordination of the rhizobial community composition in different invasiveness categories (Jaccard similarity) in the (A) native and (B) introduced experiments based on the 16S rRNA gene from different soil treatments derived from nonmetric multidimensional scaling. Invasiveness categories more similar in rhizobial community composition are closer together in ordination space.

the naturalized category) (Fig. 4B **[see Supporting Information—Table S6B]**).

Survival for individual species was also high across soils. We observed over 50 % survival in a minimum of six soils (*A. baileyana* [N]) and a maximum of all 10 soils (*A. longifolia* [N], *A. melanoxylon* [I] and *A. verticillata* [C]) **[see Supporting Information—Fig. S3 and Table S7B]**.

The model with the best support for nodulation presence included only soil inoculation as a main effect with species as a random variable (ΔAIC = 3.99, w_i = 0.88) **[see Supporting Information—Table S8B]**. In contrast to the native experiment, nodule presence across soils was low for the introduced experiment (>50 % in six soils for the casual category, three soils for the invasive category and one soil for the naturalized category) (Fig. 5B **[see Supporting Information—Table S9B]**).

Nodulation presence was generally low across soils for individual species as well, with over 50 % nodulation presence for each species in a maximum of six soils (*A. longifolia* [N], *A. pycnantha* [C] and *A. verticillata* [C]) and a minimum of zero soils (*A. baileyana* [N] and *A. cultriformis* [C]) **[see Supporting Information—Fig. S5 and Table S10B]**.

We found a significant interaction between soil and invasiveness category for nodulation index of effectiveness (ΔAIC = 22.32, w_i = 1.00) **[see Supporting Information—Table S11B]**. We, therefore, examined soils individually using MCMs with Tukey contrasts and found no significant difference in nodulation index among invasiveness categories (Fig. 6B).

Visual assessment of ordination diagrams indicated no significant difference in rhizobial community composition

among acacia invasiveness categories (Fig. 7B). PerManova results lent further support to this conclusion, with no significant difference in rhizobial community composition found among invasiveness categories (ADONIS, R = 0.08, P = 0.21). In addition, we found no significant difference in rhizobial richness among categories of invasiveness (F = 1.224, P = 0.31).

Summary of results: native and introduced experiments

Biomass results from the native experiment showed no significant difference in aboveground biomass among acacia invasiveness categories except in one soil. Survival did not differ among categories in both the native and introduced experiments. Nodule presence and index of effectiveness was generally high across all invasiveness categories for the native experiment, but low for the introduced experiment. We found no circumstances in which multiple models were equally likely for individual response variables (i.e. differed by <2, see above) for either the native or introduced experiments. Rhizobial composition differed slightly among invasiveness categories in the native experiment only; richness did not vary among categories for either the native or introduced experiment.

Discussion

The goal of this study was to examine whether variation in host promiscuity with rhizobial symbionts plays a role in the differential invasion of *Acacia* species in California.

We found that host promiscuity as measured by plant growth in the native experiment, and survival and nodulation response in both native and introduced experiments did not differ among acacia invasiveness categories. However, acacias in the native experiment (regardless of invasive status) were able to develop nodules in a greater number of soils than in the introduced range experiment. We found limited variation in rhizobial associations among acacias that vary in invasiveness in California. While rhizobial community composition differed slightly among acacia invasiveness categories in the native experiment, rhizobial richness or the number of strains with which host species in these groups formed an association was not significantly different. Results from the introduced experiment showed no difference in community composition or richness of rhizobia associating with *Acacia* species in different invasiveness categories. Plant growth response, paired with belowground rhizobial richness results, suggests that variation in host promiscuity may not be a major determinant of invasiveness of Australian *Acacia* species in California.

Results from T-RFLP analyses indicated a slight difference in the rhizobial communities acacias in different invasiveness categories associated with, when paired with Australian soils. However, no such differences were evident with Californian soils, perhaps reflecting a greater diversity of compatible rhizobial strains in Australian soils. We found no difference in rhizobial richness among invasiveness categories for either set of experiments. Together, these results suggest that partner choice as opposed to partner breadth may be more important in explaining how interactions with rhizobia influence potential for invasiveness in this set of *Acacia* species. However, more work is required to generalize these observations. Birnbaum *et al.* (2012) found similar results when examining acacias that have become invasive within their native continent; species examined associated with the same abundance of rhizobial strains in both native and novel ranges, and for two species tested (*A. longifolia* and *A. melanoxylon*), they associated with similar rhizobial communities between ranges.

In the native range experiment, rates of nodulation and survival were similarly high across almost all soils. Although we paired acacias with soils in which they did not occur in their native range, effective rhizobial strains may be broadly distributed, as has been previously found with rhizobial (Barrett *et al.* 2012) and mycorrhizal fungal symbionts of acacias in Australia (Birnbaum *et al.* 2014). In their introduced range, acacias may be more likely to encounter rhizobial strains that are more distantly related to those with which they have co-evolved, or appropriate strains may be completely absent, such that it

is more difficult to find suitable partners (perhaps partially explaining the generally lower nodulation rates we observed in the introduced experiment).

In addition to rhizobia, other organisms in soil communities may have influenced plant performance. We used whole-soil inoculation treatments, which may host multiple rhizobial symbionts as well as pathogens and other mutualistic microorganisms, and plant response may be influenced by the presence of such organisms (Thrall *et al.* 2007). The presence of non-rhizobial mutualists may have had a greater effect on plant performance in the experiment utilizing Australian soils, because acacias native to Australia likely have higher compatibility with the resident microorganisms. The presence of pathogenic organisms such as fungi or nematodes as well as interactions among co-occurring soil biota may also affect *Acacia* species growth response, influencing the potential positive benefit of being a promiscuous rhizobial host. However, whether pathogenic interactions are more likely to have influenced plant growth in Australian or Californian soils is difficult to assess. Other, more complex synergistic or antagonistic interactions may also occur when using whole soil inoculations. For example, rhizobial competition arising from the presence of multiple rhizobial genotypes within soils may have influenced mutualistic outcomes. Barrett *et al.* (2014) found evidence that acacias paired with multiple rhizobial strains suffered diminished plant growth response, likely due to altered patterns of rhizobial association. Hence, an important caveat is that we are unable to tease apart the complex species interactions that may occur among the myriad organisms occurring in natural soil communities, and which may have influenced plant performance in this study.

A previous study has shown that more invasive *Acacia* species are more promiscuous rhizobial hosts. Klock *et al.* (2015) paired 12 rhizobial strains ranging in effectiveness with 12 *Acacia* species differing in global invasiveness (four invasive, four naturalized and four casual species). In regard to plant growth, invasive acacias were generally more promiscuous hosts, able to associate and have a positive growth response with more rhizobial strains than naturalized and casual acacias. However, in this previous study, acacias were paired with single rhizobial genotypes rather than whole soil inoculations and acacia invasiveness was categorized on a global, rather than regional scale (Klock *et al.* 2015). *Acacia* species tested in this study vary in invasiveness in California; however, all except for one are invasive in at least one region of the world (Richardson and Rejmánek 2011; Rejmánek and Richardson 2013). We were interested in what drives differences in invasiveness on a regional scale; however, since all *Acacia* species tested here are invasive at least somewhere in the world, they may very well all be promiscuous rhizobial hosts, constrained by mechanisms other than host selectivity for rhizobia from becoming invasive in California. Host promiscuity with rhizobia may indeed influence the ability of acacias to invade novel regions, but other biotic and abiotic factors likely contribute to the establishment and colonization of these species, limiting some species from invading particular regions, and promoting the invasiveness of others.

The lack of aboveground biomass data in the experiment using soils collected in California reduced our ability to determine whether patterns of plant performance are consistent between native and introduced experiments. While we were able to assess the ability of acacias in different invasiveness categories to nodulate with rhizobia and their subsequent survival, we do not know whether this resulted in a beneficial growth response in the introduced range experiment. This limits our ability to assess whether acacias in the introduced range responded in a beneficial manner as a result of being paired with unfamiliar rhizobial symbionts. Future studies would benefit from assessing aboveground biomass of acacias paired with soils from their introduced range. Still, results from our nodulation, survival and molecular analyses provide strong evidence that acacias in different invasiveness categories tested here do not vary in host promiscuity with rhizobial symbionts.

Rhizobia-related mechanisms other than host promiscuity may influence the invasiveness of acacias introduced to novel regions. There is increasing evidence that some legumes have been introduced abroad with their native rhizobial symbionts (Rodríguez-Echeverría 2010; Crisóstomo et al. 2013; Ndlovu et al. 2013) and for similarity in associated rhizobial strains across native and novel ranges (Birnbaum et al. 2016). The introduction of both invasive species and their co-evolved beneficial symbionts may circumvent any need for introduced species to develop novel mutualistic rhizobial associations. *Acacia pycnantha*, a native Australian species that has become invasive in South Africa (Ndlovu et al. 2013), has been found to associate with rhizobial strains more closely related to those of Australian origin (Ndlovu et al. 2013). Both *A. longifolia* and *A. saligna* associate with rhizobia of Australian origin in Portugal (Rodríguez-Echeverría 2010; Crisóstomo et al. 2013). Legumes native to Portugal were also found to form associations with rhizobial strains of Australian origin in areas where *A. longifolia* occurred (Rodríguez-Echeverría 2010). Birnbaum et al. (2016) found evidence for three *Acacia* species associating with the same rhizobial strains between native and novel ranges within their native continent. Dual invasion of symbiotic plant and microbial species may thus be occurring in regions where acacias have been introduced, or certain rhizobial strains may be particularly widespread, potentially contributing to both above and belowground structural changes in native habitat composition.

Acacias that become invasive in California may benefit from mutualistic interactions other than the legume–rhizobia symbiosis that aid in their establishment and colonization. As indicated here, host promiscuity with rhizobia alone does not appear to delineate invasiveness of acacias in California. However, as a general trait promoting invasiveness, host interactions with other taxa may be important to the establishment, colonization and survival of these species. Ant mutualists may aid in seed dispersal and seed bank accumulation as well as protection from herbivores for *Acacia* species that become invasive in their novel range (Holmes 1990; Montesinos and Castro 2012). *Acacia* species that have become invasive in California may also develop successful mutualisms with avian seed dispersers (Glyphis et al. 1981; Underhill and Hofmeyr 2007; Aslan and Rejmánek 2010). Being hosts for a variety of mutualistic organisms may increase the opportunity for *Acacia* species to develop self-sustaining, spreading populations that invade novel ranges.

Conclusions

Species that have become invasive in multiple areas of the world may be constrained from establishing and colonizing all regions where they are introduced. Identifying as well as ruling out potential mechanisms influencing expansion of species that have become invasive globally but are constrained regionally can inform management of species introduced abroad. We found that acacias varying in invasiveness in California do not differ in their ability to form symbioses with nitrogen-fixing bacteria, as evidenced by a lack of difference in plant performance and rhizobial richness when paired with diverse soil inoculants. Invasive status of introduced acacias in California, therefore, does not appear to be determined solely by the ability to associate with larger numbers of rhizobial symbionts.

Due to the demonstrated capacity of almost all *Acacia* species introduced to California to invade at least one other region of the world, and previous research showing that globally invasive acacias are promiscuous hosts, all *Acacia* species, whether currently invasive or not in California should be monitored closely for further colonization and expansion in their introduced range. Just as species differentially establish in their native ranges, the levels of invasiveness that species accomplish when introduced abroad may also vary. Our results suggest that taking scale into account when examining the factors that drive invasion of species is important; those species that are deemed invasive on a global scale may not be so on a regional scale, and different mechanisms may be influencing their capacity to invade novel regions. By identifying the mechanisms that both promote and constrain acacia invasion in particular regions, we can better inform management and future introduction of these species abroad, thereby mitigating their potential to cause negative impacts on native communities.

Sources of Funding

Our work was funded by a Louisiana Board of Regents EPSCoR grant, funding from the National Science Foundation (DEB-1311290) (United States), a Sigma Xi Grant-in-Aid of Research (United States) and a Louisiana Environmental Education Commission Research grant.

Contributions by the Authors

M.M.K., L.G.B., P.H.T. and K.E.H. conceived of and developed the idea. M.M.K. and L.G.B. analysed the data; and M.M.K. led the writing with input from all authors.

Acknowledgements

We thank Caritta Eliasson, Mohammad S. Hoque, Kristy Lam and Alexandre de Menenzes, for help and guidance in the laboratory, glasshouse, and with data analysis. We thank Meredith Blackwell, James T. Cronin, Hallie Dozier, Bret Elderd and Richard Stevens for support and insight throughout the development of the project. We thank Sandra P. Galeano, Katherine Hovanes and Hector Urbina for discussion in manuscript preparation. We thank Matthew Ritter for discussion and invaluable insight. We thank the guest editor and two anonymous referees for helpful comments that improved the manuscript.

Supporting Information

The following additional information is available in the online version of this article —

Table S1. Soil collection sites for *Acacia* species inoculants in Australia and California.

Table S2. GLMM Models predicting difference in aboveground biomass (g) among California invasive rankings.

Table S3. Average aboveground biomass (g) and SEM of all *Acacia* species replicates in each invasiveness category when grown with each soil in the native experiment.

Table S4. Average aboveground biomass (g) and SEM of each *Acacia* species when grown with each soil in the native experiment.

Table S5. GLMM Models predicting difference in survival among California invasive rankings in the native and introduced experiments.

Table S6. Average percent survival (in decimal percentage) and SEM of all *Acacia* species replicates in each invasiveness category when grown with each soil in the native and introduced experiments.

Table S7. Average percent survival (in decimal percentage) and SEM of each *Acacia* species when grown with each soil in the native and introduced experiments.

Table S8. GLMM models predicting difference in nodulation presence among California invasive rankings in the native and introduced experiments.

Table S9. Average percent nodulation (in decimal percentage) and SEM of all *Acacia* species replicates in each invasiveness category when grown with each soil in the native and introduced experiments.

Table S10. Average percent nodulation (in decimal percentage) and SEM of each *Acacia* species when grown with each soil in the native and introduced experiments.

Table S11. GLMM models predicting difference in nodulation index among California invasive rankings in the native and introduced experiments.

Figure S1. Aboveground biomass (g) response of individual host species to different soil treatments.

Figure S2. Percent survival for each host species × soil treatment combination in the native experiment.

Figure S3. Percent survival for each host species × soil treatment combination in the introduced experiment.

Figure S4. Percent nodulation for each host species × soil treatment combination in the native experiment.

Figure S5. Percent nodulation for each host species × soil treatment combination in the introduced experiment.

References

Aslan CE, Rejmánek M. 2010. Avian use of introduced plants: ornithologist records illuminate interspecific associations and research needs. *Ecological Applications* **20**:1005–1020.

AVH. 2015. Australia's Virtual Herbarium, Council of Heads of Australasian Herbaria, http://avh.chah.org.au, accessed 03 Nov 2014.

Barrett LG, Bever JD, Bissett A, Thrall PH. 2014. Partner diversity and identity impacts on plant productivity in Acacia–rhizobial interactions. *Journal of Ecology* **103**:130–142.

Barrett LG, Broadhurst LM, Thrall PH. 2012. Geographic adaptation in plant–soil mutualism: tests using *Acacia* spp. and rhizobial bacteria. *Functional Ecology* **26**:457–468.

Bates D, Maechler M, Bolker B. 2012. lme4: linear mixed-effects models using S4 classes. R package version 0.999999-0. http://CRAN.R-project.org/package=lme4.

Bauer WD. 1981. Infection of legumes by rhizobia. *Annual Review of Plant Physiology* **32**:407–449.

Bever JD, Broadhurst LM, Thrall PH. 2013. Microbial phylotype composition and diversity predicts plant productivity and plant–soil feedbacks. *Ecology Letters* **16**:167–174.

Birnbaum C, Barret LG, Thrall PH, Leishman MR. 2012. Mutualisms are not constraining cross-continental invasion success of *Acacia* species within Australia. *Diversity and Distributions* **18**:962–976.

Birnbaum C, Bissett A, Thrall PH, Leishman MR. 2014. Invasive legumes encounter similar soil fungal communities in their non-native and native ranges in Australia. *Soil Biology and Biochemistry* **76**:210–217.

Birnbaum C, Bissett A, Thrall PH, Leishman MR. 2016. Nitrogen-fixing bacterial communities in invasive legume nodules and associated soils are similar across introduced and native range populations in Australia. *Journal of Biogeography* **43**:1631–1644.

Blossey B, Notzold R. 1995. Evolution of increased competitive ability in invasive nonindigenous plants: a hypothesis. *Journal of Ecology* **83**:887–889.

Bolker BM, Brooks ME, Clark CJ, Geange SW, Poulsen JR, Stevens MHH, White J-SS. 2009. Generalized linear mixed models: a practical guide for ecology and evolution. **24**:127–135.

Brooks ML, D'Antonio CM, Richardson DM, Grace JB, Keeley JE, DiTomaso JM, Hobbs RJ, Pellant M, Pyke D. 2004. Effects of invasive alien plants on fire regimes. *BioScience* **54**:677–688.

Burnham KP, Anderson DR. 2002. *Model selection and multimodel inference: a practical information-theoretic approach*, 2nd edn. New York: Springer.

Butterfield HM. 1938. The introduction of acacias into California. *Madrono* **4**:177–187.

Calflora: Information on California plants for education, research and conservation. [web application]. 2015. Berkely, California. The Calflora Database [a non-profit organization], http://www.calflora.org/, accessed 20 July 2015.

Cal-IPC. 2006. *California invasive plant inventory*. Berkeley, CA: Cal-IPC Publication 2006-02.

Callaway RM, Aschehoug ET. 2000. Invasive plants versus their new and old neighbors: a mechanism for exotic invasion. *Science* **290**:521–523.

Carruthers J, Robin L, Hattingh JP, Kull CA, Rangan H, van Wilgen BW. 2011. A native at home and abroad: the history, politics, ethics and aesthetics of acacias. *Diversity and Distributions* **17**:810–821.

Crisóstomo JA, Rodríguez-Echeverría S, Freitas H. 2013. Co-introduction of exotic rhizobia to the rhizosphere of the invasive legume *Acacia saligna*, an intercontinental study. *Applied Soil Ecology* **64**:118–126.

D'Antonio CM, Vitousek PM. 1992. Biological invasions by exotic grasses, the grass/fire cycle, and global change. *Annual Review of Ecology and Systematics* **23**:63–87.

De Mendiburu F. (2009). Una herramienta de analisis estadistico para la investigacion agricola. Tesis. Universidad Nacional de Ingenieria (UNI-PERU).

Funk JL, Vitousek PM. 2007. Resource-use efficiency and plant invasion in low resource systems. *Nature* **446**:1079–1081.

Glyphis JP, Milton SJ, Siegfried WR. 1981. Dispersal of Acacia cyclops by birds. *Oecologia* **48**:138–141.

Hierro JL, Maron JL, Callaway RM. 2005. A biogeographical approach to plant invasions: the importance of studying exotics in their introduced *and* native range. *Journal of Ecology* **93**:5–15.

Holmes PM. 1990. Dispersal and predation in alien Acacia. *Oecologia* **83**:288–290.

Hothorn T, Bretz F, Westfall P. 2008. Simultaneous inference in general parametric models. *Biometric Journal* **50**:346–363.

Jepson Flora Project (eds.). 2015. *Jepson eFlora*, http://ucjeps.berkeley.edu/IJM.html, accessed on 01 July 2015.

Kahle D, Wickham H. 2013. ggmap: spatial visualization with ggplot2. *The R Journal* **5**:144–161. URL http://journal.r-project.org/archive/2013-1/kahle-wickham.pdf.

Keane RM, Crawley MJ. 2002. Exotic plant invasions and the enemy release hypothesis. *Trends in Ecology & Evolution* **1**:164–170

Klock MM, Barrett LG, Thrall PH, Harms KE. 2015. Host-promiscuity in symbiont associations can influence exotic legume establishment and colonization of novel ranges. *Diversity and Distributions* **21**:1193–1203.

Kull CA, Rangan H. 2008. Acacia exchanges: Wattles, thorn trees, and the study of plant movements. *Geoforum* **39**:1258–1272.

Levine JM, Vila M, Antonio CMD, Dukes JS, Grigulis K, Lavorel S. 2003. Mechanisms underlying the impacts of exotic plant invasions. *Proceedings of the Royal Society of London* **270**:775–781.

Liu WT, Marsh TL, Cheng H, Forney LJ. 1997. Characterization of microbial diversity by determining terminal restriction fragment length polymorphisms of genes encoding 16S rRNA. *Applied and Environmental Microbiology* **63**:4516–4522.

Mack MC, D'Antonio CM. 1998. Impacts of biological invasions on disturbance regimes. *Trends in Ecology & Evolution* **13**:195–198.

Mack RN, Simberloff D, Mark Lonsdale W, Evans H, Clout M, Bazzaz FA. 2000. Biotic invasions: causes, epidemiology, global consequences, and control. *Ecological Applications* **10**:689–710.

Martyniuk S, Oroń J. 2008. Survival of rhizobia in two soils as influenced by storage conditions. *Polish Journal of Microbiology* **57**:257–260.

Masclaux-Daubresse C, Daniel-Vedele F, Dechorgnat J, Chardon F, Gaufichon L, Suzuki A. 2010. Nitrogen uptake, assimilation and remobilization in plants: challenges for sustainable and productive agriculture. *Annals of Botany* **105**:1141–1157.

McKnight T. Queensland Division of Plant I. 1949. *Efficiency of isolates of rhizobium in the cowpea group, with proposed additions to this group*. Brisbane: Division of Plant Industry.

Miller JT, Murphy DJ, Brown GK, Richardson DM, González-Orozco CE. 2011. The evolution and phylogenetic placement of invasive Australian *Acacia* species. *Diversity and Distributions* **17**:848–860.

Montesinos D, Castro S. 2012. Invasive acacias experience higher ant seed removal rates at the invasion edges. **12**:33–37.

Ndlovu J, Richardson DM, Wilson JRU, Le Roux JJ. 2013. Co-invasion of South African ecosystems by an Australian legume and its rhizobial symbionts. *Journal of Biogeography* **40**:1240–1251.

Oksanen J, Blanchet FG, Kindt R, Legendre P, Minchin PR, OHara RB, Simpson GL, Solymos P, Henry M, Stevens H, Wagner H. 2015. Vegan: Community ecology package.

Pimentel D, Lach L, Zuniga R, Morrison D. 2000. Environmental and economic costs of nonindigenous species in the United States. *BioScience* **50**:53–65.

R Core Team. 2015. R: a language and environment for statistical computing. R Foundation for Statistical Computing. Vienna, Austria. URL http://www.R-project.org/.

Ramette A. 2009. Quantitative community fingerprinting methods for estimating the abundance of operational taxonomic units in natural microbial communities. *Applied and Environmental Microbiology* **75**:2495–2505.

Rejmánek M, Richardson DM. 2013. Trees and shrubs as invasive alien species - 2013 update of the global database. *Diversity and Distributions* **19**:1093–1094.

Richardson DM, Allsopp N, D'Antonio CM, Milton SJ, Rejmánek M. 2000. Plant invasions - the role of mutualisms. *Biological Reviews* **75**:65–93.

Richardson DM, Carruthers J, Hui C, Impson FAC, Miller JT, Robertson MP, Rouget M, Le Roux JJ, Wilson JRU. 2011. Human-mediated introductions of Australian acacias – a global experiment in biogeography. *Diversity and Distributions* **17**:771–787.

Richardson DM, Pysek P, Rejmánek M, Barbour MG, Panetta FD, West CJ. 2000. Naturalization and invasion of alien plants: concepts and definitions. *Diversity and Distributions* **6**:93–107.

Richardson DM, Rejmánek M. 2011. Trees and shrubs as invasive alien species – a global review. *Diversity and Distributions* **17**:788–809.

Rodríguez-Echeverría S. 2010. Rhizobial hitchhikers from Down Under: invasional meltdown in a plant-bacteria mutualism? *Journal of Biogeography* **37**:1611–1622.

Shea K, Kelly D, Sheppard AW, Woodburn TL. 2005. Context-dependent biological control of an invasive thistle. *Ecology* **86**: 3174–3181.

Southern EM. 1979. Measurement of DNA length by gel electrophoresis. *Analytical Biochemistry* **100**:319–323.

Sprent JI. 2001. *Nodulation in legumes.* In: Dickerson S, ed. Great Britain: The Cromwell Press Ltd.

Sprent JI, Sprent P. 1990. *Nitrogen fixing organisms: pure and applied aspects.* Cambridge: University Press.

Thrall PH, Burdon JJ, Woods MJ. 2000. Variation in the effectiveness of symbiotic associations between native rhizobia and termperate Australian legumes: interactions within and between genera. *Journal of Applied Ecology* **37**:52–65.

Thrall PH, Hochberg ME, Burdon JJ, Bever JD. 2007. Coevolution of symbiotic mutualists and parasites in a community context. *TRENDS in Ecology and Evolution* **22**:120–126.

Thrall PH, Laine A-L, Broadhurst LM, Bagnall DJ, Brockwell J. 2011. Symbiotic effectiveness of rhizobial mutualists varies in interactions with native Australian legume genera. *PLoS One* **6**: 1–11.

Thrall PH, Millsom DA, Jeavons AC, Waayers M, Harvey GR, Bagnall DJ, Brockwell J. 2005. Seed inoculation with effective root-nodule bacteria enhances revegetation success. *Journal of Applied Ecology* **42**:740–751.

Underhill LG, Hofmeyr JH. 2007. Barn swallows *Hirundo rustica* disperse seeds of Rooikrans *Acacia cyclops*, an invasive alien plant in the Fynbos Biome. *IBIS* **149**:468–471.

Vitousek PM, D'Antonio CM, Loope LL, Westbrooks R. 1996. Biological invasions as global environmental change. *American Scientist* **84**:468–478.

Wandrag EM. 2012. Do mutualists matter? The role of pollinators, seed dispersers and belowground symbionts in the invasion success of Acacia. Doctoral Dissertation, Lincoln University, New Zealand.

Permissions

List of Contributors

Christine N. Wheaton
Department of Biology, Bryn Mawr College, Bryn Mawr, PA, USA

Joshua S. Caplan and Thomas J. Mozdzer
Department of Biology, Bryn Mawr College, Bryn Mawr, PA, USA
Smithsonian Environmental Research Center, Edgewater, MD, USA

Aigar Niglas, Priit Kupper and Arne Sellin
Institute of Ecology and Earth Sciences, University of Tartu, Lai 40, 51005 Tartu, Estonia

Arvo Tullus
Institute of Ecology and Earth Sciences, University of Tartu, Lai 40, 51005 Tartu, Estonia
Institute of Forestry and Rural Engineering, Estonian University of Life Sciences, Kreutzwaldi 5, 51014 Tartu, Estonia

Gordon G. McNickle
Department of Biology, Wilfrid Laurier University, 75 University Avenue West, Waterloo, ON N2L 3C5, USA

Joel S. Brown
Department of Biological Sciences, University of Illinois at Chicago, 845 W. Taylor St. (MC066), Chicago, IL 6060, USA

Loralee Larios
Department of Environmental Science, Policy and Management, University of California Berkeley, 137 Mulford Hall, Berkeley, CA 94720-3114, USA
Division of Biological Sciences, University of Montana, 32 Campus Dr HS104, Missoula, MT 59812, USA

Katharine N. Suding
Department of Environmental Science, Policy and Management, University of California Berkeley, 137 Mulford Hall, Berkeley, CA 94720-3114, USA
EBIO, University of Colorado, Ramaley N122, Campus Box 334, Boulder, CO 80309- 0334, USA

Sara Jo M. Dickens, Edith B. Allen and Louis S. Santiago
Department of Botany and Plant Sciences, University of California Riverside, Riverside, CA 92521, USA

David Crowley
Department of Environmental Sciences, University of California Riverside, Riverside, CA 92521, USA

Thomas E. Miller
Department of Biological Science, Florida State University, Tallahassee, FL 32306, USA

Robert R. Blank, Tye Morgan and Fay Allen
Great Basin Rangelands Research Unit, USDA-Agricultural Research Service, Reno, NV, USA

Nathan L. Brouwer, Alison N. Hale and Susan Kalisz
Department of Biological Sciences, University of Pittsburgh, Pittsburgh, PA 15260, USA

Lambertus A.P. Lotz
Plant Research International, Wageningen University and Research Centre, Droevendaalsesteeg 1, 6708 PB Wageningen, The Netherlands

Tanja A.A. Speek
Plant Research International, Wageningen University and Research Centre, Droevendaalsesteeg 1, 6708 PB Wageningen, The Netherlands
Laboratory of Nematology, Wageningen University and Research Centre, Droevendaalsesteeg 1, 6708 PB Wageningen, The Netherlands
Department of Terrestrial Ecology, Netherlands Institute of Ecology (NIOO-KNAW), Droevendaalsesteeg 10, 6708 PB Wageningen, The Netherlands

Wim A. Ozinga
Centre for Ecosystem Studies, Wageningen University and Research Centre, Droevendaalsesteeg 3a, 6708 PB Wageningen, The Netherlands

Joop H.J. Schaminée
Centre for Ecosystem Studies, Wageningen University and Research Centre, Droevendaalsesteeg 3a, 6708 PB Wageningen, The Netherlands
Department of Ecology, Aquatic Ecology and Environmental Biology Research Group, Radboud University Nijmegen, Heyendaalseweg 135, 6525 AJ Nijmegen, The Netherlands

Jeltje M. Stam
Laboratory of Entomology,Wageningen University and Research Centre, Droevendaalsesteeg 1, 6708 PBWageningen, The Netherlands

Wim H. van der Putten
Laboratory of Nematology, Wageningen University and Research Centre, Droevendaalsesteeg 1, 6708 PB Wageningen, The Netherlands
Department of Terrestrial Ecology, Netherlands Institute of Ecology (NIOO-KNAW), Droevendaalsesteeg 10, 6708 PB Wageningen, The Netherlands

Philip A. Fay and H. Wayne Polley
Grassland, Soil, and Water Laboratory, USDA-ARS, 808 E Blackland Rd., Temple, TX 76502, USA

Beth A. Newingham
College of Natural Resources, University of Idaho, 441133, Moscow, ID 83844, USA
Great Basin Rangelands Research, USDA-ARS, 920 Valley Rd., Reno, NV 89512, USA

Jack A. Morgan and Daniel R. LeCain
Rangeland Resources Research Unit, USDA-ARS, 1701 Centre Avenue, Fort Collins, CO 80526, USA

Robert S. Nowak
Department of Natural Resources and Environmental Science/MS 186, University of Nevada Reno, 1664 North Virginia, Reno, NV 89557, USA

Stanley D. Smith
School of Life Sciences, University of Nevada, Las Vegas, 4505 S. Maryland Parkway, Las Vegas, NV 89154, USA

Ragan M. Callaway
Division of Biological Sciences and the Institute on Ecosystems, The University of Montana, Missoula, MT 59812, USA

Lixue Yang
Division of Biological Sciences and the Institute on Ecosystems, The University of Montana, Missoula, MT 59812, USA School of Forestry, Northeast Forestry University, Harbin 150040, China

Daniel Z. Atwater
Department of Plant Pathology, Physiology, and Weed Science, Virginia Tech University, Blacksburg, VA 24061, USA

Pamela R. Belter and James F. Cahill Jr
Department of Biological Sciences, University of Alberta, Edmonton, AB, Canada T6G 2E9

Stanislaus J. Schymanski
Department of Environmental Systems Science, Swiss Federal Institute of Technology Zurich, Universita¨tstrasse 16, 8092 Zurich, Switzerland

Max Planck Institute for Biogeochemistry, Jena, Germany

Michael L. Roderick
Research School of Earth Sciences and Research School of Biology, Australian National University, Canberra 2601, Australia
Australian Research Council Centre of Excellence for Climate System Science, Canberra 2601, Australia

Murugesu Sivapalan
Department of Geography and Geographic Information Science, University of Illinois at Urbana-Champaign, Urbana, Illinois, USA
Department of Civil and Environmental Engineering, University of Illinois at Urbana-Champaign, Urbana, Illinois, USA

Simon Antony Croft and Jonathan W. Pitchford
Department of Biology, University of York, Wentworth Way, York YO10 5DD, UK
York Centre for Complex Systems Analysis (YCCSA), The Ron Cooke Hub, University of York, Heslington, York YO10 5GE, UK

Angela Hodge
Department of Biology, University of York, Wentworth Way, York YO10 5DD, UK

Thalita G. Zimmermann and Antonio C. S. Andrade
Laboratório De Sementes. Instituto De Pesquisas Jardim Botânico Do Rio De Janeiro. Rua Pacheco Leão, 915, Jardim Botânico, Rio De Janeiro, RJ 22460-030, Brazil

David M. Richardson
Department of Botany and Zoology, Centre for Invasion Biology, Stellenbosch University, Matieland 7602, South Africa

Rafael Dudeque Zenni, Wanderson Lacerda da Cunha and Guilherme Sena
Department of Ecology, University of Brasília, Campus Universitário Darcy Ribeiro, Brasília CEP 70910-900, Brazil

Metha M. Klock and Kyle E. Harms
Department of Biological Sciences, Louisiana State University, Baton Rouge, LA 70803, USA

Luke G. Barrett and Peter H. Thrall
CSIRO Agriculture Flagship, Canberra, ACT 2601, Australia

Index

www.ingramcontent.com/pod-product-compliance
Lightning Source LLC
Chambersburg PA
CBHW080635200326

41458CB00013B/4633